CALAVERAS GOLD

*Wilbur S. Shepperson Series in History and Humanities*

# Calaveras Gold

*The Impact of Mining on a Mother Lode County*

RONALD H. LIMBAUGH & WILLARD P. FULLER JR.

University of Nevada Press ▲▲ Reno & Las Vegas

Wilbur S. Shepperson Series in History and Humanities
Series Editor: Jerome E. Edwards

University of Nevada Press, Reno, Nevada 89557 USA
Copyright © 2004 by University of Nevada Press
Manufactured in the United States of America
Design by Carrie House
Library of Congress Cataloging-in-Publication Data
Limbaugh, Ronald H.
Calaveras gold : the impact of mining on a mother lode county /
Ronald H. Limbaugh, Willard P. Fuller, Jr.
p. cm.— (Wilbur S. Shepperson series in history and humanities)
Includes bibliographical references.
ISBN 0-87417-546-1 (hardcover : alk. paper)
1. Gold mines and mining—California—Calaveras County—
History—19th century.   I. Fuller, Willard P., 1918–   II. Title.
III. Wilbur S. Shepperson series in history and humanities
(Unnumbered)
TN423.C2 L48   2003
338.4′76223422′0979444—dc21   2003011142
The paper used in this book meets the requirements of
American National Standard for Information Sciences—
Permanence of Paper for Printed Library Materials, ANSI
Z39.48-1984. Binding materials were selected for strength
and durability.

13 12 11 10 09 08 07 06 05 04
5  4  3  2

# CONTENTS

ILLUSTRATIONS

ACKNOWLEDGMENTS

Many people were involved in helping the authors select materials for this manuscript and complete its preparation. The staffs of the Calaveras Heritage Council and the Calaveras Museum supported the project in its beginning stages back in the 1970s, and have continued to be helpful over the years. The Segerstrom family of Sonora was particularly helpful in allowing the authors access to family archives that have subsequently been deposited at the Holt-Atherton Library at the University of the Pacific. Daryl Morrison and her staff at the Holt-Atherton Library aided materially in locating and organizing resources. Gene Gressley and Carlo de Ferrari reviewed an early version of the manuscript and provided many helpful suggestions.

Special thanks are due to the following individuals and organizations for their cooperation and assistance during the long years of preparation: William B. Clark and the California Division of Mines and Geology, Dale M. Stickney of the California Mines and Geology Library, Mike Kizer and Grandview Resources of Carson Hill, Linda Guerra and the Sonora Mining Company at Jamestown, Bonnie Hardwick and Mary Morganti of the Bancroft Library, Peter Blodgett of the Huntington Library, Mrs. Lorrayne Kennedy of the Calaveras County Archives, the Calaveras County Historical Society, the Tuolumne County Museum, the Calaveras Cement Company, the Conference of California Historical Societies, Ted Bird, Philip R. Bradley, Julia Costello, Richard Dyer, Michael Dell'Orto, Donald Dickey, Anthony Dutil, E. M. Gerick, Ella McCarty Hiatt, John W. James, Robert E. Kendall, Judith Marvin, Glen Nevens, Duane Oneto, Richard Rolleri, Carlos Schwantes, Edgar Smith, Barden Stevenot, Charles Stone and Rhoda Stone, and Howard Tower.

# Introduction

Sam Casoose slung a burlap bag over his shoulder and headed for town. At fifty cents each, his burden of eviscerated salmon commanded a ready market in Murphys, a mile due south of his cabin on the rancheria above the Oro y Plata mine. Caught and dried at Clark's Flat on the Stanislaus a few weeks before, the spring chinook had been sluggish and easily trapped as they neared the end of their life cycle after spawning. The seasonal run had brought out dozens of Central Sierra Miwok, from rancherias at Six Mile, Vallecito, Sheep Ranch, and others. But fishing had been good, and the rocky beach had been covered with drying fish.

As he trekked down the ridge path toward the dusty wagon road leading to town, his thoughts were interrupted by the sight and sound of miners at work. Two hundred yards downslope, a miner had taken an option on the Oro y Plata, sluicing the old tailings and figuring what it would cost to dewater the underground levels. The scene reminded Sam of earlier times, when the noise from the mine's stamp mill, now quiet, had echoed across the valley. Perhaps that was why Miwok elders had named the rancheria Mol-Pee-So, or "earplugs."

\* \* \*

The story of Sam Casoose and the Oro y Plata mine exemplifies the interaction of people and mines along the lower slopes of the central Sierra in what is now Calaveras County. Some inhabitants had been born in the region or had ancestors whose bones had rested there for more than a thousand years; others were recent arrivals from foreign lands; still others were Americans born in the States but inflected with the accents and attributes of their European antecedents. The newcomers came singly, in pairs, or in groups and companies at different times and by different conveyances, but the means of their arrival are not as important as the motives. To borrow an old cliché, they were hoping to get something or to get away from something, but what they found was not always what they expected, and whether they stayed or moved on often depended on economic or social forces beyond their control.

The land they came to provided an abundant resource base for social and economic development. A 660,352-acre wedge of land on the western slope of the Sierra Nevada, Calaveras is a diversified landscape with a wide range of natural resources and climatic conditions. The narrow eastern border lies

The Oro y Plata mine at Murphys. Note the rancheria buildings on the ridgeline. (Courtesy Dr. Milton B. Smith)

in rugged alpine terrain 7,200 feet above sea level, where total snowfall approaches 50 feet in an average winter. The broader western border is drawn across the low-lying Sierra foothills adjacent to the great Central Valley, a fallow but hot and rainless land from May until storms begin in the fall. The ragged lines of the flanks follow the northeasterly trend of two major tributaries of the San Joaquin River, the Mokelumne on the north and the Stanislaus on the south, almost to their sources high in the mountains. Those rivers supply much of the water for farmers in the San Joaquin Valley, and for industrial and domestic users along the eastern shore of San Francisco Bay. In between the county's northern and southern borders is the Calaveras River, actually much smaller in recent geologic times than the ancient river system that once traversed the central Sierra. It still drains much of the county and provides the principal water resource for San Joaquin County communities and farms downstream.

Gold miners first gained access to the southern Sierra foothills by following the rivers upstream from the San Joaquin plains through the Bear Moun-

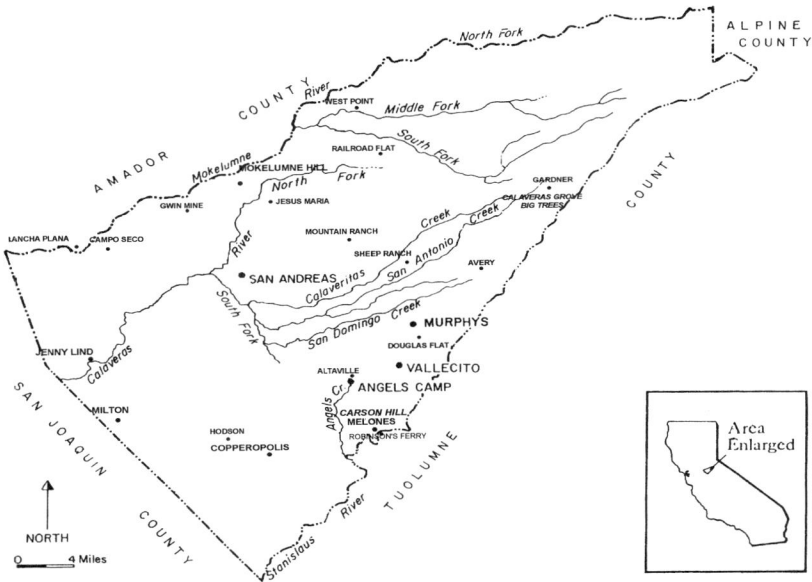

Calaveras County map. (Adapted from James Gary Maniery, *Six Mile and Murphys Rancherias* [San Diego: San Diego Museum of Man, 1987])

tain Range, a north-south trending barrier the Calaveras River breached near Toyon Flat. Gold Rush overland routes evolved into major arteries—Highways 4, 12, and 26 today—connecting the supply centers in the Central Valley to the mining camps, once the county's principal population centers. Now, new structures are sprouting along these transportation corridors all the way from Wallace and Copperopolis on the western boundary to neighboring Alpine County on the Sierra crest. Most of the newcomers, however, settle in the foothill towns and suburban developments, close to the Central Valley but still part of the Mother Lode. Angels Camp, the county's largest town with an estimated 3,150 residents as of 2001, is growing about 2.3 percent a year, slightly faster than the county as a whole.[1]

The varied life zones ranging from lower Sonoran to Canadian in this angular country support an abundance of wildlife, with many species endemic to the area, although some have been endangered by human and exotic plant intrusions over the past century. The forest canopy extends from the scattered digger pines and oaks at lower elevations through the foothills to the mixed pine, fir, and cedar forests in the higher elevations. The diversity of surface resources attracted the first humans to Calaveras shortly after the last ice age some 10,000–12,000 years ago. For thousands of years Native Americans lived quietly and successfully on the land, adapting to its rhythms and natural cycles, utilizing its animals and fibers, learning to drive game by setting fire to the grassy rangeland and the brushy understory of the upper slopes, moving higher and lower as the food supply and living conditions changed with the seasons, keeping population in tune with the resource base.[2] Environmental historians recognize the impact of early human changes on natural ecosystems, correcting distorted pioneer portrayals of starving "diggers" as well as romantic views that fostered an idyllic image of Indians living in unchanging "harmony" with the land.[3] But the pace of change accelerated rapidly with the coming of Europeans, and escalated explosively with the arrival of Americans during the Gold Rush.

Beneath the placid Calaveras foothills, the Mother Lode angles across the western part of the county in a northwestern-southeastern direction. This deep and tectonically significant structure of faulting and mineralization testifies to the long and violent geological history of the Sierra Nevada range. A relatively narrow zone stretching more than one hundred miles from Mariposa to El Dorado Counties, the Mother Lode is sliced by numerous faults, many of which were mineralized with gold-bearing quartz veins. The subsequent deep geologic erosion of the Sierra Nevada freed the gold from the grip of quartz, and concentrated it in many placer deposits in the rivers and streams. The Gold Rush followed directly from the discovery of nuggets in these deposits.

In Calaveras for the past 150 years, mining has been the most dynamic of the external forces affecting the lives of individuals and families. It is one of the counties within what was called in early days the "Southern Mines," a region that began at the Mokelumne River on the Amador-Calaveras boundary and included the counties of Tuolumne and Mariposa as well as the lower valley counties from San Joaquin to Kern. With an estimated gold production of more than twenty-five million ounces, the Southern Mines rank second in total California gold production, well below the seventy-five million ounces attributed to the "Northern Mines" along a 120-mile stretch of the Sierra foothills between the Mokelumne and Feather Rivers. The Southern Mines include a large part of the Mother Lode, a celebrated zone stretching from Georgetown in El Dorado County to Mariposa 120 miles south.[4] While the Mother Lode perhaps is better known in song and story than any other mineralized region in the world, production figures show it to be less significant economically than those districts extending into the northern Sierra Nevada. Calaveras ties with Butte and Sierra Counties for fourth place in recorded gold production, with about nine million ounces recovered between 1848 and

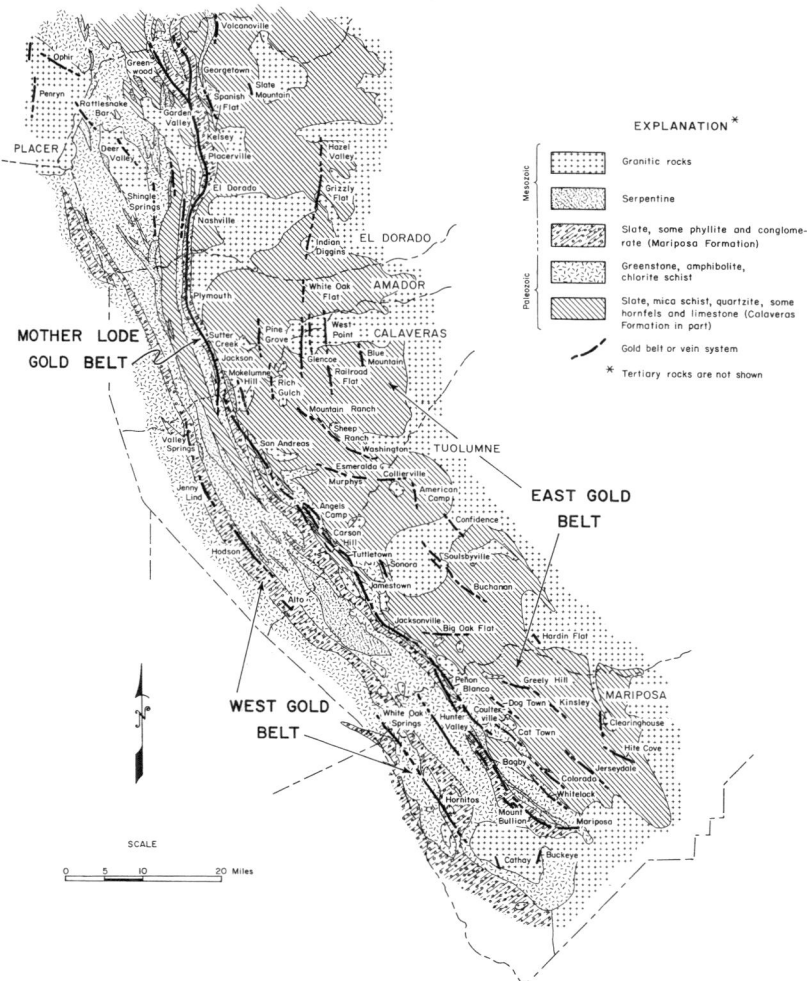

Mother Lode geology map. (William B. Clark, *Gold Districts of California,* Bulletin 193 [Sacramento: California Division of Mines and Geology, 1970])

1965. It is therefore a major California producer, but less than its neighbors Tuolumne to the south and Amador to the north, and far behind the state's number-one gold county, Nevada, with twenty-two million ounces.[5]

This is a book about mining in Calaveras and its economic impact from the Gold Rush to the present. It looks in depth at mining as an industry and the technology that shaped it in a typical county in the southern Sierra foot-

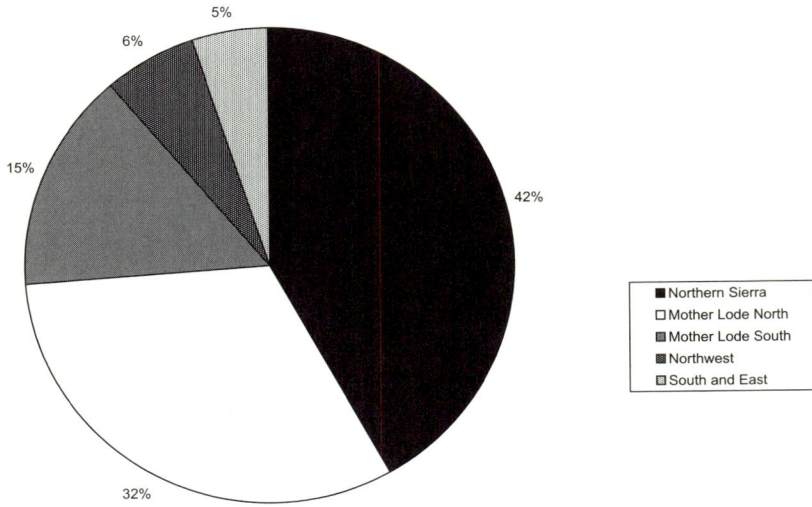

Gold Production. (Authors' database)

hills. Though technically precise and factually detailed, it is geared to general audiences, with just enough geology and technology included to help readers understand what problems miners had to contend with and how they tried to solve them. Technical terms are defined in the text and more fully explained in the glossary. An appendix provides details on county population data.

Mining in Calaveras is a complex story. Rather than use a topical approach, like many modern Gold Rush books, or hit the highlights in a broad survey, as Mary Hill did recently in an intriguing combination of geology and history, we provide a chronological case study of mining and its economic dimensions in a single region.[6] Starting with the initial discoveries, chapter 1 discusses the methods used by the pioneer placer miners, and the changes brought about by evolving placer technology. The next chapter follows mining underground with the first hardrock efforts. It explains how lode mining was financed, and explores the first base-metal developments in the copper deposits of the West Belt. Although our primary focus is economic, in chapter 3 we consider the social consequences of the Gold Rush, especially the impact on people of color and ethnic minorities suddenly juxtaposed against Euro-American immigrants. Chapter 4 shows how mining spawned supportive networks of economic activity that often evolved into separate industries. The emergence of modern technologies and subsidiary industries is covered in chapter 5. Chapter 6 surveys the major mines and districts during the prosperous era from the 1890s to World War I. How the mines and miners struggled through the twenties and thirties is the subject of chapter 7. The final chapter brings the Calaveras story down to the present with the decline of mining and its ancillary industries after World War II.

Today the pace of change in the foothill counties is again accelerating, but the main engine this time is not mining but lifestyle. Attracted by visions of cleaner air, quieter neighborhoods, and lower costs, thousands of retirees from the coastal cities and inland valleys have literally headed for the hills. Thousands more fully or semiemployed workers, with jobs in Sacramento, Stockton, Modesto, and other valley cities and towns, are moving to the lower Sierra despite a long daily commute. Along the southern Mother Lode, the county growth rate is among the highest in the nation.[7]

The impact of gold mining on regional economic development is one of the oldest themes in historical literature, so why produce another book on the subject? The simplest answer is that scholars and storytellers alike have largely overlooked Calaveras. A few monographs can be found describing Calaveras in the Gold Rush, but aside from technical publications, the region has lacked a comprehensive mining narrative. Mark Twain's venerable jumping-frog tale proves the rule.[8] Even though it virtually launched his career as a humorist and writer, and was the inspiration for an annual jubilee that attracts thousands to Angels Camp, the story has little to do with Calaveras history. Jubilee visitors leave the region knowing little more about its connection to gold mining than they did before. The primary purpose of our book is to make that connection explicit, to inform readers about the region's gold mining heritage and its economic ramifications as they changed over time.

Did those economic ramifications have a positive or a negative impact? Most contemporary observers were overwhelmingly positive: They saw the quest for wealth as the primary engine of economic growth and development that propelled California into a modern industrial state. By the mid-twentieth century, however, older progressive models were giving way to more complex interpretations. Modern historical scholarship on the meaning of the Gold Rush began with pathbreaking monographs by Rodman Paul in 1947 and John Walton Caughey a year later. With critical insight and broad perspective, Paul reformulated the classic case that California's mining legacy provided a constructive framework for western economic development. Caughey's work was more voluble and philosophical. In considering "whether the gold of California did not do more harm than good," he answered in the positive, though his explanation amounted to a critical review of the Gold Rush's darker, more troubling side.[9]

A second reason for our book is to test broad economic generalizations by applying them to a specific region. As two sociologists wrote recently, new scholarship has questioned many earlier views "about the implications of re-source extraction for socio-economic development," but "analyses of long-term historical data have been quite limited."[10] While we do not overload the text with data analysis, our story rests on empirical evidence gleaned from both primary and secondary sources. This volume is a descriptive history of events that heretofore have not been adequately chronicled, but it is also an interpretation of gold mining and its economic significance for Calaveras and its resident population.

Since the 1960s, with the advance of postmodern thinking and the "new history," scholars have rewritten much of California's golden legacy. No serious study of Calaveras or any other mining region in California can ignore the implications of new interpretations that have added social, urban, environmental, and multicultural dimensions to what was once primarily a progressive chronicle of political and economic events. Although our book concentrates on economic themes, mining was not just a matter of methods, results, and production records. The mines had a profoundly human impact. They brought people together from all ethnic backgrounds, sometimes with unhappy and even tragic consequences. They made a few rich and bankrupted many more. They gave some men their start to fame and fortune, and drove others to crime or insanity. By generating and distributing capital, they stimulated growth and development everywhere the money flowed, but they also contributed to exploitation of those who were unable to protect themselves from wealth and power. They gave rise to towns and permanent settlements, stimulated subsidiary industries like logging and hydroelectric power generation, guided the growth of transportation networks, and laid the foundations of the region's modern economy. They also despoiled the streams, denuded the landscape, and left a toxic residue for others to face. The impact of the mining industry can be traced in the region's political institutions, government, laws, courts, and law enforcement efforts. From the mining age came schools, churches, libraries, opera houses, and recreational activities, but also came the downside of growth in an age without regulatory agencies or safety nets. In short, more than any other single enterprise, and for good or ill, mining urbanized, industrialized, and modernized Calaveras County.

As most readers are aware, historically gold has been both a commodity and a monetary unit. For most of the time period covered by this book, governments rather than market forces determined the price of gold. For more than 180 years in the United States, the federal government set gold prices and generally required American gold producers to sell gold bullion directly to the U.S. Mint, although there were some exceptions. Before 1933 the official U.S. price was $20.67 per ounce for "pure" gold, that is, 1,000-fine or

24-carat gold. In California gold bullion usually varied between 750 and 980 fine. The "impure" part was almost entirely silver, which between 1792 and 1893, except for a five-year period in the 1870s, also had a fixed monetary value, set at a ratio to gold of fifteen or sixteen to one. During the Depression, President Roosevelt raised the gold price to $35 per ounce, and there it remained until 1971, when Congress dropped the gold standard and allowed the price to float on the open market.

In this book production values are given generally in numbers of ounces, although sometimes it was more expedient to use dollar amounts. In converting from dollars to ounces, to compensate for variations in actual payments received by miners, we used an average of $17 per ounce for gold before 1933, and $30 per ounce between 1933 and 1971.

For modern investors who purchase gold today at around $350 an ounce, the historic prices fixed by the government may seem incredibly low. Mining itself a century ago was quite limited and simple compared to operations of the global giants that constitute modern mining corporations. But size and value are relative terms. Variations in the size and scope of individual mines, as well as incomplete, uncertain, or undisclosed financial transactions, make comparisons difficult. Fluctuations in the price of gold over nearly two centuries reflect the inflationary forces that have affected all aspects of the U.S. economy. When gold was $20 an ounce, the daily wage of a common laborer was little more than $1 per day. A gold mine worth $50,000 in 1870 might well be the equivalent of a $5 million property or operation today. But simple price conversions do not tell the whole story. In addition to size, readers must take into consideration changes in labor costs, commodity prices and supplies, the availability of capital, and especially living standards. In short, converting yesteryear's economic and social values to modern equivalents is very difficult, not just for readers but also for historians.

Though mining today is a fading force in the U.S. economy, its legacy is still visible in the "hard places" of the western landscape, to borrow a phrase from historian Richard Francaviglia, and its political clout is felt whenever Congress tries to revise the 1872 mining law that still provides the rules for locating and acquiring mining property on federal lands. Its darker social and environmental dimensions are also with us today, whether we recognize them or not. As author and California state librarian Kevin Starr puts it, the "Gold Rush is not something back there in time. The Gold Rush is everywhere around us, even in its tragic consequences. The Gold Rush is who we are as a people."[11]

What happened in one mining region happened elsewhere, though perhaps not in the same ways.[12] With Calaveras as our main arena, and with gold our central theme, we offer in this narrative an example of how mining influenced a small geopolitical region over time. While our findings may be applicable to other mining regions, we recognize the limitations of trying to conclude too much from too little. Richard White's caveat is well worth repeating: "The Gold Rush had consequences, but specifying those consequences—and not overplaying them—that is the trick."[13] Though we offer a broader perspective than most regional histories, this book is not about present controversies, nor does it attempt to extrapolate cosmic lessons from a regional focus. Let readers draw their own conclusions about the role of mining in the making of modern America. We will be satisfied if we help put Calaveras on the mining map, explore its mining life and lore, and in the process tell a good story.

# 1: The Early Placer Era

The Gold Rush, in a technical sense, did not begin with James Marshall's discovery at Coloma on January 24, 1848. That was only the spark. The fire did not break out until May, when Sam Brannan rode through the streets of San Francisco with a pouch full of nuggets, shouting, "Gold, gold, from the American River!" The first epidemic of "gold fever" emptied San Francisco and spread quickly through the coastal countryside from Sonoma to Monterey. It started the mass movement that over the next three years poured nearly two hundred thousand people into the Sierra foothills. Out of this great convergence of newcomers came new towns and new political units even before California became a state. At the first legislative session on February 18, 1850, Calaveras was created, one of twenty-seven charter counties in California. Initially a bat-wing-shaped county stretching across the crest of the Sierra to the Nevada border and containing most of what became Amador, Alpine, and Mono Counties, it was cut down to its present boundaries by subsequent legislation over the next decade.[1]

## Calaveras in the Gold Rush

The story of gold mining in Calaveras begins curiously not in the Sierra foothills, but in the enterprising mind of Charles M. Weber, Stockton's founder and presumptive empire builder. A friend and former employee of John A. Sutter, Weber was one of northern California's most prosperous entrepreneurs. Within two years after immigrating to California in 1841 with the Bidwell party, he was a successful merchant, with a general store in San Jose that profited handsomely in the early 1840s. By 1845 he had acquired full title to a forty-four thousand–acre rancho on the lower San Joaquin River after buying out his business partner who had secured a land grant from the Mexican government two years before at Weber's instigation. Weber retained his San Jose store and his Bay Area business connections after he moved to his rancho in 1847 as land baron, town founder, and colonizer. At the head of navigation on the San Joaquin he founded "Tuleburg," but soon changed the name to Stockton to honor the boisterous commodore who commanded the successful U.S. land forces during the brief California phase of the Mexican War.

The town of Stockton was still in its infancy on January 28, 1848, when James Marshall, after a hard ride in a rainstorm, reached Sutter's fort. He gleefully showed his boss the first glittering yellow samples that he had re-covered four days earlier from the tailrace at Sutter's Mill. Despite Sutter's efforts to keep the discovery a secret, the news spread quickly among Sutter's inner circle of friends and acquaintances. One of those friends was Charles Weber, who frequently traveled to Sacramento on business. He was at Sutter's fort early in March 1848, and must have learned of the discovery at that time. However, Frank Gilbert, who interviewed Weber thirty years later and wrote the first published history of San Joaquin County, said Weber did not catch the "spark from the flame" until one of Sutter's men turned up in Stockton with a few shiny specimens later in March.[2]

As a businessman and entrepreneur in a territory desperately short of hard currency, Weber needed gold for Stockton's development, but neither he nor anyone else at the time knew much about gold or how to find it and mine it. In spite of the long history of gold mining and milling from ancient times to the nineteenth century, Americans in northern California before the sum-mer of 1848 had almost no personal mining knowledge or experience.[3] Fortu-nately for these young gold hunters, California surface placer deposits were so widespread and so easily tapped that practically anybody with a strong back had a good chance of locating pay dirt. Some went exploring with little more than a pick and a long knife for probing creek-bed crevices. Weber was more systematic: He rejected the idea of a quick scramble in favor of an orga-nized expedition. From the first he also had commercial interests in mind. Like Collis P. Huntington, Leland Stanford, and other merchant-Argonauts, Weber was perceptive enough to realize that there was more profit and less work in supplying miners than in mining himself. It took no depth of in-sight to recognize the market possibilities of selling scarce merchandise to gold seekers isolated from easy sources of supply.

Weber spent most of March and April gathering information, supplies, and men for a mining expedition. At least two members of his party later became prominent in Calaveras County. Although no record of the original party has been located, doubtless it included the two Murphy brothers John and Daniel, sons of Martin Murphy Sr., who had brought the first wagons over the Sierra in 1844. Weber had known most of the twenty-six members of the Murphy clan since their arrival, and knew them intimately as in-laws after courting and eventually marrying their younger sister Helen.[4] Late in April, two weeks before Sam Brannan's giddy gallop through San Francisco, Weber's party headed east toward the Stanislaus River. They explored the river and its tributaries, but found nothing of real interest. Then they turned north toward Coloma, prospecting as they traveled. For nearly two months they worked the foothill streams and gulches, digging closer to bedrock than had the first

"Captain" Charles M. Weber. (Courtesy Holt-Atherton Library, University of the Pacific)

hasty prospectors out of Coloma. Along the Mokelumne River their luck changed. They found gold in several rich pockets, perhaps the first from what became Calaveras County, although the exact location is unknown. But they kept moving north, leaving the river to the remnants of Colonel Jonathan D. Stevenson's New York regiment of Mexican War volunteers. These newcomers had arrived in California nearly too late to fight, but promised to remain after the war to help Americanize the country. Soon after the unit disbanded in September 1848, some of these young adventurers came to the Mokelumne, taking up claims the Weber party had probably explored briefly on their way north.[5]

Just above Coloma, on a small tributary of the American River, Weber's company found the most promising prospects yet. Here was the literal pot of gold they had been looking for! They staked claims and settled in for some serious mining. Soon Weber's Creek was a flourishing camp, with miners recovering thousands of dollars in small nuggets and fine gold at the field price of ten dollars per ounce. Weber, calmly mercantile despite the excitement, seemed to recognize where the real money was. He controlled the distribution of supplies from a hastily erected brushwood store. James H. Carson, arriving a few weeks later, reported that Weber "was daily sending out mules packed with gold."[6]

By June, all of northern California had caught gold fever, and Weber realized that he would have to reprovision and reorganize to keep ahead of the crowd. Leaving some men behind at Weber's Creek to hold their claims, he returned to his rancho and organized the Stockton Mining Company as a joint stock mining and trading venture. Among his friends and acquaintances along the Cosumnes River, at San Jose, and elsewhere, he sought out supplies and enlisted stockholders in the enterprise.

He also contracted for a group of local Native Americans to labor in the mines. Indians were cheap and exploitable, living on lands they once called home that had been appropriated by Mexican and American impresarios with government acquiescence. As historian Jim Rawls has made clear, in California "the Hispanic system of labor exploitation was transferred from the ranchos to the mines." In 1847, when Weber had taken possession of his San Joaquin rancho, scattered remnants of plains and mountain Miwok tribes were still living in the large rectangle roughly bounded by the San Joaquin, Cosumnes, and Fresno Rivers.[7] These were the survivors of the devastating European diseases that had wiped out so much of the Native population prior to the U.S. takeover. One nearby subgroup of Plains Miwok, known as Síakumne or Siyakum, inhabited a small village on the south bank of the Mokelumne, with a range that extended south to the Stanislaus. They were led by José Jesús, described as a tall, "stunning" figure clad in colorful Hispanic garb. Jesús had assumed a leadership role after the death of Estanislao, the leader of a combined Yokuts-Miwok renegade band that had been nearly wiped out in 1829 by a Mexican force under Mariano Vallejo. Estanislao's successor harbored strong resentment against the Mexican government, a feeling that played into the hands of Charles Weber. He befriended Jesús by convincing him that the development of his rancho on the San Joaquin would serve as a buffer between coastal Hispanic settlements and the interior Indians.[8]

A year later, in the spring or early summer of 1848, Weber came once again to Jesús, this time as a friend offering opportunities for Indians to work in the mines. Contracting for Native labor was not unusual in pre–Gold Rush California. Sutter used Native help in constructing the sawmill at Coloma.[9] In the northern Sacramento Valley, Pearson Reading and John Bidwell also employed Indians under labor contract, and continued to use them as miners. In Tuolumne and Mariposa Counties, James D. Savage, according to contemporaries, had "thousands" of Indians mining for him.[10] Weber and Jesús worked out an agreement by which Weber would be supplied with a Native crew to prospect ground along the Stanislaus and the Tuolumne, an area Weber's initial discovery party had skimmed over without success or had missed entirely. Brought to Weber's Creek for a short course in placer mining, the Indians departed with instructions to deliver any yellow metal they unearthed to Weber's ranch foreman at French Camp. They would receive in exchange blankets, utensils, beads, and other trade goods. Whether they got fair value in the exchange is debatable. Weber had tried to cultivate goodwill among the Miwok, but he still charged them one hundred dollars in gold per blanket.[11] Unsuspecting Indians were fair game for unscrupulous traders who weighed Indian gold with the "Digger Ounce," an overweight lead slug used instead of standard weights. As James H. Carson observes, "weighing gold for Indians and white people was a different matter," and traders saw no harm in "throwing the lead" when naive Natives were the customers.[12]

The first established gold discovery site in what is now Calaveras County is not precisely known, nor is the date recorded. George Tinkham, an early regional historian who interviewed Weber long after the event, places the discovery on the Stanislaus, but does not identify the site or the date. Others identify the date as late July or early August 1848, and place the site on the north side of the Stanislaus below Carson Hill along a steep and dusty gulch first called Dry Diggings, later Indian Gulch.[13] Regardless of location, the honor of discovery goes to Native Americans. Sometime in late July or early August, some of Weber's hired Miwok appeared before his representative at French Camp with a handsome collection of Calaveras nuggets that were the largest yet discovered in California, and the first found in what became known as the Southern Mines.

Gold from the Stanislaus created a wild excitement that summer when it reached Weber and his partners in the Stockton Mining Company, still working claims on Weber's Creek that were declining rapidly.[14] Many decided on the spot to abandon the local diggings and head for Indian Gulch, detouring through Weber's rancho in Stockton long enough to gather supplies. Joined by others who had heard the news, the miners reached the Stanislaus by the latter part of August. Soon the whole countryside was astir with prospectors spreading out to explore the gulches and streambeds throughout the region. In September, Weber also left his diggings near Coloma, and the company dissolved not long afterward. Stockton's founder recognized the importance of his town as the "Gateway to the Southern Mines," and he wanted to develop its waterfront and transportation connections.[15]

A few months earlier, while the Weber company was still at "Weberville" near Coloma, another individual, later prominent in Calaveras County mining history, arrived at Weber's Creek. James H. Carson, a Virginian who had come west with Stevenson's regiment, had initially settled in Monterey after the regiment mustered out. When the first news of the Marshall discovery reached town Carson remained skeptical, but early in May, after meeting up with an old army acquaintance with a sack of gold, Carson caught the fever and headed east. He found Weber's camp flourishing, with "Indians giving handsful of gold for a cotton handkerchief or a shirt." Apparently, this was gold both from Weber's Creek and from the Stanislaus country, for Carson noted that the Natives were coming in with news of outlying discoveries.[16]

In August 1848 Carson joined the rush to Stanislaus but split off from the main party when they reached what soon became known as Angels Creek. Aided by Native guides, he and a small group found likely placer deposits farther upstream, along a creek, later called Carson, which circled the base of a hill rising some nineteen hundred feet above the north bank of the Stanislaus River. He remembered averaging 180 ounces per man in ten days of digging. Had Carson worked up the hill to the quartz outcrops he would have discovered the source of this golden bounty, which contained some of the most productive veins on the Mother Lode. But he knew nothing about quartz, and moved on. Others located Carson Hill's best treasures two years later.[17]

Carson explored the area until October, returning to Monterey with some good specimens but eager to find more. The next spring he came back with a party of ninety-two men, including John W. Robinson, a coleader of the expedition. They explored the southern Sierra foothills near Mariposa, then worked their way north as far as Columbia with respectable results. But Carson had seen better diggings, and in May 1849 he returned to Carson Creek, along with Robinson and Stephen Mead. The latter two constructed a ferry across the Stanislaus at the mouth of the creek and set up a trading post.[18]

Carson's enthusiasm for prospecting waned with the onset of rheumatism, a common complaint among miners who often worked days on end in frigid water or wet clothes. In 1850 he left the diggings and formed a partnership to haul supplies between Stockton and Mariposa, but soon his partner defaulted and left Carson heavily in debt. He remained a prominent regional figure, however, telling his tale to the newspapers, publishing a book, and running successfully for the state legislature from Calaveras County in 1853. It turned out to be his last victory, for illness soon took his life.[19]

Another pair of prospectors from Stevenson's regiment were two brothers, Henry P. and George Angel. Natives of Rhode Island, they were at Weber's Creek when Carson arrived in August. In the southern exodus that followed the camp's abandonment, Carson and the Angel brothers traveled south together until they reached the Stanislaus watershed. Learning from Weber's experience, the Angels stopped where a minor tributary later called Dry Creek intersected a stream named for the two brothers. There they prospected and set up a primitive trading post to exchange Native goods for gold. Whether they worked for themselves or for Weber is not clear, but Angels Camp soon had more white prospectors than Indians. By the spring of 1849, the camp numbered about three hundred.[20]

The exodus from Weber's Creek also brought the Murphy brothers to the Calaveras region. Working to the east of Angels Camp along what they named Coyote Creek, they prospected for a few weeks at a place later known as "Murphy's Old Diggings." Eventually, it was renamed Vallecito. Sometime in the fall of 1848, John and his brother Daniel moved from Coyote Creek northward to more likely gravels in a higher basin drained by the upper arm of Angels Creek (soon locally known as Murphys Creek), which had been prospected earlier. Who was actually first on the scene is uncertain.[21] John M. Murphy, who had clerked for Weber both at San Jose and later at Weber's Creek, brought enough supplies from Weber's stock to set up a trading post at what was originally called "Stoutenberg," then "Murphy's New Diggings," and ultimately "Murphys Camp" or simply "Murphys."[22] As the Murphy party dug down four feet or more to the concentrations of alluvial gold along the limestone bedrock, they opened a placer field of extraordinary richness.

News of the new diggings at Murphys spread quickly in the fall of 1848. To accommodate newcomers and spread the wealth, a miner's meeting in 1849 drafted rules limiting each man to a single placer claim of eight feet per side.[23] Miners made money despite the restriction: The average take in 1849 was three thousand dollars per claim, with some claims yielding sixteen ounces to the pan in a day when one or two ounces was considered an acceptable yield. Some of the first comers became quite wealthy. John M. Murphy, it was alleged, took out between $1.5 and $2 million in gold dust after only two mining seasons. On his last pack trip out, he carried "as much gold as six mules could haul from the camp." In October 1848, Walter Colton found Murphy's tent stocked with trade goods that he exchanged with local Indians "who gather gold for him," and perhaps also for his supplier, Captain Weber. The tent also contained "refreshments that would have graced a scene less wild than this." Contemporaries claimed Murphy's marriage of convenience to a local chief's daughter provided him with plenty of Native labor to work his claims. But unlike less scrupulous traders, Murphy "allowed no liquors in the camp" to go to the Natives. White and Mexican miners, on the other hand, had no shortage of alcohol in Murphys.[24]

Hard drinking matched the hard work confronting the Argonauts at Murphys. Placer operations on the gravelly flats were especially difficult and dangerous. To reach pay dirt, miners sunk "coyote" holes, vertical shafts to bedrock that varied from five to forty feet below the surface, trying to avoid falling debris and cave-ins while in the process. Shallow holes were generally untimbered, but sensible miners in deeper holes timbered the shaft before drifting outward in a radial pattern to locate the pay streak. Using buckets attached to horse- or hand-powered windlasses, they hoisted auriferous gravel from the bottom and washed it through sluices on the surface.[25]

As they found out in the rainy season, the flat lacked good drainage. Jacob Bachman, who arrived in Murphys early in February 1850 with the John Woodhouse Audubon party, prospected the area but had "very little success, the large flat being too wet to dig as far down as necessary to get at the gold." Audubon left in March, "bitterly disappointed," but Bachman stayed on until August, still waiting for the water to recede. Heavy digging was safest in the dry months when caving was less likely. Yet, that time of year was also bad for placer pioneers because the surface water was insufficient for sluicing their gravel. "I think we can very easily make 8 or 10 hundred dollars apiece," wrote one argonaut near San Andreas in November 1850. "[W]e shall go to throwing up dirt soon as there is no water to wash with now in the ravines."[26]

Altering nature for human benefit was an unalloyed good in nineteenth-century American thinking. At Murphys, miners demonstrated their cultural conventions as well as their engineering skills in the 1850s. To extend the season for sluice boxes, in 1852–1853 the Union Water Company completed a flume that brought water from the Stanislaus River, fifteen miles away. The groundwater problem took longer to solve, however. Some of the early miners, like Bachman, gave up and left in frustration. Others resorted to ingenuity and hard work to come up with a solution. As early as 1850, Murphys miners tried draining the flat by digging a three hundred–yard trench nearly eighteen feet deep. It took nearly ten years and several later reorganizations to complete what eventually became known as the Deep Cut, or Murphys' Bedrock Flume. Ultimately reaching a depth of thirty-seven feet in some sections and a length of four thousand feet, the ditch opened in time for the 1860 mining season. Draining the basin gave miners a chance to rework their shallow claims and send the tailings downstream toward Angels Camp.[27]

Despite the difficulties and the overcrowding, Murphy's New Diggings prospered. As late as 1852, miners were reportedly averaging twenty to fifty dollars a day. By that time the population had mushroomed to five thousand.[28]

Other Calaveras camps also boomed in this pioneer period. At Vallecito, which the Murphy brothers had briefly explored but abandoned, others found paying placer ground that in one week during the early 1850s produced sixteen hundred ounces of gold. At its height the camp population reached four thousand. Upstream from Vallecito, Douglas Flat spurted to life after another of the Stevenson's regiment prospectors panned coarse gold there in the summer of 1848.[29] Wade's Flat, Humbug Hill, and many other satellite camps opened in the wake of these pioneer discoveries.

On the Mokelumne River, surface placers attracted moderate attention after Weber's spring discoveries, although real excitement did not develop until a former soldier from Stevenson's regiment unearthed a twenty-five-pound nugget near Big Bar. This may have been the same "lump of gold" a dyspeptic Scotsman told Walter Colton about in Monterey on September 9: It was "as big as his double fist."[30] A rush to the Mokelumne inaugurated a number of river camps, including Lancha Plana, Poverty Bar, Lower Rich Gulch, Middle Bar, Big Bar, and the bars along the upper Mokelumne River. By 1849 the Mokelumne was booming. Bayard Taylor, sent by Horace Greeley to describe the Gold Rush for *New York Tribune* readers, rode a mule in August from Stockton to a hill overlooking the river and saw before him a valley "dotted with the tents of the gold-hunters, whom we could see burrowing along the water." The next day, touring the diggings at Lower Bar, he watched a company of ten men complete a diverting canal that exposed about twenty yards of riverbed. They proceeded to mine the claim with shovels, a "rude cradle," and bateas. One man attacked a likely spot with his knife, "scraping up the sand from the bed . . . and throwing it into a basin, the bottom of which glittered with gold. Every knifeful brought out a quantity of grains and scales, some of which were as large as the fingernail." By noon they had recovered "nearly six pounds."[31]

On higher ground not far from the river was Mokelumne Hill, which had started by the fall of 1848 as a modest dry placer camp. It mushroomed in importance in November, after a Frenchman, working along a likely crevice with a pocketknife, recovered a nugget worth two thousand dollars. The rush of '49ers rapidly depleted the surface streams and gulches, however. Byron McKinstry, arriving in October 1850, struggled a year and a half with pick, shovel, rocker, and Long Tom, producing a daily wage that, discounting inflation, barely matched what he could have earned back in Illinois. His journal entries record a dreary daily regimen of hard work periodically interrupted by illness, claim disputes, and ethnic violence. Occasional bull and bear fights provided some diversion, but not enough to relieve his growing loneliness and desire to return home.[32]

By the early 1850s, as the nature of the Tertiary gravels that made up the richest diggings there was better understood, production boomed at nearby

Nigger Hill, Stockton Hill, and Corral Flat. These discoveries attracted thousands of miners. An estimated fifteen thousand populated the region by 1853.[33]

Mexican miners were the first to prospect the ravines and gulches along the North Fork of the Calaveras River. In the winter of 1848 they located good placer diggings in what became known as San Andreas Gulch, and established a camp with the same name in honor of Saint Andrew, their patron saint. Within a year, a sizable Hispanic community had risen on the hill above the gulch, including a number of Chileans and Mexicans driven out of the richer mines farther north. American miners at first did not consider the San Andreas diggings "worthy of any great note," but the great rush of 1849 brought new pressures on old placer ground.[34]

In October of that year, a party of Iowans staked claims on the main gulch and built winter cabins on the North Fork just below the confluence of Murray Creek. John Hovey counted about fifty Americans in the vicinity when he arrived early in November, including one familiar scout: "We found one camp occupied by Kit Carson and about twenty Sons of the Ocean at work with him all in fine spirits." The average daily take was between eight and sixteen dollars, but Hovey's claim was so poor he soon relocated near Double Springs. A month later, the Iowa Log Cabins he left behind was the scene of a bloody skirmish in the so-called Chilean War, discussed in chapter 3.[35]

By the fall of 1850, when eighteen-year-old Lucius Fairchild arrived with a company of Wisconsin gold diggers, surface gravels in and around San Andreas had been nearly worked out. Fairchild worked three months in very dry conditions, piling dirt for later sluicing. When the rains finally came, the best he and his partners could do was about five dollars a day. Then for a time his luck changed. Early in 1851, he found a new partner with more experience and a better claim. They hired three hard-up miners for two months at five dollars per day and made "good wages off of thier [sic] work, besides what we make ourselves." Every dollar he saved above expenses he placed into an "old bag" and repeated words echoed throughout the diggings: "*one more step nearer home.*"[36]

In the 1850s, with the surface placers in decline, San Andreas witnessed a Tertiary gravel boom. It began in the spring of 1852, when dry diggings were discovered on Gold Hill, a mile west of town on a segment of the main Tertiary channel. Two years later Orson Murray discovered the upstream segment of this channel under Douglas Hill. With shafts and tunnels, gravel miners followed it right through the main part of town. By 1859 there were some eighty gravel mines at work, some right in the business district.[37]

These sporadic mining booms spurred town growth, but economic stability remained elusive until the community diversified as a trade and transportation center. Voted the county seat in 1863, San Andreas has outlived many of the more spectacular Gold Rush camps.

## Geology of Gold Placer Deposits

A gold placer is essentially a river- or streambed of sand, gravel, and silt that contains particles of native gold. Because of its high density, placer gold is about sixteen to nineteen times heavier than water, and five or six times heavier than the material in the stream gravels. Pure gold has a specific gravity of 19.3, but because the gold found in the Sierra Nevada placers is a natural amalgam of gold and silver, the actual specific gravity is less, as low as 16. Pieces of sand and gravel in the streambed have a specific gravity of 2.7 to 3.0. Coarser gold "nuggets" are generally found at the bottom of the channel, often right on bedrock. Finer grains, referred to as "flour gold" or "gold dust," can be found more widely distributed in the stream gravels but concentrated in the lower part of the bed.

Sierra placer gold comes from quartz veins in bedrock. Disintegration and erosion of the bedrock over geologic time liberate the gold, which then travels downslope under the forces of gravity and running water. Because of its much greater density, gold works down through streambeds toward the bedrock. The bigger the pieces of gold, the slower they move. Very fine pieces can travel downstream for many miles, but big nuggets are found relatively close to where they were eroded from the vein outcrops. As pieces of gold move downslope and downstream, they are rounded and flattened by the continual pounding and working of stream gravels. The high malleability of gold affects the shaping of nuggets and smaller grains. Smaller particles are flattened more than larger ones as they travel farther downstream and are subjected to more stream action. As gold is nearly indestructible, and because a river is a very efficient environment for concentration, essentially all gold weathered out of veins remains in streambed gravels until recovered by placer miners.

Gold tends to become concentrated in pay streaks in the lowest part of a streambed. However, in a meandering stream the pay streak may be splayed out over the upper inner part of a meander curve rather than in the deeper outside part. Placer deposits are also richer in the flatter portions of a stream profile, where the forces of river action are relatively weaker, than in the steeper reaches, where they are stronger.

The original gold-quartz veins were emplaced in the bedrock of the ancestral Sierra Nevada about 130 million years ago (at the end of the Jurassic period of the Mesozoic era). Then followed some 65 million years (the Cretaceous period) of intense erosion and degradation, reducing the great ancestral range to a mature or even old-age topography. Cretaceous rivers and their many tributaries were very efficient. They gathered essentially all the gold that had weathered out of the veins, and concentrated it into pay streaks. By the beginning of the Tertiary period of a new geologic era (the Cenozoic), renewed geologic processes were at work on the Sierra Nevada block. Mild but repeated periods of erosion and then sedimentation were coupled with an enormous amount of volcanic debris that was spewed out over the entire Sierra Nevada range. For 60 million years or so, some of the gold placer deposits were eroded out of their original stream gravels and reconcentrated in later streambeds. Thick deposits of volcanic and other debris in turn often covered these beds.

Within the last several million years the present great Sierra Nevada range was created by strong and repeated uplifts of the eastern side of this structural block, accompanied by the rejuvenation of deep erosion that had not occurred since far back in the Cretaceous period. This new uplifting and tilting caused the cutting of the spectacular canyons we see today and the final sculpturing of the Sierra landscape, with its innumerable ravines tributary to the major streams. During the deep erosion many portions of the old Tertiary auriferous gravel channels were cut out and their gold redistributed into the local streams and into the riverbeds, creating an entirely new generation of gold placers. It was this latest generation of gold placers in the Stanislaus, Mokelumne, and Calaveras Rivers that was discovered in 1848 and mined in the early years of the Gold Rush. Not until the early 1850s were the rich remnants of these Tertiary placers identified. Curiously, a few remnants, such as those in the Murphys area and near Mokelumne Hill and Chili Gulch, were actually exposed in the upper reaches of small tributaries, and were actively mined before anyone realized their true significance.

## Mining the Surface: Early Placer Tools and Techniques

Mining is a technology rooted in antiquity and carried forward into modern times with remarkable continuity and tradition. Gold Rush argonauts used techniques and methods Jason and his mythical Greeks would have understood. To separate gold from barren material, placer and lode miners in California and the West for fifty years after 1848 followed the same principles of gravity separation used in ancient Egypt. Woodcuts in Agricola's sixteenth-century treatise, *De Re Metallica,* illustrate many of the basic tools that later reappear in mid-nineteenth-century California placer diggings and lode gold mills. Contemporary Gold Rush literature too often credited "Yankee ingenuity" for the development of methods and tools known centuries before.

National pride is behind some of this bombast, but ignorance of history was perhaps just as important.

So extensive were surface placer deposits in Gold Rush California that diggers at first used the same tools found on any nineteenth-century farm or construction site: picks, shovels, crowbars, and knives. "The first thinking anyone in California does when he is someplace new is to scrape the earth around a bit and see if anything appears that might look like gold," wrote a Chilean '49er. "This is exactly what I did. I took out my knife and made a small hole, and in a second I saw two or three bits of gold." Picks and shovels were, of course, universal tools dating to at least the Middle Ages. But pragmatic miners soon needed more specialized implements, adapted to particular jobs. By the early 1850s, California began to import or manufacture digging or trenching tools especially built for mining, and as the industry advanced so did the specialties. Within a decade one San Francisco machine shop alone was turning out more than thirty different sizes and varieties of mining picks. One was the poll pick, a Cornish import with a pick on one end and a hammer on the other, popular with both placer and "hardrock" miners, or those working vein or lode deposits. In Calaveras, Wade Johnston claimed credit for introducing in 1855 the "Wisconsin drifting pick," an improved flat-eyed pick.[38] For the most part, Gold Rush miners in Calaveras and the rest of California were not innovators but imitators, improving upon but not inventing mining techniques or tools that were already in use by miners of Hispanic or European origin.

Many placer gold–washing techniques used in the Gold Rush had been brought to the New World by the Spanish and had been in use for nearly three centuries in Mexico and South America.[39] The miner's pan may have developed directly from the wooden "batea" long familiar to Hispanic miners, although Isaac Humphrey, an experienced Georgia miner, panned gold near Sutter's Mill—probably with an iron frying pan—when he arrived in March 1848, months ahead of the Hispanic influx. Americans used bateas or even baskets woven by Native Americans in the early months of the Gold Rush, but preferred the sheet-metal pan when it began to appear in large numbers by the late summer and fall of 1848.[40] The pan method rapidly gave way to the "rocker," an ancient washing machine also called a "cradle" because of its resemblance to an infant's bed. It allowed a miner to wash a much larger amount of dirt than he could with just a pan. Theodore H. Hittell, California's first historian, claimed that Humphrey also introduced the rocker to California in 1848, though Mexicans may have been the first to use rockers and quicksilver in Calaveras County. Walter Colton found them in use when he visited Murphys in October of that year.[41] The origins of the rocker are

Miner with Cornish poll pick. (Courtesy Holt-Atherton Library, University of the Pacific)

obscure, but it may have come from China, a nation with more than a thousand years of gold mining experience. For expediency as well as for economic reasons, Chinese miners clung to their rockers long after others had abandoned the machine. It was inexpensive and easy to construct, could be built on site if necessary, and in emergencies could be dismantled and packed out by

Ah See Wahn on Latimer's Gulch near North Branch, a tributary of the North Fork
of the Calaveras River. (Courtesy Calaveras County Historical Society)

miners who often had to move quickly because of the anti-Chinese terrorism
practiced by Americans as competition increased after 1850.

Where water was inadequate for the simplest washing machines and methods, California miners often resorted to winnowing, a technique adapted
from Hispanic gold seekers brought north by the discovery news. Winnowing had been in use south of Mexico for two hundred or more years, and in
southern California at least since 1842, when Californios had discovered the
placer deposits at Placerita Canyon. In the southern Mother Lode in '48 and
'49, Hispanics used dry-washing methods extensively. If water was scarce, as
in the late summer and fall along the gulches and ravines of the central California foothills, miners hand-sorted auriferous gravel, discarded the coarse,
piled the fines on a blanket, and tossed them repeatedly in a breeze. That got
rid of the dust as well as moved the gold downward through the barren material. Finer winnowing required lung power, as Walter Colton witnessed at
Mokelumne Hill in the late summer of 1849. Observing a party of Mexicans
from Sonora coyoting near the crest of a gulch, he said they took out "lumps
large enough to pay for the labor," then turned to the "refuse sand" that was

Dry washing or winnowing. (William V. Wells, "How We Get Gold in California,"
*Harper's Weekly Magazine* [1860])

"full of fine gold, and only needed washing out." After blanket winnowing
and handpicking to remove the bulk of the barren dirt and rock, they poured
the residue into a batea:

> [T]hen, balancing the bowl on one hand, by a quick, dexterous mo-
> tion of the other they cause it to revolve, at the same time throwing
> its contents into the air and catching them as they fall. In this man-
> ner everything is finally winnowed away except the heavier grains of
> sand mixed with gold, which is carefully separated by the breath. It is
> a laborious occupation, and one which, fortunately, the American dig-
> gers have not attempted. This breathing the fine dust from day to day,
> under a more than torrid sun, would soon impair the strongest lungs.[42]

Captain Joseph L. Folsom in 1848 observed winnowers at work on the
American River, but thought air separation highly inefficient. "Much of the
finest of the gold is . . . blown off with the sand and lost," he wrote. But such
losses did not bother the earliest California gold diggers, who sought only the
visible nuggets. In late 1848, Walter Colton watched a Sonoran woman, proba-

bly on the Calaveras, who washed a bowl, found only about half a dollar's worth of gold, then "hurled it back again into the water, and straightening herself up to her full height, strode off with the indignant air of one who feels himself insulted."[43]

The great rush of 1849 crowded the goldfields and accelerated the pace of technological change. With the richest surface placers already claimed, '49ers had to work lower-grade deposits and needed more refined technology than their predecessors to make money at gold mining. As Philip May has argued, the motive behind technological advance was not to make placer mining techniques more efficient, but to increase the volume of production. Hence the need for improved gold-washing devices.[44]

Using "quicksilver," or mercury, to capture fine gold was an ancient technology, but expensive and little used in Mother Lode placer camps before the 1850s. Unaware of the values in microscopic "fines," most early argonauts looked for coarse, visible gold. But the rush of '49, and the rapid depletion of surface deposits, provided new incentives for efficiency. The result was expanded use of quicksilver after 1849, even though it cost four dollars a pound or more until the price dropped in the 1850s.[45] Though use of the liquid metal increased recovery in a riffle box, a significant part of the values still washed away in tailings, along with some of the quicksilver. Cleanup amounted to shutting off the water supply, pulling the riffles and scraping the amalgam into a buckskin bag or chamois skin, and squeezing out the excess quicksilver. The final step was to heat the remainder in a retort—or on the end of a shovel if no retort was available. This boiled off the quicksilver, leaving a gold sponge that was melted into bars or ingots. A British observer, watching this procedure in 1852, thought it "must have caused considerable waste," but frugal miners recovered and reused as much quicksilver as possible.[46]

Gold Rush miners, of course, did not understand the long-term problems associated with mercury use, although they learned quickly to avoid breathing mercury fumes. The "single-minded pursuit of wealth and its consequences" did not worry many argonauts. They rationalized greed as both patriotic and beneficial, their version of "winning the West."[47] Staking a claim and working it successfully required a willingness to take risks. It also required ingenuity and pragmatic adaptation of old technologies whenever they promised better results. Thus, from the rocker evolved the "Long Tom," a much more powerful tool that required more water but could keep a company of several miners busy washing four or five times the volume of gravel a rocker turned out.[48] By the early 1850s the rocker and the Long Tom were both giving way to the sluice box as the most widely used method of gold washing, at least among Euro-American miners. Sluice technology was ancient as well as simple, consisting of a long series of wooden troughs with riffles, designed to fit together end upon end downslope for thorough washing. The sluice was not used much in California prior to 1850 primarily because its massive washing power was not needed to produce a respectable return. With a good head of water, only two or three miners could operate a string of boxes.[49] With the advent of mass mining, the sluice became the washing instrument of choice for large placer companies where water and topography made hydraulic mining possible. As will be seen later, water diversion became a major ancillary industry that laid the foundations of California's future growth long after mining receded.

Water was supplied to sluice boxes by short ditches from upstream diversions. "Dry diggings" required longer diversions through ditches generally controlled by water companies. With the advent of sluicing, miners now had to develop earth-moving innovations to keep pace with the washing machines and methods. If adequate water was available, particularly in dry diggings, miners often resorted to a practice called "ground sluicing" to wash larger areas without having to dig the dirt and shovel it into sluice boxes. By this method, water washed topsoil and pay dirt downslope where it was channeled into large sluices that were often simply ditches dug into bedrock and lined with riffles. This method could be used where the ground was not rich enough to justify hand shoveling. A variation was called "booming" or "gouging," where a sudden release of large volumes of water washed ground containing boulders or dirt that would not respond to normal ground sluicing.

The 1850s also saw the expansion of river mining, the technique of diverting the flow so that the streambed could be mined down to bedrock. This was done in a variety of ways, generally with flumes or by diking. When the topography was favorable, sometimes tunnels were driven to divert the stream past large bends. Current wheels, mounted in the flumes, were often used to power pumps for dewatering the workings. Masts and booms or derricks were rigged to lift out boulders and pile them on the banks. Smaller boulders and cobbles were stacked in walls along the stream edge, and overburden piled behind the walls or washed downstream. The water level and the amount of flow in the stream were thus very important. In the wintertime miners worked the higher bars and benches, and in the summer they placered the deeper parts of the streambed, especially the bedrock bottom. This type of labor-intensive mining was particularly suited to Chinese companies who took over many failed or abandoned Euro-American efforts.[50]

Frequently, miners found the accumulation of sand and gravel on the bars and benches quite deep, giving rise to "coyote mining" or "coyoting," by which a shaft was sunk to bedrock and drifts were driven outward from the shaft on the richer gravel. The rewards were uncertain, but a few were lucky. John Steele, a Wisconsin immigrant with lead-mining experience, helped a coyoter sink a forty-foot shaft near Nevada City. Within three weeks they hit bedrock, with gratifying results. "Many of the [first] tubs of gravel," he exclaimed, "with a capacity of less than half a barrel, contained over one thousand dollars worth of gold."[51]

Fortunate were the few, but the risks were great. Many a miner lost his life to a caving coyote hole. Steele faced death in another deep mine on Coyote Hill in Nevada County in the fall of 1850. Working at the bottom of a sixty-foot shaft with five other miners, loose gravel at the collar began to give way. The men barely escaped by protecting themselves under the windlass bucket, then worked their way up over the debris. The next day they cleared the collar of loose boulders and descended to find the shaft timbers had "leaned a little towards a worked out and abandoned mine, which had caved in; the immense weight of earth was indicated by the heavy timbers being bent over the posts in the form of ox yokes; and the great mass of earth overhead had settled eight or ten inches." At Owlsboro in the same decade, Leonard Noyes barely escaped death at the bottom of a shaft when his partner accidentally kicked the bucket back down the hole without hooking it to the windlass. As another observer dryly remarked while watching coyoters at work: "It is rather dangerous looking to one not accustomed to working under ground." Emphasizing the risks as well as the frustrations, he concluded: "A man might dig wells for six months and never see the color of gold."[52]

The gulches and ravines that dried up in summer had to be worked in winter when water was available. Some miners built reservoirs in ravines upstream from their claims or on hillsides close by with ditches to collect the immediate runoff during cloudbursts and heavy rains, then used the water until it was gone and waited for the next storm. To extend mining beyond the rainy season, others constructed reservoirs on hilltops and filled them with ditch water flumed in over trestles.

The older gold districts throughout Calaveras still clearly show the results of these mining methods. To an experienced eye it takes little imagination to visualize argonauts hard at work. Large trees are now growing in the "diggings," and the flumes and sluices have long since rotted away. But the piles of cobbles are there, and the ditches and banks, too, although they are sloughed and overgrown.

The success of surface placer mining was limited by the shortage of development capital, the primitive state of geologic and geographic knowledge, the rather elementary technology discussed above, and the lack of control over the water supply. Small mining companies were particularly handicapped, as

Leonard Kip learned after a week of hard work near Mokelumne Hill in 1850. His party of six earned just enough the first week to cover expenses. After another week of poor returns, with provisions depleted and dysentery breaking out, the company gathered for a final meeting: "The decisive vote was passed; the chairman jumped up, kicking the wash-pan over the tent and dissolved the meeting; and so ended all our experience in mining."[53]

While some companies quit early, others hung on despite the odds against them. Most of the early technical problems were overcome to a surprising degree by brute strength, sheer determination, stubbornness, and, above all, the ever present golden hope of a quick fortune. Unschooled but observant miners soon became very adept at locating pay dirt. Those who lost hope and gave up left the diggings to the better qualified. Doubtless some discoveries came by accident, although romantic legends such as the story of a drunken prospector discovering Rich Gulch by rolling down it and landing on pay dirt, or the yarn that greenhorn black miners sent to high ground as a joke discovered the rich Tertiary gravels of Nigger Hill, must be used with caution if not discounted altogether.[54] Since placer gold could at first be found in the majority of foothill streams, gulches, and ravines, neither experience nor luck was really necessary before 1851, and even for a few years after that. Indeed, even while new placer technology was making pick-and-pan mining old-fashioned if not obsolete, individual placer operators continued to produce well. The aggregate annual total for California placer mining between 1851 and 1854 was 3.5 to 4 million ounces.

The adventures of Josiah Foster Flagg, a Pennsylvania dentist who caught "gold fever" in 1849, provides a typical example of both the simplicity of early placer mining and the gullibility of the miners. Arriving in San Francisco on December 18, 1849, by the Cape Horn route, Flagg and his companions reached "Woods Diggings" in Tuolumne County by mule and began work April 1, 1850, with immediate results:

> Chester seized a pick and assaulted the bowels of the earth bringing to light a piece of gold weighing about 1.50 oz. You should have seen us four miners, but you probably, had you been there, would have only noticed the eight eyes, as this realization of the truth of Gold Digging bust upon us—we stood "aghast" and passed around the treasure in solemn silence—but altho' we all worked as tho' impelled by an engine of one horse power each, we found no more "big hunks" on that day.[55]

Flagg mined intermittently for the next year in both Tuolumne and Calaveras, but had to supplement his income by practicing dentistry on the side. By 1851 he had earned twelve hundred dollars, but paid out seven hundred dollars in expenses. Soon he gave up mining and turned to farming along the Mokelumne and then merchandising in Sacramento—all with little success. In 1855, chastened by his experiences, he returned to Pennsylvania to pick up the pieces of his dental practice.[56]

## Mining the Deeper Placers: Drift and Hydraulic Mining

In the earliest days of California mining, argonauts worked only the placers in the streambeds and bars of the present-day river systems and their tributaries.[57] Locating these surface deposits was easy, as one '49er explained in a letter back home: "The mere finding of gold mines here is nothing, but the labor of getting it is the grand item, therefore scientific geologists are worthless and laughed at, many are desponding and going off and many more wish themselves away. . . . On the other hand thousands are in high spirits and are making money fast."[58] By the very early fifties, however, it was becoming apparent to many miners that they were dealing with two types or rather ages of placer deposits. As the rivers and ravines became overcrowded, the more adventurous gold seekers spread out and soon found "dry diggings" in many different settings.

The Murphy's New Diggings early in 1848, and the discoveries at Mokelumne Hill later that year, could hardly be explained by the classical placer theory of origin. These could have resulted only from previously formed placers of an earlier geologic epoch that had been mysteriously exhumed by processes of nature. By 1851 astute miners, helped no doubt by more educated curious visitors, were distinguishing between the present-day (Quaternary age) placers and the auriferous gravels of ancient or fossil rivers (Tertiary age). The full extent and detail of these Tertiary stream systems took years to work out, but the immediate effect of their discoveries, often on ridges and hilltops, was spectacular. In contrast to present-day stream placers, which were rapidly being depleted under the intensive attack of hordes of miners, the new "dry diggings" required a more organized approach, often with slower but richer returns.

As the new discoveries accumulated and new methods of mining were designed to develop them, the patterns of these ancient rivers began to unfold. In some places all that was found of the Tertiary channels were occasional remnants on ridges and hilltops, literally right on top of the ground. Other deposits, discovered on the slopes, proved to be river channels "entering" or "breaking out" of the solid ground. If upon digging into these channels they were found to be going downgrade, the channels were entering, and if going upgrade, they were breaking out. The channels were often buried by hundreds of feet of gravels, sands, tuffs, indurated mud flows, and other rock types that soon visiting geologists identified as being of volcanic origin. In many places the old stream channels had been "cut out" by modern streams that contained much richer placers at those points (and just downstream) than elsewhere. They had mined out the Tertiary gold and reconcentrated it in a second generation of placers in the river bars and benches.

The origin of the rich Murphy's New Diggings may be explained as a portion of a Tertiary stream that had cut across the limestone bedrock, making an unusually fine trap to capture large amounts of gold nuggets. Although it had once been deeply buried, modern erosion by the vigorous headwaters of Coyote and Angels Creeks had stripped off the cover of later sedimentary debris right down to the Eocene placer, readily accessible to the argonauts. Similarly, many otherwise isolated and inexplicable "dry placers" were vestiges of eroded Tertiary channels, leaving only a "vestigial" or "ghost" placer, where the gold was too heavy and indestructible to have been completely removed or washed down into the lower drainages. There were a number of ghost placers mined in Calaveras County, and from their locations many details of the Tertiary drainage can be reconstructed.

One of the largest of the late Cretaceous and early Tertiary stream systems of the Sierra Nevada flowed down through lower Calaveras and into the Ione or Western Sea that occupied the present site of the Great Valley floor and the Coast Range. Only Nevada County had a greater stream, the mighty Tertiary Yuba. The tributaries of the large Calaveras system drained much of Amador, Tuolumne, and Alpine Counties in addition to all of Calaveras. The spectacular placers of Columbia and Shaws Flat in Tuolumne County, as well as most of the Tertiary placers of Amador, belonged to this system.

The two main forks of the Tertiary Calaveras River traversed much of the county's Mother Lode region before they converged below San Andreas at Central Hill (not to be confused with Central Hill just south of Murphys) and then flowed westward to the sea. These two forks and their tributaries enriched themselves as they accumulated gold from the outcropping Mother Lode veins, adding this to the extensive amounts of gold they had already garnered from the East Belt veins.

Further complicating the early Tertiary auriferous gravel channels was their additional erosion and cutting by the Miocene river system (named by geologists the Valley Springs system) and later in the Pliocene epoch by the Mehrten River system. These later stream gravels generally contained gold only where they cut out the Eocene channel deposits. The final sculpturing of the landscape occurred in post-Mehrten time (within the past few million years), extending right up to the present, during the great canyon-cutting period that produced the Mokelumne and Stanislaus canyons, and reduced the ancient Calaveras River to its very minor condition today.[59]

### Drift Mining

Recognition of the nature of Tertiary placers soon led to new methods of exploitation. The exposed gravels could be worked like present-day placers when water was made available to these "dry diggings" by using sluice-box and ground-sluicing techniques, as long as there was little overburden. But many of these remnants of Tertiary placers had literally mountains of overburden or, at best, many feet of overlying gravels that were lean in gold values. In this case, to follow the pay streak very close to bedrock, miners had to drift in from the outcrop.

Operators opened a drift mine either through a tunnel (hence "tunnel" mine) or through a vertical or inclined shaft, depending upon topographic conditions. A shaft mine had to be pumped to keep the workings free of water. A tunnel or drift mine, if it was developing upstream in a Tertiary channel, was self-draining, but a mine developed in the downgrade direction of a channel had to be pumped unless topographic conditions permitted an upgrade tunnel to be driven underneath the channel and the gravel reached by "raises" from the tunnel. In some cases it was necessary to drive long crosscut tunnels instead of a shaft to reach a buried channel.

Once into the channel, drifts and crosscuts explored and developed the gravel. That material rich enough to mine was "breasted" out, or, as we say today in underground mines, "stoped," and transported to the surface for washing. This was all handwork, mostly by pick and shovel, far more expensive than surface placering. Miners were hired as "pickmen." Tramming the gravel out to the surface was accomplished in mine cars running on wooden rails capped by iron straps. In the breasts, however, much of the "muck" was moved by wheelbarrow to where it could be loaded into mine cars, then "propelled outwardly easily by man power," as the *Miner's Own Book* described it in 1858.[60]

The main entries and drifts had to be timbered for ground support, but often the gravel would stand well enough in the breasts to allow the miners to excavate the pay dirt with almost no timbering. Sometimes the coarser gravels were forked out and used for "gobbing" or backfilling the breasts, along with other mine waste material. Ventilation was generally provided by an occasional "raise" to the surface to create air circulation. With sufficient water, a downdraft of fresh air was created by a "water blast" by dropping water down the raise.

Development of drift mining in Calaveras began as early as 1851, when newspapers reported that miners at Mokelumne Hill had reached bedrock by sinking a shaft 150 feet below the surface to search for a channel. However, the practice of mining ancient channels by drifting along the bedrock evidently did not become widespread in Calaveras until after the Murray company introduced the technique at Douglas Hill just southeast of San Andreas in 1855. Within three years the San Andreas area had more than eighty drift gravel mines.[61]

The most extensive development of Calaveras drift mining came in the 1860s and 1870s. The Stockton Ridge channel at Mokelumne Hill was one of the first to be actively opened. Initially developed by shallow shafts on the south side of the ridge, almost in town, it was then reached farther downstream, under Stockton Ridge, by tunnels driven in from upper Chili Gulch. As more channels were recognized, they were opened up in Happy Valley and in Upper Spring Gulch in the late 1850s. The extension of drift mining to Chili Gulch was already begun by late 1860. Channels were explored in Old Woman's Gulch the next summer.[62] Rapid development of these complex Tertiary channels in the Mokelumne Hill–Chili Gulch region soon eclipsed the activity around San Andreas, and kept this district busy for nearly twenty years.

In the pioneer period, extensive drift mining was concentrated in the San Andreas and Mokelumne region. Development of other Tertiary channel deposits in the county proceeded more slowly because of the more isolated exposures of these gravels and because in many places they either had been totally removed by later erosion or were still buried by thick late-Tertiary cover. The Central Hill and Fort Mountain channels were opened up at a few localities, such as the Central Hill mine above Murphys, the San Domingo (Jupiter) at Dogtown, the McElroy near Altaville, the Wild Goose above Vallecito, and at more isolated localities near Mountain Ranch and Railroad Flat.

All of these early developments did not lead to sustained activity, however. By 1880, the California Mining Bureau concluded that drift mining in Calaveras had declined to near extinction. Later there was considerable revival of interest in drift mining in the Vallecito area and elsewhere, which resulted in a large amount of shaft mining, activity that continued well into the first half of the twentieth century. The richer and more easily mined channels had been worked out before 1880, however, and the days of cheap labor and intolerable working conditions were gone. Many of the later operators attempted to work in the intervolcanic channels or in channels that had been reworked by the later Tertiary streams that had disseminated the gold and made the gravel much lower in grade.[63]

In the twentieth century, statements by observers such as C. E. Julihn and F. W. Horton (1938) that only half of the Tertiary channels had been mined out and that those remaining "undoubtedly contain many millions of dollars in gold" were obviously made in the enthusiasm of the moment. More pragmatic persons doubt that it is feasible any longer to mine the remaining auriferous gravels by underground methods with the present-day labor and environmental constraints.

*Hydraulic Mining*

Early surface placering on the "outcrops" of the Tertiary auriferous gravel channels and the follow-up with drift or tunnel mining demonstrated that the ancient placers were often very rich. Although drift miners found and efficiently mined out the pay streaks for some distance from the outcrop, it was not profitable to continue to mine the gravel unless it had sufficient gold values to pay for underground operations. Large amounts of auriferous gravel had a much lower grade, but to mine them inexpensively required removal of enormous amounts of overburden. Hydraulic mining proved to be a very effective method for this work.

Breaking down high Tertiary gravel banks with large nozzles or "monitors" fed by very high–pressure water, and then washing the "muck" or "dirt" into large sluices, was a distinctly California contribution to mining technology in the best traditions of American initiative. As Philip May has observed, both the hose and nozzle were prior inventions, but Californians combined them for "a novel purpose." The result was "an imaginative transfer" that may qualify as an invention, since it had not been done before.[64]

Within six months after hydraulic mining began near Nevada City in March 1853, it had spread as far south as Mokelumne Hill. Here versatile Major Case introduced this method on his claim at the point where the Stockton Ridge channel entered the ridge on the southwest edge of town. Not until 1855 was it used with much success locally, however, and by 1857 there were already signs of decline. Later it was revived, but not extensively used until the 1870s or early 1880s.[65]

The first requirement for successful hydraulic mining was a good supply of water at a high head of pressure. At the zenith of hydraulicking, individual "monitors" operated under a head of 400 to 500 hundred feet at 175 to 220 pounds per square inch. They consumed up to 1,500 miners' inches, equal to approximately 17,500 gallons per minute. This enormous discharge delivered an unbelievable force, eating away thousands of tons of debris every day. Smaller pressure and volume meant less efficiency and poorer results. The water was collected in large ditch systems at higher elevations and then intro-

Hydraulicking near Mountain Ranch on the Cave City road in the 1880s. (Courtesy Calaveras County Historical Society)

duced into pressure pipes of riveted wrought-iron construction. We see pieces of this pipe lying around the old hydraulic diggings even today, a century and more later, rusting away.[66]

Second, the hydraulic method also depended on a very cleverly designed and constructed nozzle capable of delivering high-volume, high-pressure water in a concentrated stream that could be easily controlled by one operator.[67] Anyone fortunate enough to operate or even see such a piece of equipment in operation would readily agree that a monitor is truly a remarkable device. In recent years there was one still in operation near Santa Cruz, California, used for mining sand. In the 1880s these "giants" were seen throughout the Sierra gold districts.

The next requirement for a successful hydraulic operation was a means of recovering the gold in the auriferous gravel washed down by the monitor. In large pits this generally involved long bedrock tunnels into which the gravel was sluiced and then run downstream. In smaller pits deep ditches in the bedrock performed the same function. To capture the finer gold as amalgam, hydraulic miners added quicksilver in the upper sluices. The higher volume of processed material required much larger doses of quicksilver than in stream placering, although efficient operators tried to recapture and reuse as much of the liquid metal as possible. Nevertheless, quicksilver loss rose as its use increased in both placer and hardrock milling operations, a troubling legacy for western waters. California health officials have posted warnings on six reservoirs and streams where dangerous levels of mercury have been found. Scientists investigating the downstream impact of hydraulic mining at Dutch Flat have concluded that this one district alone released "many thousands of pounds" of mercury into the environment.[68]

Fourth, hydraulic mining needed a sizable Tertiary deposit that could provide sustained production over a long period. The tremendous efficiency of hydraulicking enabled miners to operate successfully in extremely low–grade deposits, as low as those containing only five cents to the yard of material (at the prevailing price of gold). The distribution of the gold, however, was rarely if ever uniform throughout the deposit. Often, much totally barren material had to be washed away to get at the underlying productive ground.

Finally, hydraulic mining required space downstream to accommodate the tailings of gravel, sand, and fines (or "slickens" as they were sometimes called), debris that eventually choked the drainage systems in the lower valleys, crippled river navigation, caused floods, and devastated farmlands. Miners at first ignored farmer complaints and paid little attention to the tailings. It took nearly thirty years before mounting opposition culminated in a series of famous court battles that pitted miner against farmer, merchant, shipper, and flood victim. Calaveras residents, as might be expected, sided with the miners in this controversy, although both hydraulicking and its termination had less economic impact locally than in the more important hydraulic districts.

California was the birthplace of hydraulic technology. Where geographic conditions were favorable, the industry spread throughout the West and abroad, often by experienced Californians who roamed widely after the Gold Rush. Within California, Nevada County was the center of the hydraulic industry. There the massive North Bloomfield pit at Malakoff State Park, a man-made canyon a mile long and 350 feet deep, where 40 million cubic yards of debris washed through hydraulic sluice boxes, can still be seen today. Nevada surpassed all other counties in both the quality of the Tertiary gravel and the amount of water available to work it. Calaveras County probably ranked about fifth in the production of gold by hydraulicking. The pits in Calaveras were generally small, but there were many of them.[69] A number of these contained only low banks, and the use of the monitor in such pits could be termed merely a special category of ground sluicing.

The most abundant local pits were in the Mokelumne Hill–Chili Gulch region. Several can still be seen from Highway 49. There were also a number from the Murphys area downstream to San Andreas on the Central Hill channel. Outlying districts upstream, such as Mountain Ranch and Railroad Flat, on the upper reaches of the Fort Mountain channel, had numerous small hydraulic pits. Downstream, hydraulicking was practiced near Campo Seco and Camanche and above Jenny Lind at Brushville. Even the main street of Murphys experienced the blast of hydraulic monitors in the 1860s.[70]

This type of placer mining reached its peak of development statewide in the late 1870s and early 1880s, when it was producing some $15 million a year. The *Sawyer* and *Field* court decisions in 1884 abruptly terminated most large-scale hydraulicking by requiring the impoundment of hydraulic debris.[71] The practice was revived briefly in the 1890s when the federal Caminetti Act allowed its continuation under the strict supervision of the Anti-Debris Commission, but it never reached its former scale and steadily declined to insignificance during the first decade or two of the twentieth century. Calaveras County was not hit nearly as hard as the bigger hydraulicking counties to the north, because in Calaveras the pits were smaller and the topography was more favorable to their continuation by constructing debris dams. However, the exhaustion of many of the deposits and the steadily increasing overburden at others, as the pits were advanced, inevitably put them out of business. A brief revival in the 1930s was short-lived.

* * *

The Gold Rush in Calaveras County and other parts of the Mother Lode was a haphazard affair mostly by raw amateurs with little knowledge of geology and no mining experience. Using tools that had changed little from Roman times, they shoveled, gouged, and creviced in gulches, stream banks, and bars, looking for shiny "nuggets" of a mineral most had never seen in its native state. The most likely result of these crude conditions and limited knowledge for many miners was failure, or at least very meager returns, and more than a small amount of discouragement. Everywhere they worked they left behind a degraded environment, but before the 1850s the lack of efficient tools for denuding the trees and ground cover, removing topsoil and gravel, and exposing bedrock limited the damage. Nature has reclaimed much of the old workings, so it is difficult today to determine just where the '49ers worked and how much destruction they caused. Hydraulic mining by the 1860s made all other forms of gold digging environmental child's play by comparison. But whatever the means or methods, waste, inefficiency, and environmental degradation were characteristic patterns of all early California mining. As Raymond F. Dasmann has observed, the Gold Rush ended any thought of rational planning and "gained for the miners unlimited license to create environmental havoc." Yet the promise of instant wealth was an overpowering motivating force in nineteenth-century America, where the average working wage for men hovered around $1 to $1.50 per day throughout the century. Equally important was what Stuart Udall has labeled the "myth of superabundance," the prevailing American belief that the nation's natural resources were virtually inexhaustible. Religion also played a role in explaining the fervent, even reckless, assault on the landscape by Americans in the late Jacksonian era. A Protestant nation had a "manifest destiny" to conquer the continent, chase out the Indians, defeat the Mexicans, exploit the resources, and convert the wilderness to farms and cities. As David Stiller has recently argued, early miners were also guided by raw "business instincts." If the federal government "placed no value on the land, why should they?" They were given the resource for almost nothing, paid no royalties and few taxes, and walked away without consequences "because it cost them less then repairing the damage. No one made them do otherwise." To not take advantage of the situation, in Stiller's opinion, would have made them "fools" in the eyes of their peers.[72]

Thus, Calaveras argonauts, like their counterparts swarming over the gulches and streams all along the Sierra foothills from Rich Bar to Coarsegold, participated in the march of progress. Not all benefited equally, however. Legal and cultural obstacles in the form of discriminatory laws and attitudes confronted some. Fires or mine closures or other calamities, natural or synthetic, defeated or discouraged others. Yet for most denizens of early Calaveras, progress, as defined by American cultural norms from that day to modern times, came in the form of towns, transportation, industrial development, jobs and rising living standards, and the growth of farms and ranches. With frenetic energy and massive movements of men and earth, the Gold Rush and its aftermath were inevitable consequences of the forces of modernism in the Western world.

# 2: The Beginnings of Lode Mining, 1850–1885

For the first couple of years after Marshall's discovery at Sutter's Mill, the argonauts centered their attention on placer gold and where to find it. Apparently, no one bothered to investigate Marshall's alleged observation before discovery day that he knew there was gold in the hills because he saw the "blossom of gold" in the white quartz outcrops. Americans in California had little or no experience with quartz, and so long as the placer bonanza made news, few took time to consider the source. Some of the early speculations were far-fetched. In the summer of 1848, Walter Colton, otherwise a sensible journalist, dismissed any efforts to find the origins of California's placer gold. Those who seek the source, he wrote, "might as well hunt the fleeting rainbow. The gold was thrown up from the bed of the ocean with the rocks and sands in which it is found; and still bears, where it has escaped the action of the elements, vivid traces of volcanic fire." Edward Buffum was not much closer to the mark in the fall of 1848. After two weeks of exploring and prospecting, he was "fully satisfied that at some early date in the world's history, by some tremendous volcanic eruption, or by a succession of them, gold, which was existing in the form of ore, mixed with quartz rock, was fused and separated from its surrounding substances, and scattered through every plain, hill, and valley, over an immense territory." The volcanic theory also appealed to a well-educated Chilean, Ramon Gil Navarro, who bought an unusually shaped nugget in Stockton after his arrival in 1849. He wanted it as "as proof of volcanic eruption. How else could something as strange and prodigious as this have been formed?"[1]

Even though geology was still a young discipline, trained scientists knew better. An English geologist, David Ansted, while the great trek of 1849 was just getting under way, published a treatise on California gold. Basing his opinions on the fieldwork of earlier observers, he described the erosive forces shaping the Sierra Nevada, and concluded that the foothill "alluvia" causing so much gold fever had its origins in the "altered schists, diorites and metamorphic limestones . . . [that] are traversed by auriferous veins of quartz often containing also much silver."[2] In short, if you wanted to find the source, look for outcropping quartz veins in the bedrock.

Those who looked for quartz in California had not far to seek. By the end of 1848 prospectors were pressing forward, working upstream along all the major tributaries of the Sacramento and San Joaquin, and most of the minor.

Quartz outcrop at Virginia Mine, Mariposa County. (Courtesy Holt-Atherton Library, University of the Pacific)

A favorite tool of these early mineral explorers was a gad, or short crowbar, handy for probing crevices and cracks in the fractured slates and schists of narrow ravines. Walter Colton wrote from a ravine near Murphys in October 1848 that the "sounds of the crowbar and pick, as they shake or shiver the rock, are echoed from a thousand cliffs." As the explorers worked their way upslope, they found numerous vein outcrops, but most of them were barren or "bull" quartz, creamy white and devoid of metallic mineralization. Yet there were enticing discoveries of "float," mineralized rock broken away from eroding veins, that led the prospectors on. Colton's party, camped in an unidentified Calaveras canyon, found "a boulder of trap and quartz which had evidently traveled some distance, as nothing of the kind existed in the ravine." Breaking off a hunk with a pick, Colton "found [it] veined with gold. But it is the only specimen of this combination with which I have met. Where the fellow came from I know not." A few days later, moving south toward Carson Hill, the party crossed over a "steep mountain spur," and along the ridgeline saw "a vein of white quartz run along the ridge, like a line of unmelted snow, with here and there spangles of gold glittering in the sun." With a hunting knife he vainly tried to pry out the gold, then took out his pistol and fired at it with a lead ball. That "knocked out the gold-drop, but jewel and lead went over the steep verge together." Colton's experience never brought him great personal wealth but considerable wisdom in the quartz business. A concluding chapter of his narrative—one of the best in Gold Rush literature—ends on a philosophical note: "The great Hebrew proverbialist says there are three things about which there is no certainty,—the way of an eagle in the air, the way of a serpent upon a rock, the way of a ship in the midst of the sea; and he might have added—the way of a thread of gold in a vein of California quartz."[3]

Mariposa County, rather than Calaveras, claimed the honor of identifying the first significant gold-bearing quartz veins in California. Within the vast "floating land grant" acquired by John C. Frémont and later confirmed after a long and protracted dispute, prospectors found an ore body that assayed at 160 ounces of gold to the ton.[4]

Frémont's Mariposa grant lay within a one hundred–mile-long mineralized zone that soon became known as the Mother Lode. The name probably originated with Mexicans who came to the Southern Mines from Sonora. Familiar with vein mining near Alamos, they explained the mysteries of the *veta madre* to Americans. Trending northwest-southeast and extending through five counties, the Mother Lode gold belt is primarily a continuous series of parallel *en echelon* quartz veins in metamorphic strata. In Calaveras County the gold belt has three divisions: the Main Belt or true Mother Lode running through Carson Hill, Angels Camp, and the Gwin mine; the West Belt through Copperopolis and Campo Seco; and the East Belt, a broad zone extending from five to twenty miles east of the Main Belt, where Sheep Ranch and West Point are located.[5]

Prospectors working these three belts located outcrops in the mineralized zone that varied greatly in depth and width. A few quartz leads began at the surface but pinched out within a hundred feet. Others were large disseminated bodies in the country rock adjoining vein structures (the so-called gray ore) that could be worked only by open pits or glory holes. Often, Mother Lode miners did not encounter major ore bodies until their workings reached five hundred or more feet in depth, although many mines, especially in Calaveras, had outcropping ore. Some ore shoots extended to a depth of more than six thousand feet. Widths of major veins sometimes exceeded fifty feet, and gray ore zones were often wider.[6]

Values also varied widely. Some high-grade gold ores in the pioneer era ran nearly six ounces per ton or more. Especially rich were the intersections of small stringer veins in both Calaveras and Tuolumne Counties. Through most of its productive years the Kennedy mine in Amador County averaged a half-ounce per ton, while at Carson Hill, major production in the 1930s depended in large part on a massive ore body averaging only eight hundredths of an ounce per ton. Complex or refractory ores might assay as high-grade but could not be efficiently processed before 1900, when flotation and cyanidation modernized milling and increased its efficiency.[7]

Popular literature today has often confused the Mother Lode with the more productive foothill deposits farther north in Nevada County, where California hardrock mining had its longest and most lucrative stand. The heart of Mother Lode mining was at Jackson in Amador County, where the largest concentration of lode gold was found, and where mines remained open the longest and reached the deepest levels. Both to the north and to the south of the Jackson district, productive gold mines were widely separated. Across the Mokelumne River, six miles south of the Jackson district, was the Gwin mine, and sixteen miles farther were the Angels Camp mines. About four miles from Angels was the Carson Hill district, and beyond the Stanislaus River some five miles were productive mines in Tuolumne County. Even though in the mineralized zone, all the ground between these clusters of good mines generally proved unproductive.

Although Calaveras veins have much in common with Mother Lode deposits elsewhere, each district has its own geological characteristics. Along the Main Belt, the Gwin vein resembled those in Amador, but the ore bodies in the Angels mines were wider, richer, closer to the surface, and higher in sulfide content—conditions that made mining and milling at Angels more difficult than elsewhere. Carson Hill, too, had hardrock difficulties. There, gold mineralization was concentrated around a steeply dipping, deep pipelike structure, with the gray ore in the wall rock of the quartz veins carrying much of the values. While noteworthy for some spectacular small high-grade shoots in flat stringer veins, Carson Hill was primarily a low-grade deposit that contained more sulfide than typical Mother Lode deposits. The presence of gold telluride minerals, rare in California, added to Carson Hill's complexities and contributed to its metallurgical problems.

Mines in the East and West Belts of the Mother Lode in Calaveras were less productive than those along the Main Belt, but still they made sizable contributions to county economic development. East Belt mines were small and widely separated with varying strikes and dips, and with much more sulfide minerals than elsewhere along the Mother Lode. Only one mine had significant gold production, but in the aggregate they supplied the extensive placers of the Tertiary Calaveras River with huge amounts of gold. In the West Belt, the Madam Felix–Hodson district was the only significant center of gold mining, but West Belt copper deposits were extensive and provided Calaveras with two of its more important mines.

In the flush of quartz excitement following the Mariposa strike, enthusiastic mine promoters along the southern Mother Lode imported European
advisers and machinery at great expense, overbuilding the surface plant before understanding the size and nature of the ore body—a management mistake all too common in early American hardrock mining. British investors
poured at least two million pounds sterling into these early California ventures. Samuel F. Marryat's experience at Tuttletown was typical of the times.
In the fall of 1851, this English seaman and adventurer traveled through the
southern Mother Lode with several friends. Crossing the Stanislaus River into
Tuolumne County from Carson Hill, he spotted an outcrop covered with
"massive veins of rock, peppered with specks of gold." After taking samples
to an "engineer" whose opinion was "highly satisfactory," Marryat and his
party staked a claim and worked for three months to bring the prospect into
production. Hiring a Mexican labor force to produce the ore, they went to
San Francisco, bought a steam-powered stamp mill, and hauled it back with
great difficulty. It failed miserably, but Marryat's company tried again with
another mill. When it also failed, the Englishmen threw up their hands, paid
off their crew, and abandoned the claim, poorer but wiser.[8]

For a few months in 1851, Peter Cool, a Methodist lay minister, and a
few pious colleagues operated one of these early stamp mills at the Spring
Hill Quartz Mine near Amador City, then in Calaveras County. Evidently of
European design with wooden stems and camshaft and square nonrotating
stamp heads made of wrought iron, it was a flimsy affair given to frequent
breakdowns. When it was assembled, the results were far from satisfactory.
"Ran for about an hour when the band [belt] flew off and was seriously damaged," Cool reported late in September. Before it could be repaired, one of
Cool's preacher partners sold his share for sixty-eight hundred dollars and
returned to Ohio. The next week the remaining partners managed to repair
the machine and run "about one hundred buckets of decomposed quartz
thorough our pessels. When we got it amalgamated and retorted, found we
had about seven ounces or one hundred and twelve dollars." Three days later,
they were again "[r]epairing our engine, or rather our pessels." Within a
month, the mill had to be completely rebuilt, this time with an iron camshaft
and stronger stems. Before the end of the year Cool had also sold out, but
with better machinery and management the claim would pay handsomely.
Eventually, the Keystone Company, one of the most important producers on
the Mother Lode, absorbed Cool's original claim.[9]

After two or three years of effort, underwritten primarily by naive British
capitalists relying on dubious field reports prepared by inexperienced and self-
serving operators, this first Mother Lode quartz bubble burst. "Many early

companies had far more capital than financial management or technical skill," concludes one historian of the Cornish miners, who were often called on to work these fledgling promotions. A contemporary London observer, sent out in 1852 to investigate, wrote that the Mariposa quartz mines were a "humbug," and that Carson Hill quartz showed "specks" of gold, "but without metal enough to pay half the expenses of reduction." That was very different from reports a year earlier that "big machines" were at work on Carson Hill veins that were "flush with as much gold per square block of terrain as all of the placers where gold is washed." The highly promoted mines shut down, the surface plants abandoned or scrapped, and investment capital dried up, leaving a sour taste for California hardrock mining in U.S. and European financial markets that took years to overcome.[10]

The lode mining industry got a second start in the later 1850s, primarily with regional capital, but for thirty years it struggled with crude technology, inefficient methods, and the dubious advice of untrained or inexperienced "experts." In this period only relatively high–grade ore could be mined, and only at shallow depths, especially in the hard metamorphic country rock of the Mother Lode. Ores from below the weathered zone resisted treatment by arrastras. Most miners, rather than struggle with unoxidized deep deposits, abandoned the claim and looked for another outcrop.[11] By the 1880s, however, with improvements in explosives, drilling techniques, and equipment, along with increases in capital investment and professional education, modern mining came into being. Only the problem of effectively recovering the gold content of refractory ores remained unsolved.

## Early Discoveries and Developments

As a microcosm of the industry and its early troubles, Calaveras County's hardrock history is instructive. Both the promise and the pitfalls of quartz mining can be studied by surveying the major mines and districts in Calaveras as they grappled with the fundamental technical, legal, and economic problems of locating, mining, and milling large gold ore bodies.

### The Paloma (Gwin) Mine
One of the first lode mines in Calaveras was the Paloma, known after 1867 as the Gwin mine. Considered in its heyday "one of the most interesting of the Mother Lode mines," its early history was marked by intermittent development and lack of capital.[12] In late 1849 or early 1850 prospectors located rich quartz outcrops on what they called the Paloma Ledge in Lower Rich Gulch. It lay unworked until 1853, when Dr. Hugh Toland, later founder of the

Senator William M. Gwin, M.D.
(Courtesy Calaveras County
Historical Society)

University of California Medical School, started an inclined shaft to follow
the vein. Lacking capital for a full-scale operation, he soon turned it over to
J. H. Alexander, who organized a company to develop the ledge north of the
original works. By the early 1860s a twenty-four-stamp mill had been erected,
but below two hundred feet the values diminished and the work stopped.
After the Civil War part of the Paloma vein fell into the hands of William M.
Gwin, who had visited Lower Rich Gulch on his initial trip to the diggings
and perhaps had a lingering interest in mine speculation.[13]

Known as "the California Machiavelli" during the state Constitutional
Convention of 1849, Gwin was one of California's most prominent early poli-
ticians. A southern Democrat who had practiced medicine before launching a
political career with the help of an influential friend, President Andrew Jackson, Gwin served a single term in Congress as delegate from Tennessee. After
the Democrats were defeated in the 1840 election, he returned to private practice but joined the California Gold Rush in 1849, hoping for new political
and economic opportunities. In the rough-and-tumble of frontier politics he
rose quickly, leading the statehood fight and fulfilling his own prediction in
1850 when the newly formed California state legislature sent him to Washington as a U.S. senator. For a brief interlude Gwin was a rising star in national
politics, but the sectional troubles that rocked the nation in the 1850s rever-

Sinking a new shaft at the Gwin mine at Paloma, 1894. (Courtesy Calaveras County Historical Society)

berated in California, divided his party, and ultimately destroyed his political career.[14]

Turning from politics to business after the Civil War, Gwin bought the old Paloma mine adjacent to Alexander's shaft in 1867 for fifteen hundred dollars. It was a smart move. Within ten years the renamed Gwin mine became the most productive lode mine in Calaveras County, thanks to the excellent ore grade and to the Gwin's manager, the senator's son William M. Gwin Jr. He greatly expanded production in 1867–1868 by erecting a thirty-six-stamp mill at the old Paloma plant, and in 1872 he absorbed other Lower Rich Gulch claims to monopolize the Paloma vein. The mine's productivity was a marvel to the senior Gwin, who could not help boasting to relatives. A letter dated 1871 said the mine was "yielding from $15,000 to $20,000 per month in gold bullion, and during the next year the yield will be double that amount, more than half of which is clear profit." In 1870 Gwin and his son listed their individual real and personal property values at twenty-five thousand and fifty thousand dollars each, probably a gross underestimate but by far the largest declared personal assets in the county. By 1882, when it ran out of profitable ore above nine hundred feet, the mine had yielded more than one hundred thousand ounces.[15]

With the help of professional engineers it might have produced even more.

As the physical plant and equipment grew old and obsolescent, and pumping and mining costs increased, the Gwins faced the dilemma of undertaking a deep development project and replacing expensive mining and milling machinery, or shutting down. They decided to take their profits and leave, rather than pump money back into the mine to keep it operating efficiently. W. H. Storms, a reputable mining engineer highly critical of many Mother Lode operators, wrote early in the century that the Gwin management had been, "if not incompetent, at lease [sic] improvident."[16] This, of course, was hindsight, prompted perhaps by an earlier critical report by the state Mining Bureau, but also reflecting the professional engineer's disdain for amateurs who "pick the eyes out" of a mine by selectively mining the high-grade ore and leaving a degraded property that lacked systematic development.

### Carson Hill

Better known but also much more chaotic were the early mining efforts at Carson Hill, the county's first lode mining district on the Mother Lode. Complicating quartz or hardrock mining development in the pioneer era was the frequency of litigation over claim boundaries, ownership rights, contract disputes, water fights, damage suits, and other issues. Court battles tied up mining properties for years and cost thousands of dollars in legal fees. Unwary investors, instead of making quick profits, often received stock assessment notices instead.[17]

In Calaveras County, Carson Hill suffered the most from legal disputes. Eventually ranking as the county's second-most productive gold mining district (after Angels Camp), it made little progress before 1889, when the first large-scale low-grade operation began.[18] Before that time mining on the hill was unproductive and disorganized at best and anarchic at worst.

In the remote camps during the Gold Rush, miners held impromptu meetings to draft and enforce their own claim laws. They regulated the size and shape of a claim, stipulated the method and manner of marking its boundaries and the amount of work required to hold it, and established the rules for recording it with a designated local official. In the surface placer camps, where both geology and technology were relatively simple, these ad hoc claim laws worked well so long as the majority identified with both the lawmakers and the law enforcers. However, lode mining was a much more complex process. Even before Congress in 1866 adopted the first uniform federal mining law, idiosyncrasies in both geologic formations and local mining laws kept batteries of attorneys busy, especially in important districts where the stakes were high. The 1866 federal law, largely written by mining representatives in Con-

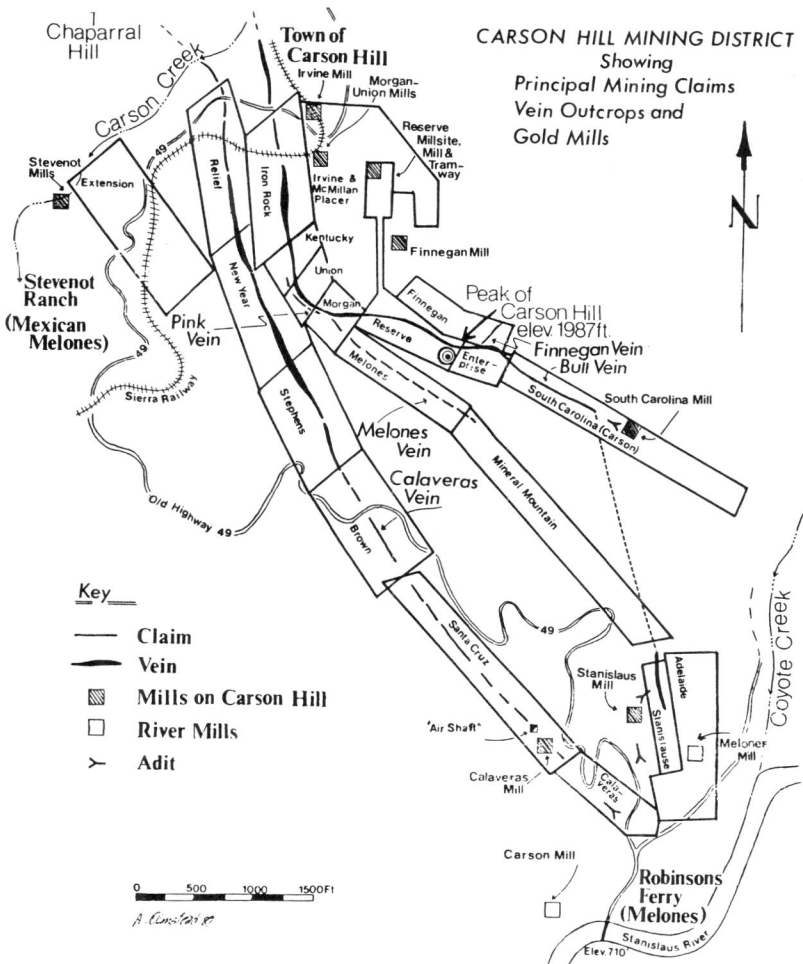

Carson Hill Mining District map. (Courtesy Calaveras County Historical Society)

gress, adopted the basic rules of the lode mining districts. In 1870 Congress expanded the law to include placer mining. Finally, in 1872, a federal uniform mining code was adopted for the entire industry, a law that still governs the industry today despite intense efforts by environmentalists, tax reformers, and others in recent years to modify it. Adopting standardized mining procedures after 1866 helped reduce litigation in the lode mining industry to some extent, although trying to reconcile the apex law governing extralateral rights with the complex geology of lode deposits resulted in many prominent mining lawsuits for at least a half century after it was adopted in the 1872 revised code.[19]

The principal legal dispute on Carson Hill grew out of the conflict that began a few months after the original discovery. In October 1850, William Hance and a group of miners working high up on the hill from Carson Creek discovered the outcrops of a highly enriched mineralized zone of weathered quartz that subsequently proved to be the apex of the main deposit.[20] Within a month of the discovery, Hance, Alfred Morgan, and several others organized the Carson Creek Consolidated Mining Company, claiming about seventeen hundred feet along what they called the Morgan Lode.[21] The Morgan company worked the claim for nine months without contest, but the richness of the discovery attracted a crowd of newcomers who protested the Morgan claim's boundaries. To miners used to smaller claims, the Morgan claim seemed excessive. In the winter of 1851–1852 the newcomers held a meeting, organized a new mining district, and established a set of district rules and regulations that limited claims to thirty feet square. Whether it was democracy or greed that motivated Morgan's protagonists is not clear. At least one shocked editor thought the effort to revise the rules after the first party had recorded its claim was "communism with a vengeance and in its worst form."[22]

The Morgan party resisted at every turn, even taking the matter to the nearest court in Murphys, but Jacksonian democracy had its day. The judge ruled in favor of the new limits. Anticipating the good news—or perhaps an adverse decision—a party of well-armed miners invaded the Morgan and divided it. Soon there were at least thirteen claimants to the same ground.[23]

By the winter of 1851 a full-scale rush was on to grab what was left of Carson Hill. A newspaper correspondent visiting the area recorded an example of some of the claim notices he found. One "curious" posting just below the hill got the traveler's attention: "Notes We due clame from this ere tre to the won above for our watter digens P S it is five akers mor or les which we will defend threwe life and threwe deth."[24]

Lacking the manpower to physically protect their claim, the Morgan defenders sought legal redress. They obtained an injunction from a different judge who recognized the Morgan company's right to one thousand feet along the vein and ordered the usurpers to withdraw. At first they refused, then left temporarily when the sheriff arrived to take possession of the claim for the Morgan company. Almost as soon as the authorities were out of sight, wrote one correspondent, "the boys took possession again, and swear if Morgan and his friends come up again, they will shoot the whole of them."[25]

Over the next few weeks, according to Leonard Noyes, who came to Carson Hill the next spring, the Morgan ground changed hands at least five more times, with each side gathering men from surrounding regions to take part. More than two hundred armed miners were on the hill by the winter of 1851–1852.[26] Noyes said the trespassing miners had possession when he arrived in the spring of 1852 to take up a claim alongside the Morgan shaft. The dispute had racial as well as legal overtones, for Mexicans made up a large part of the Morgan workforce, and a Mexican named Pacheco may have been part of the original discovery party.[27] As J. Ross Browne remarked in his 1868 review of Carson Hill's early history, American miners held the Mexicans in contempt, but "they had had some experience in this kind of mining and their services were indispensable."[28]

Basing his report on an article in a San Francisco paper the year before, Browne went on to describe the Morgan dispute. He wrote that a notorious San Francisco gang leader, Billy Mulligan, had led the trespassers.[29] Soon after Leonard Noyes and his partners purchased the Block and Tackle mine alongside the Morgan, James Finnegan, Mulligan, and some others from San Francisco invaded the Morgan claim, chased the Mexicans out, and took possession. They held the mine for several weeks, tried to take over others by force, and were finally chased out themselves by angry miners, including Noyes, who organized resistance and helped secure a court injunction against the jumpers.[30] Other boundary disputes and court fights hindered any systematic mining on Carson Hill for the next fifteen years.

Even though systematic development was stymied by boundary disputes and litigation, there was plenty of unsystematic mining in the interim. Trespassing pocket hunters riddled the surface of the Morgan claim with shallow shafts and pits, excavating or "high-grading" the best ore and leaving the rest. What they sought was a rich ore pocket, described by John David Borthwick as "a spot not more than a few feet in extent, where lumps of gold in unusual quantities lie imbedded in the rock." Some simply blasted the surface with black powder charges, and then scoured the area with baskets to pick up the pieces. Others used picks and shovels and knives to extract gold in rich stringers near the surface. The hard work brought fabulous rewards to some who came down from the mountains with saddle packs filled with high-grade ore specimens. Stockton papers pounced on reports of gold discovered by the pound.[31] The effect of such stories converted skeptics who had previously denied that the Southern Mines contained paying quantities of "metaliferous [sic] quartz."[32]

Pocket miners swarmed on and around the Morgan like gnats to nectar. Their efforts occasionally paid big dividends. Out of one twenty-by-thirty-foot excavation on the summit in 1851, pocket miners blasted, picked, chiseled, and hammered out an alleged three million dollars in gold.[33] Three years later, in a highly weathered quartz pocket on what was later called the Pink vein of the Morgan Lode, miners discovered a 195-pound lump of nearly pure gold, the largest single mass ever recorded in California. Lesser Carson Hill gold pockets brought to Carson Hill a bonanza reputation and swarms of pocket hunters that pockmarked the landscape like a battlefield. One pocket in the Pacheco shaft was so rich that miners used sledgehammers to break it up. A single golden lump from the mass weighed 108 pounds—worth thirty-five thousand dollars in 1854, but one-half million dollars today.[34]

The Morgan was not the only rich early claim on the hill. Southeast of the Morgan, the South Carolina was a mine with a history nearly as colorful and chaotic as that of its bigger neighbor. Discovered in 1850 and recorded in 1851 by a sixteen-man party, the claim was a high grader's dream. Pocket miners dug shallow pits with picks and shovels, used chisels to extract "very rich rock" from stringers, then crushed the ore in hand mortars and washed it in rockers or sluices to recover the gold. Most of the lower-grade ore was left behind or tossed aside. Writing thirty years later, William Jeffrey, who claimed to have a stake in the original discovery, said that in the 1850s "thousands of tons of very rich mill rock were thrown from the mouths of these pits over the banks" that "I am confident will pay from seven to ten dollars per ton with machinery."[35]

In 1852 the South Carolina locators sought easier and faster wealth by leasing the mine to a remarkable woman of French Basque extraction, Madame Eliza Martine or Martinez, perhaps the first female mine operator on the Mother Lode. Working only the richest ore for 80 percent of the gross proceeds, she produced large amounts of gold and paid the company "handsome dividends." Her crew consisted of forty to fifty laborers from Sonora, Mexico, who were paid up to $8 a day to extract and haul ore by pack mule to Coyote Creek, where she had constructed four arrastras for milling.[36] Not content with this daily wage, the crews concealed some of the richest ore in their clothes, crushing and panning it later. "I myself have seen them wash as high as three ounces in one of these batteaus or basins, with probably not over fifteen pounds of ore," wrote Jeffrey in the 1880s when he was trying to sell his interest in the property. His testimony is suspect, but even heavily discounted it is indicative of the times. In seven months the Mexicans took out $119,000, but rumors that the mine was much more productive dissatisfied the owners, who canceled the lease in 1853.[37]

The high-grading started again in earnest after the owners negotiated a new lease with Don Gabriel Metro under more favorable terms. Under his supervision the miners worked more systematically, but after he came down with an illness and was carted away, his crew "commenced a wholesale plunder of the rich ores," wrote Jeffrey.

Next on the scene was a British investment company headed by John Sadler, who bought the South Carolina in 1854 for $92,000 in cash and 20 percent in paid-up stock. Jeffrey claimed that Sadler was so unscrupulous that he misappropriated company funds, "caused delays and kept his agents traveling between England and California on the most flimsey pretenses." Caught at last, he committed suicide, and the mine eventually reverted to its original owners. During the interim Richard Inch of Tuolumne County, a mining investor and father-in-law of Charles H. Segerstrom, later a principal promoter of Carson Hill, examined the mine and said that "this claim is second to none in California."[38]

In 1855 the mine was leased again, this time to eleven Cornishmen who were better miners than mill men, according to Jeffrey. They sank a shaft below the shallow workings and ran drifts into sulfide ore that they trammed on rails three-fourths of a mile to Coyote Creek. There they built a primitive five-stamp mill, but the "rude machinery would not save the gold," said Jeffrey, although he noted that they still managed to recover about $15 to $20 per ton. Heavily in debt and set back by a brushfire that burned their "trainway," the Cornishmen "quarreled among themselves" and dissolved the partnership. They sold the mill to a shrewd miner who dismantled it and placer-mined the tailings. In three days, said Jeffrey, his rocker had "washed more than enough to pay for it."[39]

Jeffrey became president of the South Carolina Company soon after this fiasco. Just before he took control, he wrote that the South Carolina Company officials, "being now fully satisfied that they had a valuable mine," leased it to two of their members for seven years on terms that required them to construct a mill and pay the company 15 percent of the net proceeds for the first two years and 20 percent thereafter. The mill was completed in 1860, but things soon went awry. The leaseholders, two San Francisco "dry goods merchants," tried to jointly manage, but "neither of them knew anything about mining," said Jeffrey. In a matter of months they "began accusing each other of dishonesty" and stopped all development. Company officials stepped in to break the lease, but did not keep up the assessment work. Jeffrey said he was out of state at the time, and when he returned in 1863 the mine had been relocated. He sued to recover but lost in court. Undeterred, he "then bought several interests from those in the new Company, who were not so well aware of the value of the mine."[40]

Jeffrey remained financially connected to the South Carolina until the 1880s. His summary of the mine's early history reads like a rejected movie plot for a grade-B western, yet it illustrates the financial and managerial short-comings that plagued early lode mines. His conclusion is worth repeating, for it applies not only to the South Carolina but also to many other Mother Lode mines in the pioneer era: "One great evil is the giving of short leases, the parties working on short options having no regard for the protection of the workings of the mine properly, their only object being to get out the greatest amount of the richest ore with the least expense."[41]

The legal disputes that frequently flared into violence in the Southern Mines did little to disrupt the individualized scramble for high-grade ore that characterized much of the early mining efforts on Carson Hill. At Robinson's Ferry in 1863, a correspondent to a Bay Area newspaper described the type of pocket hunting carried on at the time: "The hills around the Ferry all contain numerous quartz leads, all of which contain more or less gold. It is quite common, whenever any of the miners want a little change, that they go to the first unoccupied lead, get out a sack full of rock and pound out the gold."[42]

In the 1860s two independent developers made some effort to bring order to this early Carson Hill chaos. One was James G. Fair, later one of the "bonanza kings" on the Comstock. He was a Scotch-Irish immigrant from Belfast who had arrived on American shores in 1843 at the tender age of twelve. His family settled in Geneva, Illinois, and put him in the public schools. He was eighteen when gold fever struck the town, claiming many of its young fathers and sons. With an immigrant party he set out early for California, and by August 1849 he was working a river claim in the Northern Mines. After drift gravel mining opened farther south in Tuolumne County, Fair found work in the mines at Shaws Flat and Table Mountain. At Poor Man's Creek he met two storekeepers, James Flood and William S. O'Brien, whose later careers intimately intertwined with his, both in California and in Nevada. After a brief interlude as a farmer in 1853, Fair returned to the Tuolumne mines. He worked underground until the Fraser River rush in British Columbia five years later. That excitement drained off hundreds of northern California miners and forced many marginal mines to shut down, including most of those at Table Mountain. In Calaveras the Fraser River news left some mining communities practically deserted. San Andreas was especially hard hit, since the new excitement came just after a disastrous fire that destroyed most of the town. On the South Fork of the Calaveras, where Irish miners predominated, men sold good-paying claims for "whatever they could get," said Wade Johnston, and rushed off to Fraser River.[43] But the Canadian boom, like many in the American West, turned sour quickly, and most who had left came back

James G. Fair. (Courtesy Nevada
Historical Society)

grousing about the "humbug." That was an overstatement. A few of the early
birds did find gold, but the mineralized zone was too small and isolated to
support the thousands of latecomers who wanted to stake claims.

Whether Fair rushed off to Canada during the Fraser River excitement
or stayed in California remains unclear, but after the Table Mountain mines
closed he came to Calaveras County. His initial venture was at Angels Camp
in 1857, discussed later. Sometime in the late 1850s the Carson Hill mines
caught his eye. Just prior to leaving the county for the Comstock in 1860,
Fair purchased the Morgan claim and hired William Irvine, an Irish immigrant like Fair, to clear the title and start up the mine. Soon the big bonanza
in Nevada kept Fair so busy he had little time for Calaveras County. That left
Irvine with plenty of opportunity for a double-cross. By some dubious procedure he succeeded in obtaining tax title to the Morgan property, then in
1875 he transferred it to his newly incorporated Morgan Mining Company.
First using the existing Union-Kentuck mill, then constructing his own mill

Gabriel K. Stevenot. (Courtesy Barden Stevenot)

in 1884, he mined the Morgan almost continuously until Fair stopped him with a court injunction. A subsequent suit lasted for many years during which the Morgan was idle or prey to pocket hunters.[44]

More successful than Fair as a Calaveras County mine developer was Gabriel K. Stevenot, who with his son, Emile, pioneered the consolidation of mining properties on Carson Hill. A mine promoter as well as developer and operator, Gabriel was an Alsacian attorney who reached San Francisco in 1849 and sold general merchandise a year before moving to Carson Hill. He soon began acquiring property, including the Morgan briefly in 1856. That same year he established one of the first stamp mills in the area. In 1863 Emile joined him after graduating with an engineering degree from the University of Strasbourg. Together over the next fifteen years, except for a nine-year hiatus in the 1870s when Emile was in the borax business, they consolidated the Melones, Reserve, Enterprise, and Stanislaus claims into the Melones Consolidated Mining Company, and took a major interest in the Calaveras and Santa Cruz claims.[45]

By the time of Gabriel's death in 1885, the Stevenots had come a long way toward modernizing the mining industry on the hill, but they were still ahead of their time. Lacking the technology to work complex and refractory ore, and short of financial resources to mobilize large amounts of capital, they achieved only a modicum of success. In 1883 they sold the Melones Consolidated, and after Gabriel died Emile disposed of most of the remaining Stevenot mining interests. It was not the end of the Stevenot influence in the area, but only the end of the initial phase of consolidation. Elsewhere in Calaveras County, by the mid-1880s individual developers were giving way to corporate promoters, developers, and managers who brought finance and technology together.

### Angels Camp

The lode mines of Angels Camp opened several years after Carson Hill. Originally a placer region, Angels Camp had lost most of its miners by the early 1850s. One of those who remained was a prospector by the unlikely name of Benneger Rasberry, a Georgian in his thirties, who allegedly discovered the principal vein by accident. In the process of removing a stuck ramrod from his muzzle-loading rifle, so the story goes, he pulled the trigger and drove the rod into a quartz stringer filled with gold. Like so many others in Gold Rush literature, this colorful tale is probably fictional, but it continues to circulate no matter how many times it has been discredited by scholars. Rasberry remained in Angels for more than thirty years, settling in with his English wife, Maria, and raising four children, including two boys who joined him in the mines.[46]

Better candidates for the role of discoverers are two brothers from England, Edmund C. Winter and his older brother Augustus.[47] They unearthed gold-bearing quartz outcrops while ground sluicing northwest of Angels Creek in the early 1850s. The exact date is unknown, but it may have been as late as 1854, when they opened the Winter Brothers mine.[48] Whether they realized it or not, the brothers had exposed the northwestern end of a broad vein ten to ninety feet wide that strikes southeasterly in slaty strata and runs from Altaville nearly a mile to Angels Creek. Extracting oxidized ore from small open pits or shallow shafts along the vein outcrop, and milling by arrastras, they were said to have taken between twenty and fifty thousand ounces in gold from their claim by the end of the Civil War.[49]

The Winters' mine was one of three adjoining claims on the north end of the lode that were eventually consolidated into the Sultana, one of the four principal mining companies of Angels Camp. But consolidation and systematic development were not possible with the limited technological and financial resources of the 1850s. Between 1854 and 1857 dozens of claims were recorded on what became known as the Davis-Winter lode, but only eleven ultimately proved productive. Most pioneer miners at Angels Camp were not very lucky. Disappointed by low-grade ore in the oxidized weathered zone, and stymied by "sulfurets," or sulfide ore at deeper levels that they did not know how to process effectively, many sold out for what they could get. Some simply abandoned their claims. By the end of the Civil War most of the early claims were in the hands of secondary and tertiary developers.[50]

In this transitional period William H. Bovee, a coffee merchant in San Francisco who entered the mining business after selling his Pioneer Steam Coffee and Spice Mills to J. A. Folger, took the first step toward consolidating the mines in the Angels district.[51] In 1865 he bought the Winters' mine from the two brothers, and at the same time acquired the Fritz, another old claim dating from the pioneer days. Like many other Mother Lode enthusiasts, however, Bovee was undercapitalized and lacked the know-how to process refractory ore. He replaced the arrastras with a big stamp mill in 1866, but revenues never caught up with expenses. When the mill burned in a spectacular fire six years later, he left the mining business and became a real estate broker in San Francisco.[52]

Another transitional figure in the Angels district was James G. Fair, whose activities at Carson Hill have already been described. He had been lured to Angels Camp in 1857 by A. P. Bonton, who had shown him samples of quartz ore from his mill on what was first known as the Calaveras mine, then the Utica. Angels was then in the third year of a four-year boom that would end when miners and millers on the Davis-Winter lode encountered gold-bearing sulfides below the free-milling weathered ore at the surface. Twenty-six and still unmarried, Fair lived for a time in a miner's cabin with H. L. Lightner, a Vermont miner who held an interest in claims later consolidated as the Lightner, one of the key mines in the Angels district. How long he held that interest is not clear. In the 1860 census Lightner listed no property assets, but Fair was already prominent with total assets of $13,500.[53] Sometime in the late 1850s Fair and two partners, Messrs. Caldwell and Patton, bought out Bonton's mine adjacent to Lightner's and thereby acquired a major stake in one of the best potential properties on the Mother Lode, although no one knew it at the time.

Their first objective in Angels was to improve milling. Fair and his partners rebuilt the mill, installed a ten-stamp battery, and added copper plates for amalgamation. Caldwell and Patton, perhaps unhappy with the poor results, soon sold their interests to Irwin Davis, but Fair held on even after moving to Virginia City. He was still young but well seasoned in the mining business. Portly and heavy-bearded, he looked older than his years, and his experience gave him an edge over other developers frustrated by unfamiliar technical and financial problems.

For eight years Fair was a principal investor at Angels Camp, struggling all the while to come up with an improved milling process for complex Calaveras ore. The struggle earned him a reputation for persistence if not for honesty in

dealing with local businessmen. Cornelius B. Demarest, brother of the proprietor of the Altaville Iron Works, wrote that Fair once came to the shop
asking for detailed plans and costs in writing for some extensive alterations
to his mill. The next day, without placing an order, he took the estimates
with him and left for San Francisco. Later it was learned that he had made a
deal with San Quentin officials to have the parts made with convict labor at
a fraction of the price.[54]

When the Comstock excitement swept California in 1860, Fair rushed to
Nevada along with hundreds of others from the Southern Mines. He invested
heavily and rose rapidly on the Comstock, but at the same time kept his
hand in Calaveras mining. After expanding the Utica mill to forty stamps, he
and his partner, Davis, sold out in 1865 to two naive San Francisco investors,
James T. Boyd and Judge Delos Lake. The new owners innocently turned it
over to a manager who worked it for two months before quitting in disgust.
He claimed the previous owners had salted it. No case was ever filed, but Fair
was known thereafter as "Slippery Jim." Boyd and Lake then tried leasing the
property to another operator, but the lessee walked out after a few months.
When the assessment work lapsed the mine was declared abandoned.[55]

Fair made one last effort to cash in on the Davis-Winter lode in 1866,
when his syndicate relocated the Crystal mine, an old property that had been
worked intermittently, then abandoned. In the early days it was said to have
paid well but suffered from incompetent management. According to D. C.
Demarest, the mine superintendent suspected his Cornish miners of sleep-
ing on shift, but he "was too big a man to risk himself to go underground,
on the mine bucket, so he kept tab on the miners' work by listening at the
collar of the shaft . . . and by counting the number of blasts that were the requirement for their shift." To fool the boss, said Demarest, "the Cousin Jacks
had dummy holes in the wall rock that they loaded for blasts, to make up for
the deficiency in the shift's work."[56] Fair's syndicate changed managers but
did not do much better. The new owners held the Crystal until the late 1880s
with indifferent results. Eventually, it was consolidated into the Angels mine
by another set of developers.

For the first twenty years after discovery of the Davis-Winter lode, the
hardrock pioneers of the Angels district struggled to find the right formula for
successful development. That they failed was due in part to lack of managerial
skill and experience, and in part to lack of technology. Without consolidation
they were unable to raise the necessary capital or to implement a sustained
program of development. Without adequate technology they could not efficiently process refractory ore or work the mines at depth. As at Carson Hill,
the key to successful mining at Angels was to find and systematically pro-

cess low-grade ore. William W. Maltman, a Louisiana-born miner who had gained experience from the early quartz struggles at Angels Camp, profitably mined low grade in the district as early as 1866.[57] But the major development of massive deeper ore did not come for another twenty years.

At Angels Camp the 1870s witnessed a continuation of the embryonic attempts to organize and consolidate that had characterized the late 1860s. The Utica mine got the most attention. From the end of the Civil War to the mid-1880s it changed names and owners twice. After several locators first claimed and then abandoned the property, Robert Leeper, a young Italian-born bachelor, and five others relocated it as the Invincible in 1869 and worked it by a series of shallow open cuts in the weathered zone, stamping the ore in a small mill that they supplemented with arrastras.[58] Unable to extract the values from complex ore at deeper levels, they sold it in 1884 to a partnership headed by Charles D. Lane, who restored the old name. Under Lane's management, and with improved milling, the Utica became one of the most productive properties on the Mother Lode.

The Fritz, Bovee, and Winter Brothers mines also passed through several hands before 1883, when the Marshall Mining Company took over and reorganized as the Sultana. Adjoining the Sultana to the south was the Angels mine, a collection of separate claims consolidated and reorganized in 1884 by J. V. Coleman, a nephew of William S. O'Brien, the Comstock magnate.[59] Eventually, nearly all the major claims at Angels Camp were brought into the corporate structure of four leading companies: the Sultana, Angels, Lightner, and Utica.

### East Belt Districts and Mines

The East Belt in Calaveras County is a broad zone extending some five to twenty miles east of the Main Lode, with many scattered mines, some of them clustered in vaguely defined districts. All but one of these were small operations, many hardly more than prospects. Only about eight had production exceeding fifteen thousand ounces. The principal districts were Sheep Ranch, West Point (including Glencoe and Railroad Flat), Mokelumne Hill, Murphys, Washington (Indian Creek), and Esmeralda.

The East Belt's most important lode mine was the Sheep Ranch–Chavanne, which began operations in 1868. It was the district's only substantial producer, although it was far below Carson Hill and Angels Camp in total production. In 1875, A. P. "Cap" Ferguson, one of the original locators, and William A. Wallace, a partner, sold the Sheep Ranch claim to George Hearst, James Ben Ali Haggin, and Lloyd Tevis, three of the most important developers in western mining history. With the acquisition of the adjacent Chavanne mine from André Chavanne in 1882, Hearst and company held nearly total control of lode mining activity in the district.[60] Sheep Ranch production after the Hearst takeover will be discussed in a subsequent section.

The 1880 census provides a demographic glimpse at Sheep Ranch that demonstrates the importance of mining to this isolated community of 542 residents. Though most people lived in single family residences, four board-inghouses served the unmarried miners. French and Swiss names were numerous among the mining personnel—doubtless a reflection of the recruiting efforts of local landowners André Chavanne and Jules Fricot. Among the boarders, who were primarily European, nationality and language seemed the main criteria for selecting rooms. One of the hotels, operated by a Swiss couple, served men largely from French-speaking countries; another took mostly clientele from the British Commonwealth. Miners made up 23 percent of the total town residents. The rest were family members or service personnel dependent on mining for a living.[61]

In the West Point district, surface placering during the Gold Rush led to discovery and location of several veins, but not until the postwar boom of the late 1860s did active mining begin. In the 1870s and 1880s the district blossomed, with significant production from the Blackstone, Champion, Continental, Keltz, Woodhouse, Yellow Aster, Fine Gold, and Petticoat claims. More than one hundred smaller mines were probably active in this period, none of which amounted to much. Most of the veins were narrow and short along the strike, with small ore shoots that pinched out quickly. The ore itself was spotty and high in sulfides, frustrating mill men with significant loss of millhead values. Little wonder, then, that in spite of the large number of mines in operation, the district's total production was probably not much more than one hundred thousand ounces.

Among the dozen or so small lode mines in the Murphys district, only the Oro y Plata produced much ore. As early as 1853 prospectors had investigated the ground later covered by the Oro y Plata, Blue Wing, and Red Wing claims. Selim Woodworth started the first serious hardrock development in 1861. He organized a company, sunk a shaft on the Blue Wing, and constructed a mill. Within a year he was shipping gold bullion, but the mine's uniquely complex ore stumped the metallurgists. Apparently, Woodworth sold out a few years later, and not until 1867 did another operator by the name of Bouglinval get the mine running again.[62] This period of activity was short-lived, however; others also tried to reopen the mine but without much success.

By the late 1870s, J. J. Jerome and Page Cutting acquired control of the Oro y Plata, but later sold out to Zabdiel Adams Willard, a wealthy Boston metallurgist and entrepreneur, grandson of Simon Willard of the famous clock-making family in Massachusetts. Z. A. Willard had left the family business in 1870 and headed west, employing his talents as a chemist, assayer, inventor, and practical physician in mining camps from Colorado to California. At Carson Hill he briefly operated the Stanislaus mine, but shifted over to the Murphys district after his Stanislaus operation foundered when he could not process the telluride ores effectively. Organizing the Willard Mining Company with Boston friends, he operated the Oro y Plata for six years with fair production results. His California activities may have stimulated some of his Boston friends to invest when the Melones Consolidated Mining Company on Carson Hill went looking for eastern capital. That turned out to be a better prospect than the Oro y Plata, which closed in 1887 after a disastrous flood and cave-in. A financial failure, it did not reopen for another forty years.[63]

The remaining districts in the East Belt were only moderately productive in this early period. In the Washington and Esmeralda districts, located along Indian Creek near Murphys, the most notable producers were the Washington and Mar John mines near Sheep Ranch, and the Bence and Esmeralda mines near the little town of Esmeralda. In the Mokelumne Hill district, the most important mines were the Lamphear, the Easyz [sic] Bird, and the Boston, the latter the biggest producer.

### West Belt Mines

Although the West Belt in Calaveras County is chiefly important for its copper production, there were two gold districts, Madam Felix and Alto, the latter not active in this pioneer period. In Salt Spring Valley, gold-bearing veins were first prospected in the late 1850s, but it was not until after the Civil War copper boom that attention turned to gold in the Madam Felix district. Several companies were organized in the late 1860s, but because of the highly refractory ores in the district's mineralized zones, none was successful in developing a producing mine.

In 1876 Henry Botcher began the first serious attempt at mining at the Pine Log claim. Utilizing San Francisco capital, he eventually incorporated the Pine Log Gold Mining Company. In 1881 he sold out to Isaac Wilbur and the Castle brothers of Stockton. They worked this little mine until it ran out of ore in 1884, then they took over the Royal claim nearby, turning it into a very successful though small producer.[64]

Along the Mokelumne River, the dry diggings of the Hispanic camps gave way in the 1860s to lode mining of copper. As we will see later, farther to the southeast, important copper mines also opened at Copperopolis and Telegraph City just in time to aid the Union during the Civil War.

## Pioneer Lode Technology

Despite early discoveries, hardrock miners faced both technical and financial obstacles that made lode mining before the 1880s a dubious venture. Until the 1870s many of California's quartz mines consisted essentially of shallow workings excavated by a laborious combination of hand drills, black powder, and primitive tramming and hoisting equipment. Milling methods were crude and often not efficient, losing a significant amount of the free-milling gold values, and saving essentially none of the gold locked up in the refractory sulfides. For example, the 1854–1858 boom at Angels Camp described earlier, when the Davis-Winter lode was discovered and first mined, collapsed as soon as the mines had worked out the weathered zone ore and descended into the primary ores, heavy in iron sulfide minerals.

Ore deposits at the surface in the Angels district assayed significantly higher in gold than at deeper levels. Much of the rock volume at the outcrops of the deposit had eroded away, leaving the gold behind. Long exposure to air and water had also oxidized sulfide minerals in the deposit, thus freeing the gold at or near the surface. Less complicated metallurgically were ores from the Sheep Ranch mine and from the Gwin on the Paloma Ledge. Those mines survived simply because of their relatively high–grade free-milling ores below the weathered zones.

The typical hardrock miners of the 1850s were in reality "pocket hunters" working in the weathered zone with shafts that rarely reached below one hundred feet. Many were of Mexican origin, at least in the Southern Mines. Welcomed at first for the experience in vein mining that they brought to the American camps, they worked both as employees and as employers. At Carson Hill and in some drift mining areas, Mexicans leased some mines from white locators and worked as hired hands in others.[65] Under Latin managers Mexican or other Hispanic miners employed traditional Mexican and South American techniques to follow veins underground. In European engineering schools the Spanish method was known as "rat-hole" mining because of the cramped and narrow shafts, the disorderly levels, and the serpentine drifts and raises that wandered wherever the vein led without regard to classical surveying or engineering principles. But science and technology mattered little at Carson Hill in the early 1850s. Rich ore was hauled out in leather sacks carried by hand up a notched pole that served in place of a ladder or windlass. Pack mules transported the ore from the shaft collar to nearby arrastras on Carson Creek. The system was slow and labor intensive, but satisfied the immediate needs.[66]

Remains of a twentieth-century arrastra near Mokelumne Hill. Note the concrete lining, used in place of stone liners. (Courtesy Holt-Atherton Library, University of the Pacific)

Racial animosities rapidly changed the picture at Carson Hill. Most of the Hispanics had been driven out of the district by 1854. Yet Latin tools and milling techniques, if not rat-hole mining, continued to influence developments in the American Southwest. The Mexican patio process provided essential ingredients for the technology that helped unlock the metallurgical mysteries of American silver ores, especially on the Comstock and in Colorado. The arrastra was so firmly established as a simple and inexpensive ore grinder and amalgamator that it could still be found at work in some smaller Mother Lode mines after 1900. By then, use of arrastras was confined mostly to secondary grinding after ore had passed through stamp mills.[67] The Lavagnino arrastras in Angels Creek, for example, worked many years regrinding the sand tailings washed down from the Utica-Stickle mill. They were still squeezing out the last recoverable gold that had escaped the Utica's "blanket" mill right down to its close in World War I.

In most lode districts the primary pulverizer by the mid-1850s was the stamp mill. Used by German miners as early as the sixteenth century and imported to the New World by the Spanish, stamp mills were first used in California at Grass Valley in 1850. But their use was very limited at first; in the Southern Mines they were unknown before at least 1851. William Perkins, a Tuolumne forty-niner who joined a partnership with Ramon Gil Navarro and invested in several Carson Hill lode mines near Melones, evidently relied on Mexican arrastras to treat the ore. The venture started with high expectations but ended in 1852 on a despondent note. Perkins "lost everything" when Sonora burned to the ground, and Navarro returned to Chile. His parting words to a friend: "Total sum: zero net profit divided by 'hope.' That is all I have to say to you about business: 'hope.'"[68]

In nearby Amador County, Peter Cool's Spring Hill mine used an experimental "quartz machien" in the fall of 1851 that could handle up to one hundred buckets of free-milling gold ore per shift. Using a mortar and pestle principle, it "worked rather poorly" and had to be completely rebuilt by a Georgia quartz miner after just two weeks of intermittent operation. By 1852 a sixteen-stamp mill was in operation at the Boston mine northeast of Mokelumne Hill—perhaps the first working stamp mill in Calaveras County. Four years later William Higby, a Mokelumne Hill physician and mining investor, wrote his father, "A great many cheap machines are being made this spring for the purpose of Quartz mining. Although fortunes were lost in that business in '50, '51 and '52, yet men are persevering and now fortunes are being made at it." The county assessor in 1858 identified fifty-three quartz mills in Calaveras, at least thirteen located at Angels Camp alone. Most of these early mills were built in San Francisco, which played a leading role in mining development because of its iron foundries and machine shops. However, after 1860 the Altaville Iron Works, founded in 1854 by J. M. Wooster and taken over six years later by David Durie Demarest, built many smaller mills and serviced most of the others in Calaveras and in adjacent counties. The presence of a reliable local foundry and machine shop added materially to the success of Calaveras lode mining.[69]

Stamp mills (or stampers as they were often called in the 1850s) were extremely well adapted to typical Mother Lode gold ore. Most shallow ore was free-milling, or not locked up in complex mineral compounds and coarse enough to be fairly readily caught by amalgamation with "quicksilver," the popular term for mercury. Before it could be ground fine enough for amalgamation, mine-run ore first had to be crushed dry in a "rock breaker," generally a jaw crusher. Then it was fed, first by hand, then mechanically, to the stampers for secondary crushing and grinding. In its simplest form, the stamp mill battery consisted of a cast-iron mortar box below a group of heavy iron rods mounted vertically next to a horizontal camshaft. A large belt-driven "bull wheel" powered by water or steam turned the shaft eighty to one hundred times a minute. The cams lifted and dropped each stamp five or six inches. Specially shaped and heat-treated shoes mounted on the bottom of

A ten-stamp battery in a mill near Angels Camp. (Courtesy Calaveras County Historical Society)

the stamps protected the rods and also provided a better crushing surface against the dies attached to the insides of the mortar. In California mills water was added to the mortar mix to cut down dust, form a pulp of the pulverized ore, and help keep it flowing through the screens onto the amalgamation tables. Mill men could control the degree of fine grinding by the rate of feeding and the amount of water injected with the ore, aided by the shape of the discharge side of the mortar and by the size of the discharge screen openings.

The battery was used as the initial point of amalgamation, by feeding in a controlled amount of quicksilver. Then the overflow of the pulp was passed across the amalgamating plates to catch as much as possible of the free gold remaining. Later many unique devices were designed for improving the amalgamation process, from plates to barrels, open pans, and enclosed mills. The basic idea behind all these variations was to use mercury to catch any free gold as soon in the milling process as possible, and to continue its use at each later stage in the milling circuit to minimize the loss of any free gold still left in the sulfide concentrates. The amalgam was then retorted to drive off the mercury, and the resulting gold "sponge" was shipped to the mint or smelter. For easier handling, many producers melted the sponges into doré bars or ingots before shipping.

Quicksilver worked reasonably well with free-milling gold, but not with refractory ores. They were also reduced by stamping, but recovering the gold after grinding proved much more difficult. Early engineers tried everything from alchemy to incantation in an effort to break down complex ores, but without formal training in mining, mineralogy, or chemistry, these pioneers could not solve the milling and smelting problems that took trained technicians many years of experimentation to overcome. Little wonder that refractory ores baffled untrained miners like William Perkins whose account of quartz operations at Carson Hill in December 1851 provides a firsthand glimpse of the situation:

> I am more and more convinced that Quartz mining will prove a failure. The vein in our mine is rich; the gold is visible in all parts of it, and yet when it is crushed, with great trouble, the powder mixed with quicksilver, the amalgam does not contain a tenth part of the gold in the stone. The work does not pay, and will not for some years until labor be cheap, and machinery, worked by steam be introduced. I brought away some choice chispas, in which the bulk of gold bears a proportion of thirty and forty per cent to the quartz.[70]

What made Perkins and other Calaveras miners so discouraged was the presence of sulfides (or "sulfurets") and arsenides in lode gold deposits. At Angels Camp in the late 1850s mining men finally recognized that much of the gold was locked up in these metallic minerals. Instead of tossing out complex ores they made crude efforts to roast off the sulfur and arsenic. David Strosberger erected a plant for this purpose in 1857–1858 just south of Angels, but this and similar early attempts failed.

At Carson Hill ores from one of the veins also contained tellurides, especially calaverite, a valuable gold and silver mineral named for the county where it was first discovered. Other high-grade telluride minerals, including hessite, petzite, and sylvanite, although rare along the Mother Lode, were first identified at Carson Hill in 1864.[71] They were even more frustrating to mill men. In the Stanislaus mine on the lower south side of Carson Hill, telluride concentrates assayed as high as seventeen thousand dollars to the ton, but no amount of "massaging" could free the gold. In the nineteenth century most telluride ores were either lost in tailings or left unmined.[72]

Chlorination was the first major breakthrough in treating complex ores. Initially tried in the West at Grass Valley, the process spread gradually.[73] By the 1870s it could be found in most big western precious-metal districts, but it was no panacea. Complicated, expensive, and dangerous, with relatively low gold recovery, it found limited use in larger Mother Lode mining operations such as the Gwin, but smaller operators could not afford it. Evidently, the Gwin introduced the process to Calaveras County in the mid-1870s, but even at this prominent mine the lack of sufficient concentrates kept the chlorination works shut down most of the time.[74]

Before the 1880s only a few of the county's larger and better-established operations were able to develop mines to any significant depth. Hardrock mining in the pioneer period was thus largely confined to small pockets of high-grade ore near the surface. The golden age of mining in Calaveras could not begin until all the essential components — technology, finances, and organization — were available to successfully exploit low-grade and complex ore bodies under difficult underground conditions.

## Financial and Corporate Evolution

Mining as a business enterprise involved a number of separate organizational and financial procedures. Before major investors could be attracted to raise money for actual operations, claims had to be prospected, located, promoted, and developed. In small mines one individual might undertake these steps. Usually, the larger the prospect, the more complex the organizational process and the more people involved in its development.

Early-day prospectors who discovered the lode mines of Calaveras were typical of their breed. Uneducated, hard-drinking, and fiercely independent, scouring the West often on a grubstake, they characteristically failed to reap the rewards of their labor, usually selling out to a speculator or locator long before sufficient development had taken place to ascertain the true character of the ore body. As in the placer camps, the pattern of quick sales, inadequate development, and subsequent resale or abandonment was common in lode mines in this early period.

Prospectors were less important than claim locators in the financial frenzy that often accompanied a mining boom. As soon as news of a discovery spread, locators rushed in, staked claims themselves on whatever likely ground was left, purchased options or claims to better ground from the discoverers, and awaited customers in the form of developers or promoters. As speculators rather than producers, almost all locators lacked sufficient capital and contacts to finance a major mining operation themselves. Most were fortunate simply to be able to pay for the annual assessment work that mining law required to hold a claim. In a sense, locators helped upgrade the industry by culling out poor claims. Most of the outstanding claims passed quickly into the hands of developers, but some of the more professional locators recognized and held the second best for possible sale to bigger fish in the chain of speculation.[75]

A prime target for locators who had claims for sale was the promoter. Even more important than the locator, promoters essentially selected the most promising properties and thus played a crucial role in developing a major mine.[76] Usually a partner or agent of a syndicate organized for investment purposes, promoters scoured the mining world for good prospects. After investigating an available claim, perhaps calling in an independent consultant who prepared a thorough report on the property, the promoter negotiated with the vendor, or locator's agent. Usually, an agreement was reached by which the promoter took an option on the claim or "bonded" it; in either case he promised to purchase the claim at a stipulated price within a certain amount of time, provided a development company could be organized and stocks or bonds floated to raise capital. In the meantime he leased the claim until he could meet the bonding terms. If the terms were not met in the stipulated time, the promoter was freed from his "lease and bond" contract.[77] Smart promoters thus had time to systematically investigate the production potential of a likely prospect before committing major resources to its development. By this means the best promoters improved chances of success by concentrating on the best claims and avoiding the less worthy prospects.

By the time a claim was bonded or purchased, the financial structure for development had been established. Small or intermediate mining properties might suffice with a partnership, syndicate, or closed corporation. Most of the major companies developed a full-scale corporate structure headed by a president or chairman of the board.

The top of this financial pyramid contained the most prominent men in the mining industry. Richard Peterson studied fifty of the most successful mining entrepreneurs in the late-nineteenth-century American West—the "bonanza kings" of the industry—in an effort to develop a social profile that could be compared with eastern counterparts of the same period. He found western capitalists usually of British ancestry, mostly from upper- or middle-class families (although they trailed eastern industrialists in this characteristic), often from a rural background, and frequently starting in the business at the age of fifteen or younger. Only 21 percent of western industrialists in his sample were college educated, while comparative eastern samples show a much higher percentage of college training.[78]

While these findings were based on a sampling of mining capitalists throughout the West, they apply equally well to Calaveras County entrepreneurs. Almost 20 percent of the names on the Peterson list got their start or were active at one time or another in Calaveras County. The names include Alvinza Hayward, Thomas Selby, James G. Fair, William S. O'Brien, John W. Mackay, George Hearst, and Lloyd Tevis, all with Calaveras backgrounds or connections. Most of these mining celebrities had modest although not poor beginnings; all were of British or Irish descent; few had college training; and all were hard-driving, ambitious men not afraid to step on toes as they struggled to first rank in the industrial order.

Many American mining entrepreneurs rose to the top because of their ability to raise money. Sources of domestic capital were limited in America's formative years, but Britain was a prime source of overseas development capital, especially after 1856, when Parliament established limited corporate liability. Prior to that time the doctrine of unlimited liability required stockholders to be personally responsible to creditors if a company failed. Limited liability thus was a great liberalizing force in freeing up capital and bringing the small investor into the financial marketplace, though it did not protect shareholders from financial loss due to lawsuits, assessments, mismanagement, or fraud.[79]

As the center of international mining finance and the metals trade, London provided a lucrative market for American promoters. Before 1870 the amount of British capital in the American West was minimal for a number of reasons, including the high failure rate of many early lode and hydraulic mines, promotional frauds that attracted some gullible investors and scared away many others, land title problems and claim disputes, and the Civil War and its aftermath. By 1870, however, a new era of optimism dawned, interrupted only briefly by the panic of 1873. Encouraged by glowing reports in the financial press and lured by promises of high returns, British investors pumped nearly three hundred million pounds sterling into mining ventures throughout the world between 1870 and World War I. More than half of that amount went into gold mining, primarily in South Africa, Australia, and the United States.[80]

California absorbed much of the British export capital pouring into American mines, providing a welcome overseas infusion to offset the drain of domestic capital going into Comstock speculation in the 1860s and 1870s. Between 1870 and 1888 some seventy-three British companies were operating in the California goldfields. The failure rate remained high, partly because of fraudulent promotion that was not tightened up significantly until after the turn of the century, and partly due to the lack of adequate development and technology to work complex ore bodies. Lacking good information from the field, many companies simply ran out of capital after spending far more than the property was worth.[81]

Calaveras County was not immune to promotional frauds. Perhaps the most notorious local promoter was Windsor A. Keefer. For more than two decades, using a string of interlocking corporations, he obtained control of the Union Water Company's water rights and thus was in a position to monopolize the water supply for hydraulic operations between Altaville and Dogtown. For a time even part of Big Trees Grove was a target of his financial shell game, an early-day version of the infamous Ponzi scheme of the 1920s. When gullible stockholders grew wary or even initiated lawsuits, he cleverly shifted assets to other companies, obtained more funds from new stockholders, and left investors holding a few worthless shares. In 1897, running out of customers and just ahead of the law, he absconded with a pocketful of loot. He covered his trail by leaving his San Francisco apartment in tatters and clues suggesting that robbery and murder explained his "strange disappearance." Years later he turned up in Paris, living lavishly but beyond the reach of U.S. authorities.[82]

Ironically, some Calaveras lode mines might not have been developed at all without spurious promotions that first attracted investment capital. In the early days of the Utica, for example, "Slippery Jim" Fair allegedly salted and sold the claim long before its major ore bodies had been located. It took dedicated developers like Robert Leeper and Charles Lane to bring it into production. The Utica was an exceptional property compared to most Calaveras mines, regardless of how vigorously they were promoted. Thomas A. Rickard, an experienced engineer as well as a talented writer and editor, describes one lamentable experience in the late 1880s, when he took charge of the Union mine, on the Mother Lode a few miles from San Andreas. It had been purchased at an inflated price by a gullible London investment firm upon the recommendation of a "smart American promoter" who had "pose[d] as a mining engineer." Rickard later found that his engineering experience consisted of painting "cuspidors and coal-scuttles for the Union Pacific Railroad." With unbounded confidence, the firm sent the promoter back to manage its new property. He promptly "built a twenty-stamp mill so that part of it was off the property of the company, but on another claim upon which he had obtained an option, from which he expected to gain further financial loot." He also bought cordwood from fraudulent woodcutters who padded their income by stacking the cords in such a way as to inflate the quantity of wood available. By the time Rickard took charge there were seven lawsuits pending against the mine, which was only three hundred feet deep on a minor quartz vein assaying about $3 per ton in gold. Under the best circumstances Rickard estimated the production costs at $2.50, but lawsuits and other problems cut into that slim margin. "We were doomed to be sunk," he concluded hopelessly.[83]

Even honest promoters in Rickard's day could not be fully trusted by the investing public. Many were inexperienced and ignorant of the rudiments of geology or mining, or were influenced by unscrupulous operators looking for a job. All too often they set up mining propositions on claims of little or no potential and disposed of them in good faith to equally ignorant or credulous investors. More than a few had little regard for the interests of their clients, as historian Lewis Atherton suggests in characterizing the dubious morality of one promoter: "In general, he followed the common rule of exacting the highest possible commissions and upped the asking price for property placed in his hands as much as possible with the idea of retaining such advances in the selling price for himself."[84] Under these circumstances, very few promotions resulted in a producing mine.

Despite all these pitfalls, mine promoters operated openly and legally, going from district to district peddling their dubious wares. Without a regulatory regime it was a Barnum-and-Bailey's world for the fund-raiser in the late-nineteenth-century mining West. Even a promoter with a well-known name and a credible mining background was no guarantee of honesty or success. The celebrated Philipp Deidesheimer, a mining engineer whose early success on the Comstock blinded him to some basic facts in mining development, came on hard times late in life. Turning to promotion, at the turn of the century he tried to dispose of a number of worthless claims in Calaveras and elsewhere.[85]

Promoting a mine was relatively simple compared to making it pay, especially in the promotional waves that often followed the occasional big success or that were set off by major economic trends. The most troublesome problem in Calaveras and in most other western mining regions was not lack of financing but inadequate development of unproved prospects. Promoters, investors, and managers almost always underestimated the cost of establishing a profitable lode operation. Most of the initial investment capital went into the purchase price, leaving little for mine development and operations. Inexperienced shareholders usually assumed that these costs could be paid for by the proceeds from bullion sales. Once their initial investment was consumed, they became suspicious of management appeals for more money. The lack of working capital was all too often the harbinger of failure. Local suppliers extended liberal credit to local managers, but this generosity invariably weakened the financial position of the mine and left it vulnerable to foreclosure.[86]

The nature of Mother Lode ore bodies added to the financial difficulties of early mine managers. All along the broad zone of gold mineralization from Mariposa to Georgetown, and even into the Northern Mines along the Sierra foothills, ore shoots, in the parlance of mining engineers, were "far greater in down-dip dimensions than strike length."[87] To find, develop, and mine

deep ore took considerable experience and more than a little luck, two attributes often in short supply along the Mother Lode. Too often in mining
history, managers overbuilt the surface plant for the amount of ore in sight,
and then lacked the capital for extensive underground exploration and development. Part of this was due to a lack of professional training. In the pioneer
era, when trained geologists were both scarce and distrusted, lode operators
relied mostly on visual examination at the mine's face, personal experience,
and a certain amount of guesswork to find pay dirt. Predictably, the results
often disappointed shareholders.

But lack of rational planning was also a consequence of the financial exi-
gencies inherent in precious metals mining. Committing capital to a mining
prospect is a risky gamble even to the most experienced investor. Gold fever
can strike just as hard today as in 1848; so can the lure of phony promotions.
In the words of a recent historian: "The mine swindle is part of our national
heritage."[88] The difference today is that gullible investors are reasonably well
protected by regulatory agencies and by the courts from outright fraud, but
not from their own foolish speculations.

In the nineteenth century most investors had little recourse, regardless of
how or why they lost their money. Despite the risks enough individuals with
money could usually be found to start up a new mine that sounded good on
paper. As Patricia Limerick remarks, "[P]ersuading eastern fools to invest in
an unseen mine two thousand miles away was, for some, a much more agree-
able way to make money than actually working the mine."[89] Often, what the
fools did not understand were the complexities involved in bringing a promising mine into full production. The result was that mine managers were under
constant pressure from owners and shareholders, who wanted quick returns
on their money.

The normal response under these circumstances was to maximize production regardless of consequences. Hard-pressed operators took whatever ore
they had in sight without attempting to find and block out additional ore
reserves. Wise businessmen termed this "picking the eyes out" of a mine.
The usual outcome was to shorten the mine's operating life, use up the operating capital in productive spurts rather than systematic development, and
condemn the mine to inevitable closure when the backers ran out of money
rather than out of ore.[90] As we shall see, the Carson Hill experience is a good
example of this unfortunate cycle that could repeat over and over again.

When hard-pressed for development capital, management before 1900
often resorted, in desperation, to "Irish dividends"—mining slang for stockholder assessments. Later the larger companies turned to bond sales as an
alternative financing strategy. Neither method, obviously, tended to inspire

confidence in the investment public. Lode mines thus easily fell victim to a downward spiral that started with inadequate capital for development. Like a chain reaction the result was often production delays, the decline of confidence, reduced stock sales and market values, the drying up of credit locally, and, finally, failure. Usually, the biggest losers were the small speculators and the local creditors, both of whom had too much at stake to bail out early. On the other hand, the promoters and the syndicate partners or developers—the top rung of the investment ladder—could usually write off losses without severe financial strain or interruption.

Some mines never recovered once sucked into this financial whirlpool; others came to life again only after years of dormancy and reorganization. The ups and downs of Mother Lode mining kept promoters constantly on the lookout for claims or prospects that might be worth reexamining. Sometimes mine owners or discouraged shareholders sought the aid of promoters to unload their property; at other times speculators sent promoters to well-known districts looking for prospects or properties that had good potential. The result was a constant turnover, with old mines changing owners or reverting to original owners and heirs when leases ran out or operators shut down. When a mine reopened under new owners or leasers, the cycle began all over again.

This financial pattern is well illustrated by the promotional efforts at Angels Camp in the mid-1860s. In 1865 W. A. Williams, a San Francisco promoter, discreetly tried to inspect the main prospects on the Davis-Winters lode, where quartz mining had fumbled along without much success below three hundred feet due to the sulfides that had defeated most early milling efforts. Representing a syndicate promoting the Stow and Macdonald (chlorination?) process, he hoped to tie up the Hill, Maltman, and Utica mines in a favorable lease and bond before competitors arrived to force up the price. But his efforts were defeated by a sudden boomlet in Mother Lode quartz, initiated by rival interests who arrived just as Williams was about to close a deal. He registered his disgust in a letter to his sponsor, Joseph W. Stow:

> There is no loose things lying around here in the shape of quartz ledges, for there is too many buyers here, or parties in search of sulphurets leads. There is quite an excitement up here about sulphurets leads, and parties that own quartz mines, who but a short time ago would have been glad to have sold for almost any price, now look a man right in the face and ask $50,000. Last night some ten or a doz quartz sharps arrived in town, among them . . . a lot of Grass Valley men. This morning bright and early they rushed off up in the mountains some of them bauld [sic]

headed. Who ever has had a hand in getting up this excitement certainly has done it well.[91]

The financial and corporate evolution of the larger Calaveras lode mines followed the general pattern described above. Calaveras had its share of lo-cators, promoters, British investors, and developers as well as a rather spotty profit-and-loss record. Indeed, in the pioneer period before 1885, except for the Gwin and the Sheep Ranch, Calaveras lode mines usually failed to re-turn the initial investment, much less earn a profit for their owners or share-holders. The record in the golden years of lode mining, from the late 1880s to World War I, is somewhat better, but only a few major mines even in this bonanza period could be said to be really profitable. It must be remembered that production is a far cry from profit. Although many Calaveras mines were productive, most of the big producers ate up revenue faster than they could generate it. As we shall see, the most notable example is the Carson Hill group of mines that produced many millions but apparently never paid a dividend before the 1930s.

Despite this rather sobering financial picture, Calaveras residents may take some comfort in the fact that their county was typical of the mining enter-prise as a whole. In lode mining, as in placer mining, the majority of investors came away wiser but poorer. On the other hand, whether a mine made money for its backers was only incidentally important to most Calaveras residents, who had little money invested but were directly affected by the amount of money spent locally on mining promotions. Thus, Calaveras boomed in the golden years regardless of the red ink on corporate balance sheets.

## Early Copper Mining

During the Civil War the Calaveras copper boom caught the imagination of capitalist and prospector alike at a time when California gold production had drastically declined from the spectacular days of the Gold Rush. Copper metallurgy dates back to the beginning of the Bronze Age. It is thus hardly surprising that copper fever in Civil War California cast a wide-ranging spell that infected nearly every mining region and promoter from the northern Siskiyous to the Mexican border. Surface gossan, the highly oxidized, rusty outcrop of many metal deposits, was an attractive but deceptive lure to the mining speculator. It sent copper shares soaring on the San Francisco ex-change and caught up unwary investors. Even Henry George, a decade before he led the single-tax movement against land monopoly in California, lost at least five hundred dollars in a dubious copper venture near Marysville.[92] By the early sixties, however, the mining world had its eye on Calaveras County, considered by the "experts" to be the most promising copper district in the state.

Calaveras accounted for more than 19 million pounds of copper in the 1860s. Although tiny by modern mining standards, and little more than 10 percent of the total Calaveras copper production of some 155 million pounds, this output in the sixties gave Calaveras second place in the nation. The most important producers at that time were on upper Michigan's Keweenaw Peninsula, contributing three-fifths of the country's copper supply. Significantly, Calaveras and the Keweenaw together gave the Union a monopoly of U.S. copper output, strategic material that helped Union armies defeat the South, which had to rely on copper and munitions from foreign suppliers willing to risk the Union blockade.

### Early Discoveries

As early as 1851, Mexican miners in the Campo Seco district found specimens of native copper. Some of this material was put on display in Stockton. However, Hiram Hughes touched off the first real excitement in the spring of 1860. A Calaveras pioneer who had rushed to the Comstock in 1859, he returned the next year to prospect formations that resembled Washoe outcrops along Gopher Ridge southwest of Salt Spring Valley. In May 1860 he located the Quail Hill claim on a prominent gossan. He called it a gold strike, although it was soon found to have more copper than gold.[93] Later that year, at Hog Hill, some three miles south in similar gossan, Hughes and his son staked the Napoleon claim.

News of the Quail Hill strike soon brought other prospectors scrambling over the area. In August, William K. Reed, Thomas McCarty, and Dr. Allen Blatchly located claims on the big gossan about six miles east of the Hughes strike at the present site of Copperopolis, first called Copper Cañon. According to contemporary press reports, Reed was "a penniless miner" who found the lode "while returning to his home at Burns' Ferry from the Hog's Hill copper leads."[94] He took some samples of the gossan to Thomas McCarty at his "Log Cabin," a well-known stage stop on the road from Stockton to O'Byrnes Ferry. McCarty showed them to Dr. Blatchly, a Stockton physician and businessman who was visiting the area because of the Hughes discovery. According to a contemporary account, Blatchly,

> being something of a mineralogist, at once pronounced the specimens copper of most flattering indications. Reed, McCarty and Blatchly started out early next morning, and located and staked off for them-

selves and friends all the ground now embraced in the Union, Keystone, Consolidated, Empire and Calaveras claims—11,250 feet. . . . The first work was done by Dr. Blatchly, at what is now known as the Discovery shaft, upon the Keystone mine. The Doctor obtained good shipping ore at a depth of 25 feet. Mr. Reed soon afterwards commenced work upon what is now the Union Company's ground, and obtained good ore at about the same depth.[95]

Dr. Blatchly soon sold out his interest and left for Nevada. But unlike most prospectors the other two men held onto their discovery claims to reap the reward of escalating copper stock prices.

Assaying as high as 33 percent copper, or $150 per ton at the prevailing market price, the secondary enriched zone ore of the "celebrated Reed lead" ushered in the boom.[96] Within a matter of days there were many claims along the gossan, and soon the Union, Keystone, and Empire began producing ore. At Telegraph City further discoveries were made, and within a few months the Quail Hill, Napoleon, and Collier mines were hoisting ore. As the copper excitement spread along the West Belt, prospectors in 1861 staked claims on copper ore outcrops near Campo Seco where Mexicans had found native copper "nuggets" a decade earlier. The secondary enriched zone was much less pronounced than at Copperopolis, and Campo Seco did not develop as fast or as extensively as its bigger neighbor to the south. But it did have a parallel development, and it contributed significantly to Calaveras copper production during this first boom period. Here the Copper Hill mine was shipping sacked ore before the end of the year, and the Lancha Plana and Campo Seco mines were close behind.

### The Copper Boom

The copper excitement promised much more than it ultimately delivered, but for a five-year period copper was "king" in Calaveras County. It is astonishing how rapidly the new mining communities came into being, considering the remote location from market and the incredibly difficult transportation, both for supplies and equipment coming in and for ore shipped out. The very wet winter of 1861–1862 compounded these problems. Yet between February 1861 and July 1862 a total of 3,050 tons were shipped in 57,721 bags from the Union, Keystone, Hughes, Napoleon, Copper Hill, and smaller mines, all going to Stockton for transshipment.[97]

Most of this tonnage was handled by the Stockton firm of C. T. Meader and Company. Charles Meader, a forty-niner and San Francisco merchant before coming to Stockton, had been introduced to the copper industry as a boy in Massachusetts. His father, Reuben Meader, scion of a prominent Nantucket shipping family, was involved in banking as well as copper smelting and shipping in the Boston-Taunton area.[98] Now as a Stockton merchant, Charles was well positioned during the Calaveras boom to take a major role in the mining and marketing of copper ores.

The principal mine, the Union, put Calaveras County on the map in a year when gold mining along the Mother Lode was sinking to the lowest point in nearly fifteen years. The copper mines employed dozens of "penniless" miners at forty dollars a month. The boom perked up lagging business and transportation in Stockton and its satellite towns along the southern Mother Lode. The town of Copper Cañon, founded almost simultaneously with the opening of the Union and Keystone mines and renamed Copperopolis a year later, impressed a Stockton newspaper correspondent when he visited it in the summer of 1865. "This community," he wrote, "is one of the most flourishing and active towns in the interior of the States—a town containing twenty-two saloons, three blacksmith shops, six stores, two druggists, three livery stables, two wagon-maker shops, four hotels, and three restaurants, three schools, two churches, and a weekly newspaper—a town consisting of good buildings, some of them even elegant; having a population of nearly 2,000."[99]

The boom also generated a frenzy of stock speculation. At the height of the copper excitement, Union stock zoomed upward. In February 1862, for example, W. K. Reed sold his five and one-half shares to C. T. Meader for fifty-five thousand dollars. By September the price of a single share had risen to twenty-five thousand dollars. The following year Union stock skyrocketed to an astonishing value estimated at three hundred thousand dollars per share! The Union's substantial production figures added fuel to the speculation and led effusive promoters to predict that "Copperopolis will not retrograde for a century."[100]

That prediction seemed only mildly inflated during the Civil War boom years. In 1865 alone Meader's company shipped more than 23,542 tons of high-grade ore from the Union to Boston. Meader and other exporters also shipped ore to New York and to Swansea, Wales. Shipments to Swansea had begun the previous year but did not reach significant amounts until after the war. Assuming an average copper content of at least 12.75 percent per ton, the Union produced some 3 million pounds of pure copper.[101] But the Union had mined out the secondary enriched zone and was now in the leaner primary ore.

For a time the boom continued even after Appomattox, but falling copper prices clouded the rosy picture portrayed by the wartime optimists. One of the first local mines to close was the Keystone, just north of the rich Union mine. Full of promise in the early boom days, it disappointed miners and shareholders alike. Captain "Dick" Powning, an experienced operator from Cornwall, had vainly tried to make the mine pay, but the complex vein structures and prevailing low average ore grade defeated his best efforts. His successor in 1864, H. H. Sheldon, had no better luck. At its peak of production in 1865 it shipped only 1,500 tons of ore, less than 7 percent of the Union's total for that year. Only years later, after consolidation with the Union, did the Keystone achieve substantial production.[102]

## The Smelting Problem

The California copper industry had begun with the premise that crude ore had to be shipped to smelters on the East Coast. The high shipping costs required that the ore average from 15 to 20 percent in copper, a prerequisite that at first had been fairly easy to meet, especially by hand-sorting the crude ore hoisted from the mines. But most of the ore in these deposits ran much lower in grade and had to be either hoisted and discarded, gobbed underground, or simply left unmined, thus hampering orderly mining. George Stone, who worked in 1864 and 1865 at the Union mine for a monthly wage of forty dollars plus room and board, left a vivid description of the sorting process:

> The quartz containing the copper was taken out in bulk and distributed to men working on top under a large tent, who broke it up and assorted it into grades of richness. The men sat on the ground and with small hammers broke the ore into small pieces from the size of a goose egg down. This was inspected and the very rich ore thrown into one pile, the next best into another, and all the remainder was thrown aside and hauled to the dump. Weighed and loaded into 120-pound sacks, the best grade of ore was shipped by freight wagons to Stockton, transshipped by river boats to San Francisco and by sailing vessels or steamers to east coast smelters.[103]

By 1863 the costs of mining and sorting primary lower-grade ore made the future operation of these copper mines a matter of growing concern. At the Keystone, Captain Powning tried to reduce costs by building a concentrator to upgrade ore prior to shipping, but concentration of base-metal ores is more complicated than for precious metals. Powning lacked the technical knowledge, the equipment, and the experienced mill men necessary for success.[104] The better solution was smelting, preferably on site, but that posed an additional set of problems that inexperienced and isolated Calaveras operators could not effectively resolve.

In view of these limitations it is not surprising that the first approaches to smelting in California were crude and ineffective. To prepare it for blast furnace treatment, copper ore first must be roasted, or calcined, to drive off arsenic and antimony and as much of the sulfur as practical. Then the furnace must be properly charged with ore and the correct fluxes, and fired at an adequate temperature. At most California mining sites the only available fuel was charcoal, which did not provide high enough temperatures for effective smelting. The blast furnace product, called matte or regulus, contained 35 to 40 percent copper, but early smelting efforts lost a substantial amount of copper in the slag.[105]

The first serious attempt at smelting in California was in 1863 with the construction of the California Copper Smelting Works at Antioch by San Francisco investors. The site, located on the Sacramento–San Joaquin delta, had obvious transportation advantages, but it was also situated to take advantage of recently discovered coal deposits on nearby Mount Diablo. Unfortunately, the Antioch smelter was so beset with technical and other problems that it never came into full operation until after the Civil War.[106]

At Copperopolis the lack of good alternatives during the copper boom led to the first local smelting efforts. Charles Meader, with his background and financial backing from his father's bank in Massachusetts, started the most successful early California smelter at the Union mine. Despite formidable technical difficulties Meader's smelter, under the supervision of M. P. Desermeaux, a smelter man of limited experience, began operating in 1864. It produced matte from two small blast furnaces and by early 1865 was able, through a secondary smelting process, to turn out small quantities of "black pigs" or ingots that assayed more than 90 percent copper.[107]

Other Calaveras smelting experiments started soon after the Union smelter fired up. At Campo Seco, William B. West constructed the tiny Taunton Smelting Works and began operating in November 1865.[108] By 1867 smelter practice was improving, but not enough to offset increasing costs and falling prices.

Smelting ended in Calaveras without making much of a mark on the local economy, but on the local environment it left a more lasting impression. Copperopolis residents tolerated the toxic fumes better than the vegetation near the crude roasting kilns. These were little more than hot chemical sandwiches piled on the ground, left smoldering for weeks at a time. A Stockton correspondent, visiting the scene in 1864, described the environmental consequences but found reason for optimism:

> Some weeks ago the Union Copper Mining Company erected concentrating works near their mine at Copperopolis, for the purpose of rid-

Roasting ore at the Union mine in Copperopolis, ca. 1864. (Courtesy Holt-Atherton Library, University of the Pacific)

ding inferior ores of enough base and foreign matter to make them transportable. The smoke from these works was so charged with sulphur and arsenic that the grass and vegetation all around within the circle of a hundred and fifty yards was killed, and the leaves on the oaks and buckeyes withered and dropped off. A few weeks since these works were temporarily suspended and since then the oaks and buckeyes have put forth new buds which are now unfolding into young leaves, imparting to the surrounding woods all the appearance of early spring. One would suppose that the fumes this fatal to vegetation would be also deleterious to human health, if not fatal to life, but such is not the case. Experience in other countries proves that the people living within the daily influence of the poisoned vapors are about as healthy as those of other neighborhoods where they are not felt. This is a strong argument in favor of homeopathy, which recognizes arsenic as the chief of medicines, though doubtless it is best taken in homeopathic doses.[109]

Later generations would better understand the effects of toxic chemical releases on environmental quality and human health, but in the 1860s few worried about the adverse consequences of crude smelting efforts. Copperopolis and Campo Seco residents, heavily dependent on mining and accustomed to environmental degradation as a "legacy of conquest," in Patricia Limerick's poignant phrase, welcomed these two little smelters as signs of progress.

## The Copper Decline

Most accounts attribute the copper-mining decline primarily to falling world prices for the red metal after 1865. Just as war created an artificially high price for copper, war's end pricked the bubble. J. Ross Browne noted that in 1865 the price for 10 percent–grade copper ore at Swansea, Wales, was $41.50 a ton.[110] By 1866 it had dropped to $33.87, even though it recovered slightly the next year, reaching $36.50. However, in the United States, prices for the metal continued to slide from the peak of 47 cents a pound in 1864 to 25 cents in 1868.

Other factors also contributed to the depression in the copper districts after 1865. Transportation charges from port to smelter continued to rise. In the first eight months of 1865, Meader's company shipped 16,464 tons of ore on East Coast or foreign-bound vessels at a cost of $132,000, or $8 per ton. By 1868, as commercial shippers restored and then surpassed prewar shipping levels to Europe and thus caused a shortage in cargo space, copper shipping costs had risen to as much as $16 per ton.[111] Thus, only the highest-grade ore could stand the mining and shipping costs. In addition, much of the high-grade secondary enriched zones at most of the mines had been "stoped out" by 1866, and the deeper primary sulfide ores were too low grade to ship. The Keystone concentrator and the Union and Taunton smelters were not yet efficient enough or large enough, and were plagued with high costs for fuel and labor. They also had technical and pollution difficulties. Finally, winter rains increased pumping at the deepening mines, turned dirt roads into quagmires, and delayed ore reaching market. These endless problems for mine operators and exporters added substantially to the cost of financing inventories.

By April 1867 copper mining had virtually ended in Calaveras County. The Napoleon mine, once reputed to be prosperous, dismantled its corporate structure in 1866 and filed bankruptcy papers early the next year. The Hog Hill area had declined months before, devastated by a geologist's report that "there is not money enough in San Francisco to sink a shaft upon the summit of Hog Hill, sufficiently deep to produce the copper that would be necessary for the construction of a copper tea-kettle."[112] Even the great Union mine was in trouble by early 1867. Charles Meader was dangerously overextended. Despite steadily falling copper prices he had acquired controlling interest in the Newton mine in Amador County and was financing additional mine development along the West Belt. With most Calaveras copper ore passing through his hands, inventories were piling up at the mines and smelters, and on the docks in Stockton and San Francisco. In addition, because of uncertainties in ocean transport, there were long delays between the shipment of ore from the mines and the receipt of ore by refineries on the East Coast and at Swansea. Meader was thus rich in ore but cash poor, and in the wake of declining prices he could not keep afloat. In April 1867 his Stockton business suspended operations, and a few months later, failing a desperate effort to raise money by assessing stockholders, he shut down the mines when operating revenue ran out. A disastrous fire that leveled half the town of Copperopolis in September added to the slump that was climaxed early in 1868 when Meader declared bankruptcy and creditors foreclosed. The "boom" had "busted," shutting down copper mining in Calaveras for the next twenty years.[113]

Far from the economic panacea its promoters had predicted, copper mining ended in a dismal crash. But it had sustained a spectacular five-year boom that revived the lagging economies of both Calaveras and Stockton, helped the Union win the Civil War, and stimulated the development of improved transportation connections between the Mother Lode and the Central Valley. The transportation impact lasted much longer than the copper frenzy that had helped create the demand in the first place. Finally, this first Calaveras copper excitement opened two copper districts that would continue to produce the red metal off and on for nearly a century. These two districts, it should be remembered, substantially added to Calaveras County's mining economy. Only the gold mines at Angels Camp and Carson Hill would exceed the economic importance of the Campo Seco and Copperopolis operations.

* * *

Calaveras lode mining during the transitional years from the Gold Rush to the 1880s illustrates the difficulties inherent in trying to build a nascent industry without all the necessary ingredients. Hardrock mining in California could not grow without practical tools and methods, adequate understanding of Mother Lode geology and mineralogy, a grasp of the metallurgical complexities of refractory ore, sound financial backing, cheap and efficient transportation, a staff of experienced managers, trained technicians and engineers, and—perhaps most important—a large and dependable labor force. Deficient in many if not most of these fundamental components, lode mining in the immediate post–Gold Rush years stumbled along. It hit a few strikes here, struck out there, sunk prospect holes and shallow shafts, spent large sums on machines and mills that worked poorly or not at all, and used up investment capital with very few returns. But all the time it was learning the ropes, gaining experience, and developing an industrial base that would carry Calaveras into a golden age after 1885.

By the mid-1880s a new pattern was beginning to appear. Instead of the desultory, fragmented, often inept operations of local pioneers, most of the best claims had been absorbed by outside developers who sent professional engineers to take control of mining operations. Gradually in Calaveras and elsewhere, engineers assumed administrative as well as technical leadership.[114] The stage was set for an upgrading of both capital and management, which would lead to major exploration, development, and production.

The corporate and financial evolution in the mining industry during these late-nineteenth-century decades mirrored similar trends in other fields. Between 1870 and 1900 the Far West, especially California, began to grow out of its pioneer status and reach toward economic maturity. Transcontinental rail lines tied the West to eastern merchandise and markets. San Francisco shed its frontier trappings and emerged as a specialized urban-industrial metropolis, the economic and financial capital of the West. Foundries and machine shops grew into major centers of technological innovation and expansion, building and exporting the machines of industry and agriculture throughout the West and beyond. The wealth of western mines flowed into the cities, especially San Francisco, but Denver, Salt Lake, Portland, Seattle, and Los Angeles as well. This capital upsurge stimulated a host of ancillary industries and services, and provided the material basis for a cultural growth that manifested itself in theaters, magazines, newspapers, luxurious homes, and objets d'art.

The limitations of a material culture based on environmental exploitation seems obvious to most Americans today, but belief systems were different a century ago. Even in its infancy lode mining left a residue of scrap iron, denuded hills and valleys, polluted air, waste dumps, and contaminated water supplies up and down the Mother Lode. But except for downstream flooding caused by hydraulic mining farther north, mining's impact on the environment was a nonissue to nineteenth-century Americans. They were heirs of a Judeo-Christian tradition that long ago separated mind and matter. From the Enlightenment they had inherited the Idea of Progress, the notion of bettering human lives by utilizing the land and its resources. Connecting human health with land health is a relatively recent concept, and so are epidemiology and the study of environmental pathogens. Only in the last thirty years of the twentieth century did it become fashionable or feasible to weigh the costs of progress against the health of people or landscape.

If miners brushed off environmental problems, they could not ignore social issues. Despite signs of material progress in late-nineteenth-century California, social maturity did not always accompany economic maturity. In the mining regions, as well as in the supply centers and coastal cities built by the mines, social unrest found frequent expression. Just as the troubled growth of Calaveras lode mines represented a microcosm of western economic development in the transformational years from the Gold Rush to World War I, the social tensions in the mining camps mirrored a larger society struggling to reconcile its racial, ethnic, linguistic, and cultural differences.

# 3: Mining Society in the Early Years

In 1948 California celebrated the one hundredth anniversary of the Gold Rush. It was a gala affair, with parades, speeches, dances, and fireworks up and down the state. With the melting pot myth still intact; with environmental concerns still in abeyance; and with leadership positions in solid, white, Anglo-Saxon hands, the yearlong event was unmarred by controversy. Celebrants cheered the material development, the growth of cities, the industrial advance, and the high living standards, all attributed to the march of progress started by the argonauts of 1848. At a state chamber of commerce meeting held to plan the 1948 events, Earl Warren, California's progressive governor, told his audience that California could look forward to a better future, based on the heritage of its gold mining ancestors who had contributed so much to the state's well-being.[1] Coming just three years after America's triumphant victory over totalitarianism in World War II, the Gold Rush centennial reflected the national mood of optimism, prosperity, and unity.

Fifty years after this first celebration, California began another Gold Rush commemorative, but this time the mood was different. Instead of promoting victory parades, the Gold Rush Sesquicentennial Commission tried to steer clear of controversy. Determined to avoid a repeat of the contentious clamor that had marred the Columbian anniversary of 1992, the commission encouraged educational programs. In 1998 pioneer pageants were less visible than panel discussions over the meaning of the events 150 years before, but that may have been the result of lack of funding rather than lack of enthusiasm.

What had changed? Americans in 1998, after a half century of social and political foment, could no longer accept an innocent celebration of the past because the past was no longer innocent. The civil rights movement, the environmental movement, the free speech movement, the "new history," the proactive laws and legal cases since the 1950s designed to correct past problems—all had combined to alter the national perception of what the United States stood for. A dark side had emerged to cloud the happy perspective of 1948.

Among scholars, recent Gold Rush historiography has shifted significantly from the rosy perspective of older generations. Susan Johnson, for example, has argued that the Southern Mines have been neglected as a focus of scholarly study because they do not fit the traditional interpretation of the Gold Rush. They were "poorer" than the Northern Mines, and had a much larger per-

Three Calaveras old-timers pose for a Gold Rush celebration. Wade Johnston is
at the sluice, Louis Weisbach works a rocker, and Jim Waters uses a gold pan.
(Courtesy Holt-Atherton Library, University of the Pacific)

centage of foreigners and minorities. Instead of exemplifying the progress of
American capitalism, democracy, and individualism, the Southern Mines, in
her view, illustrate the inherent conflicts that underlie America's fundamental
attitudes toward race, gender, and class.[2]

Peeling back the outer layers of Calaveras Gold Rush history reveals some
of the scars that rippled across the social landscape of the mining West in
the nineteenth century. Like one strand of an intricate web, Calaveras reflects both the highlights and the shadows of the larger structure of a frontier
society in an aggressive age, filled with nationalist fervor and ethnocentric
biases. The predominance of young, unrestrained American males, heavily
armed and highly competitive, added volatility to the demographic mix.[3] For
women, foreigners, and minorities it was a troubling, sometimes tragic, era,
exacerbated by the flood of newcomers from all over the world into a frontier region lacking the normal constraints of law and custom. Though the
levels of oppression varied from mild to severe, depending on the location,
the nationality, and the circumstances, for all too many victims the confrontation was fatal, and for many survivors the Gold Rush experience left deep
psychological wounds that carried into modern times.[4]

## The Gold Rush of Immigrants

The Gold Rush in Calaveras entered a more turbulent phase with the first wave of immigrants during the great rush of 1849. Instead of perhaps a few hundred gold seekers exploring the watersheds of the Mokelumne, Calaveras, and Stanislaus Rivers, now there were thousands, mostly Americans from the Northeast, South, and Midwest. The 1850 census for Calaveras included only 9 percent Hispanic "Californios" and Native Americans. Of the 16,884 members of the county population, 24 percent were born outside the United States. The largest contingents of foreigners in Calaveras came from Britain (30.1 percent) and Mexico (29.6 percent), with the remainder from a scattering of other countries, primarily European. While the northern mining counties had a slightly smaller proportion of foreigners than Calaveras, still the county was considerably below the statistical average (34.5 percent) for foreign-born in the southern Mother Lode. Thus, Calaveras was thoroughly Americanized, though it had sizable minority populations.[5]

J. D. Borthwick, a keen observer from Scotland, found the ethnic and racial differences quite remarkable when he toured the goldfields in 1851 and 1852. The northern camps he said were "almost entirely composed of Americans," with only occasional European, Chinese, or Mexican workers, whereas the Southern Mines "were full of all sorts of people." Yet there was little sign of integration at first, either north or south: each group formed protective enclaves based on race, language, nationality, and ethnic identity, even religion. In the Southern Mines, he wrote, some villages were nearly all Mexican, while others were nearly all French. Every town had a large foreign population, and Chinese camps were "very numerous." In one Calaveras gulch he found some two hundred Chileans at work, but at San Andreas it seemed the Mexicans were predominant. The farther south he went the more diversity he experienced. Across the Stanislaus at Sonora he saw store facades covered with sentences in English, French, Spanish, and German. The town had a strong Mexican influence, but there were also Italians, Dutch, and Chinese, each with their own restaurants or boardinghouses.[6]

## The Native American Response

Caught up in the social turbulence of the Gold Rush were the Native peoples of Calaveras, the common victims of exotic microorganisms and ethnocentric aggression from 1848 to modern times. Archaeology and anthropology have combined to assist the historian in reconstructing the lives of the Central Sierra Miwok, a peaceful people living in several villages within the Stanislaus and Tuolumne River watersheds. Adjacent to them to the north, within the Mokelumne watershed, was another group, the Northern Miwok, linguistically related to the Central Sierra Miwok but speaking different dialects. Before the Gold Rush several thousand Natives occupied the undulating landscape of what is now Calaveras County. Their culture and lifeways resembled those of other isolated California Indians nestled in the Sierra foothills. Before the Gold Rush their isolation and remoteness had spared them from much of the devastation suffered by more vulnerable coastal and valley tribes, but the onrush of argonauts and their accompanying diseases cut through them like a scythe, decimating their ranks, forcing them onto marginal lands, and forever changing old patterns of living.[7]

Although disease took the greatest toll, California Natives were also victimized by the ethnocentric attitudes of the intruders. Travelers who crossed the high plains and deserts of the Great Basin en route to the West Coast traversed an unfamiliar landscape that cut across cultural as well as physical boundaries. Frightened and physically exhausted by the searing heat, the parched earth, the paucity of vegetation, the clouds of dust, and especially the lack of good water, they saw no good in either the land or the Native people who lived there. The latter they called "diggers"—a far cry from the "noble savages" they had read about as children in *The Last of the Mohicans*. In the Great Basin some Natives did dig camas roots for food, but the label was not merely descriptive. It was also a term of contempt, a badge of inferiority in an age that measured success by conventional Euro-American standards of dress and diet. The reflections of one '49er illustrate the prevailing attitude: "The Digger Indian, is, or was at that time, the lowest type of American Indian, at least he was the lowest thing I ever saw to have any pretention to humanity." Writing years afterward, he said it was hard for later American readers "to understand the feelings of hatred and loathing with which we got to regard these people. . . . I know that we classed them in our minds with the wolf and the rattlesnake, at least the adult males."[8]

Americans in Gold Rush California saw no distinction between the Native people of the Great Basin and those in the Sierra foothills. Regardless of differences in language, dress, and culture, they were all "diggers," debased and disgraceful. John De Laittre, a Murphys miner, said they were "the most filthy and degraded in their habits of any human beings I ever saw." James Carson also used familiar '49er language to describe them in his 1852 reminiscence: "The only thing that can be called *human* in the appearance of the digger Indians of the Sierra Nevada is their resemblance to the sons of Adam."[9]

The Latino view was not much better. Reflecting attitudes derived from generations of conflict between Native Americans and Hispanic intruders,

Ramon Gil Navarro, a high-caste Argentinean political exile living in Chile before he joined the Gold Rush, characterized Native women he watched cooking acorn meal on the Calaveras in 1849: "All of these Indians seem naturally pretty civilized, because they do not flee the white men. They are all still naked, both the men and the women, and only put on a bear skin to protect them from the cold, though they take it off to work. It is hard to imagine a more horrible race of men. Only in the women can you occasionally find some whiter skin and better-looking features."

Relegated to inferior status by ethnocentric Euro-Americans, Indians in the gold districts were quickly shunted aside by the flood of argonauts after 1848. Foothill tribes, previously untouched by Hispanic efforts to control the coastal areas south of San Francisco, now confronted the grim epicenter of disease and dislocation that had decimated lowland and coastal Natives. The foothill woodland in pre–Gold Rush times was an abundant and varied natural resource that sustained a large Native population.[10] As the flood of gold seekers swept through the foothill gulches and ravines — ripping the soil with their picks and crowbars; upending the streambeds; falling the timber; and leaving a denuded, pockmarked landscape in their wake — the Native population lost much of their primary food supply as well as their cultural sustenance. Some fled; others resisted and were killed outright, women and children along with the men; still others succumbed gradually to the microbiotic killers introduced by the newcomers. One Calaveras observer estimated in 1856 that in the previous six years some 60 percent of the Miwok population had died.[11]

That Miwok lives and culture were devastated by the Gold Rush is no surprise to a postmodern, multicultural generation of Americans. The "new history" has given voice to minority views long buried or ignored. Old scars that never healed have been lanced and exposed to the light of new research and reinterpretation. Since the 1960s scholars have revolutionized the field of Native American studies, with study after study underscoring the terrible social consequences of the Euro-American conquest. These views reached their apex in 1992 with the publication of David Stannard's *American Holocaust,* a book that summed up the history of European-American-Indian relations in one word: *genocide.*[12]

Stannard's book is a landmark, but certainly not the last word. The social history of the Gold Rush is still in transition, as indicated by a number of recent studies that provide more nuanced views of American minority struggles.[13] Instead of a simple story of triumph and tragedy, or a starkly contrasting picture of victors and vanquished, revisionist historians are craft-

A group of Miwok, photographed in Sonora, California, ca. 1852. (Courtesy Richard Coke Wood Collection, Holt-Atherton Library, University of the Pacific)

ing a much more complex mosaic of interacting cultures, intersecting lives, and intermingling values among people brought together in the crossroads of life.

For the mountain Miwok, as Susan Johnson has recently concluded, adaptation was an alternative to confrontation and destruction. Some of those that survived cholera, typhus, measles, smallpox, and other diseases tried to cope by adjusting to the white man's ways. On the lower Mokelumne near Lancha Plana, for instance, a mixed population of Americans, Chileans, Mexicans, Sonorans, and mission Indians in the summer of 1849 worked placer claims without fear of local Indian reprisals. "Captain Alvino," a Native chief described by one miner as sporting "a kind of uniform" and holding "a commission from the padres of the mission," agreed to "protect the camp against any outrages from his people and take care of the companys horses and Burros" in return for payment of one pound of gold to purchase the land for mining purposes.[14] Such arrangements were rare, however. White racial arrogance usually crippled Indian efforts to accommodate or assimilate. Invariably close contact led to conflict, with Natives the losers. Placer claims discovered and worked by Native families were often confiscated or "jumped," and the original owners killed, chased out, or exploited as laborers.[15] On the upper Mokelumne and along the Calaveras in the fall of 1849, for example, Indian miners became targets after an Oregon party reported the murder of eight of their members by Natives under the leadership of Polo. In preparation for hostilities against "the old Poler Chief," John Hovey's party "went to work and fixed our rifle, & revolvers and got our ammunition in readyness," but the only Indians they saw "skulked of[f] behind a large hill and appeared affraid of us," presumably because of the white miners' dog who "is a terror to all Indians." Vengeance soon extracted its toll, however. A contemporary observer wrote that Polo died at the hands of another Indian who brought in the chief's head to claim the "5,000-peso reward that had been offered for him."[16]

Unwelcome in the diggings, and left to forage for whatever was available, Indians gathered food wherever they could. Their efforts only reinforced white "digger" stereotypes. At Murphys, John De Laittre watched Miwok women prepare a meal of dried grasshoppers mixed with pine nuts and roots that they made into a nutritious paste eaten with the fingers. He said the concoction was "about as filthy a looking mess as one can imagine." Yet he had little sympathy for Indians who begged white food and clothing. Though harmless, he wrote later, "they were scavengers, eating the refuse from miners' cabins and boarding houses, such as civilized people give to swine."[17] Another miner at Mokelumne Hill, in a similar vein, described how he treated Native beggars:

Two of our Rich Gulch Indians have just entered the camp and after the usual salutation of "Walla Walla" the call for "Carne" (meat). I shall give them four pieces of corn beef which has spoiled and which satisfies them as well as if ever so good. After thanking me in Spanish ("mucho gracious Senior") and bidding me good bye ("adios Senior") they leave in high spirits. They are a harmless race, while your eye is on them, but are great thieves.[18]

Indian foraging disgusted some and amused others. When a pair of yoked oxen stumbled into an abandoned mine shaft near Mokelumne Hill and died before they could be rescued, the remains were left hanging for days before Indians arrived who did not want "to see so much 'carne' go to waste," as James Madison Grover wrote in his memoirs. They retrieved the top carcass with the help of some white miners, but a few days later came back when "they had become hungry again, and a large number approached at the shaft, casting longing eyes to the remaining bovine, now distinctly emitting an offensive smell." While they gathered up the remains, Grover observed the hide slipping off "like the skin from an old fashioned sausage filler."[19]

Occasionally, Indians stole cattle from miners or farmers, with predictable consequences. For two decades newspapers were filled with accounts of bloody raids and reprisals. When Indians fought back they usually did not survive to fight again, although there were exceptions. On Carson Creek in the spring of 1850, William Hunter purchased two mules that were stolen a few days later. He and a friend tracked the stolen stock to a rancheria and started shooting the first Natives they saw. They killed a man and knocked down a boy, then retreated under a hail of arrows. Hunter vowed to "take the price of his mule in Indian scalps," but never got the chance. A few days later he and his brother were killed in a surprise attack by an avenging war party.[20]

In September 1852, at Mormon Gulch just across the Stanislaus in Tuolumne County, John Wallis reported that a miner returned to his claim after dinner to find a barren riffle box. While he was absent an Indian who was captured soon afterward had washed it out. He escaped before punishment could be applied, which often meant the lash for first offenders. The following spring, during the height of excitement over Hispanic gang activity on the upper Calaveras, travelers reported local Natives were "getting very troublesome," supposedly having been "incited to rob and steal by the band of outlawese [sic] Mexicans & Chileans known as Joaquins band," said one observer. Four Indians accused of killing a white man on the road between Columbia and Vallecito were captured after other Indians testified against them. When Wallis heard the news, he figured they would "[p]robably [be] hung up to

Roundhouse on the Miwok rancheria at Murphys, ca. 1910. This ceremonial structure, fifty feet in diameter, was constructed in 1901 and used for dances, gambling, mourning ceremonies, and other events. It stood until the 1930s. (Courtesy Calaveras County Historical Society)

Dry." Yet that September in a separate incident near Wallis's cabin, a gambler was found under a horse blanket. He had been killed with an ax, allegedly by Indians whom he had cheated. The murderers left camp "for parts unknown," but were not pursued, for the victim "was assertained to be a Mexican or Chilleno."[21]

Before the Gold Rush ended the downward cycle of dependence and despondency had begun in the foothills, with the liquor trade adding a malevolent spin that hastened many Indians to their graves. John Doble found them in a store on the Mokelumne in 1852 "buying large quantities of beef & Cognac which most of them are fond of that is the Cognac. . . . [T]hey seemed to have plenty of the dust & payed for all they got. Some of them were getting pretty drunk."[22]

Peter Cool, a devout Methodist unable to reconcile his moral views with the mining life that enveloped him as he worked a claim near Jackson, condemned both Indians and the whites that corrupted them. "The habits of the natives of this country are very disgusting," he wrote, "and the present class of Americans make them still worse." His journal was spotted with pious slurs against the bedraggled bands of homeless Natives moving through the mining camps between the Mokelumne and the Cosumnes on their way to the high country in search of food. "Some thirty Indians and squaws passed through town," he reported from Indian Springs in July 1851, "a spectacle enough to make a toad blush, though common for Californie." That November at Amador City, an ambitious town with cultural pretensions, he joined the local lyceum in debating the topic: "Which has the greatest right to complain of the whites, the Indians or Negro?" Taking the losing side, Cool underscored his unhappiness. "The dission," he wrote, "was given to *Negros.*"[23]

In the mid-1850s federal efforts to establish substantial reservations for displaced Indians failed when the United States Senate, under pressure from protesting western officials, refused to ratify the treaties. A few federal reserves were later established, but they were wholly inadequate and poorly administered. The Indian agents in the interim herded most of the remaining homeless Natives, now greatly reduced in numbers, onto small rancherias spotted throughout the state. Vagrant adults could also be compelled to work for white masters, and homeless children could be indentured to white families who promised to raise them. Indians considered these tactics forms of slavery, but it was rationalized as a humanitarian gesture.[24]

The Native population in Calaveras continued to decline in the latter years of the nineteenth century just as it did across the United States, a reflection of the lingering effects of disease, warfare, and cultural imperialism. Yet through the hard times during and after the Gold Rush, despite periodic wars and individual atrocities, the Native survivors learned to cope, and gradually they adjusted to the newcomers, just as the newcomers adjusted to them. Immigrant families took in orphaned Indian children and raised them like their own, as the Ben Franklin Jones family did in the 1890s, or hired them as servants or janitors and occasionally as muckers and miners. The Canepa family welcomed Indian foragers during the growing season. "On several occasions there were severe outbreaks of army worms and grasshoppers that threatened the gardens," recalled Frank Canepa, a Genovese raised in the 1860s on the family farm in Vallecito. "Each time disaster was averted by the timely arrival of the local Indians. They would go through the garden, picking the insects into their baskets. Then, after dipping the baskets in boiling water, they would spread the insects in the sun to dry for winter stores."[25]

Remnant bands were relocated on several informal rancherias in the county where they lived in modest frame cottages and lived marginally on subsidies and charity, taking odd jobs, farming, or foraging as the opportunities arose. The 1880 census recorded two small rancherias near West Point, one with twenty-two inhabitants, the second with twenty-one, almost all adults. The first included a sixty-year-old "miner and doctor," Moluska, who lived alone. Another married elder, Butney, headed the second rancheria as "patriarch."[26]

John Jeff and family at
the Murphys rancheria,
ca. 1935. (Courtesy
Calaveras County
Archives)

Near Sheep Ranch was another small rancheria established soon after the
Gold Rush but not made official until 1916. Twenty-one California Natives
lived there in 1880 under the jurisdiction of sixty-seven-year old "Captain
Chips," whom census records described as "Supt. of ranch." The rancheria
population included only three youngsters: a six-year-old boy and two girls,
ages twelve and four. These were the children of Charley and his wife, Limpy,
both aged thirty-seven.[27]

Despite more tolerant attitudes and individual acts of kindness, the passing years continued to take their toll on the rancheria inhabitants as well as
their dwellings. C. Hart Merriam, an ethnologist who visited Murphys rancheria in 1900, found only six adults on the site living in "half a dozen . . .
hovels made chiefly of old waste lumber and odds and ends." The reserve
was down to one family when Merriam revisited it two years later. But some
rancherias gained new occupants after the old ones died. John Jeff, whose
father was a chief at West Point rancheria, moved to Six Mile rancheria with

Limpy, a few years before her death in 1930. (Courtesy Calaveras County Archives)

his wife, Tillie, in 1909 and remained there until 1927. As a ceremonial leader and powerful singer who had learned the traditional songs from tribal elders, Jeff organized ritual dances and social gatherings that brought together Indians from all parts of the county. He remained a leading conservator of Miwok culture until his death in 1937 at age sixty-seven.[28]

For more than a century Central Sierra Miwok have occupied the modest rancherias of Calaveras. Over the decades Native culture has adapted to external influences, just as Native bloodlines have diffused through exogamous relationships. The woman called "Limpy" by the 1880 census takers, for instance, showed remarkable adaptive skills during her long life. Before the Gold Rush she and her family lived on the verdant hills of Vallecito. When the argonauts arrived she was still a child but remembered "when the first white man came to Calaveras County," as she explained to interviewers late in life. Chased out by the gold hunters, the Natives of her village moved to a more remote area near Sheep Ranch. That region, too, soon came under the covetous eyes of Euro-American miners and settlers, but after the Civil War an informal rancheria was established for homeless remnant bands in the area. Limpy lived there for the rest of her life, marrying and raising a family, adjusting to life as a "ward" of the government, and learning the ways of the white world while at the same time keeping alive many of the rituals and lifeways of her Native culture. Despite a crippled leg she often made her way door-to-door, barefoot except in winter when she wrapped her feet in rags, begging food and cigarettes to supplement her subsidy of five dollars a month. Desiré Fricot, talented son of Jules Fricot, a partner in the Chavanne mine at Sheep Ranch before it was sold, befriended Limpy and enjoyed her conversations. A linguist as well as a photographer, Fricot took many photos of Limpy and her family, and transcribed many of her stories in a notebook, unfortunately now lost. She outlived most of her friends and relatives, and at the time of her death in 1930 was regarded as the oldest Native American in Calaveras.[29]

## Converging Cultures, Races, and Nationalities

At the height of the Gold Rush the largest foreign element in Calaveras were English speakers from Britain and the empire, including Ireland, Canada, Australia, and New Zealand. Next to the Americans they were the most assertive in the diggings, and suffered the least from American efforts to discriminate against foreign miners. Though neither Cornish nor Welsh were identified in the early censuses, their influence exceeded their numbers, for they brought to Calaveras the hardrock experience of years in the tin, copper, and coal mines of Britain. Some Cornish came to California by way of Wisconsin and upper Michigan, where they had clustered in the late 1840s when the lead and copper deposits were discovered in the upper Mississippi Valley and the Great Lakes.[30] Most "Cousin Jacks" settled first in Grass Valley, where extensive quartz deposits were located in 1850. But the quartz excitement farther south in the early 1850s brought a growing number to Calaveras, where they demonstrated their technical skills underground. Some held leadership positions as "captains" (foremen), shift bosses, even managers.

By the 1860s, in the major hardrock districts of the Far West, the majority of skilled underground jobs belonged to the Cornish. When the Calaveras copper mines opened in the 1860s the Cousin Jacks played a substantial role both underground and in management. They were used to doing things their way, however, and did not always work well under American bosses. Some ethnic scholars consider "individualistic self-confidence" a distinctive Cornish strategy that reinforced their group identity and economic importance in the mining West. As one Cornish observer remarked during efforts to introduce the Burleigh drill in Michigan, "Cornishmen are good miners, and good mine managers—they ought to be—but they are just as apt as others to conclude that what they do not know is not worth knowing."[31]

Wade Johnston, at home in later years near Willow Creek. (Courtesy Calaveras County Historical Society)

Wade Johnston, a Missourian with extensive Calaveras placer and hard-rock experience, was not impressed by the Cornish presence at the Napoleon copper mine in 1864. Under the leadership of "Capt. Dick Pound," Cornish miners worked a relaxed shift. A snap inspection by a new superintendent caught eleven men asleep. They were fired on the spot. But Johnston's view was not typical, for most Cousin Jacks did well in Calaveras. As Johnston admitted many years later, "There was a lot of Cornishmen there that I didn't get acquainted with."[32]

The Irish had a larger representation than the Cousin Jacks in Calaveras, but they were unskilled in underground mining, and initially concentrated mostly in the placer diggings. Census records before 1870 do not distinguish Irish from those classified as "British," but scholarly studies make clear that for twenty years after the Gold Rush, Ireland contributed most of those in California from the British Isles. Along the South Fork of the Calaveras near San Andreas they were in the majority, but even when their numbers were small they made their presence known by their political leadership and social organization. When a water company diverted so much water from San Antone Creek during a very dry year that downstream placer miners could not operate, the Irish took the lead in organizing a protest that turned violent when the water company tried to protect its flumes from angry strikers. The courts finally settled the dispute in favor of the company.[33]

In happier moments the Irish used social skills to advantage in bringing some levity to the camps. Wade Johnston recalled the mock funeral staged by some Irish friends on April Fool's Day in 1860 during a justice court trial in San Andreas. "The jury couldn't agree on the case being tried, so they thought they'd have a little fun to pass away the time." After laying out Jimmy Fitzsimonds as the "corpse," they lit candles, carried the body to the Catholic cemetery, conducted elaborate ceremonies, and then proceeded to have "fun nearly all night." Levity turned to tragedy ten days later in Johnston's hydraulic pit when Fitzsimonds was buried under tons of dirt and gravel from an undercut bank. He was not as lucky as Robert Briggs, a Missouri preacher's son, working another Calaveras hydraulic claim the year before. A collapsing bank crushed his partner and covered Briggs right up to his neck. He struggled ten hours before freeing himself and crawling to the cabin of a Mexican company nearby. They nursed him back to health, but it took a year to recover.[34]

Terrence McSorley was one of several thousand Irish immigrants to Calaveras in the first decade after the Gold Rush. Emigrating from Ireland to Canada in 1847 during the potato famine, he entered the United States and first found work in the textile mills at Lowell, Massachusetts. With a new bride he traveled west after the outbreak of the Civil War, arriving in Mokelumne Hill in 1862. Hardrock mining was beginning to boom in the district, and for several years McSorley worked underground at first in the Bob Paul mine, then the Gwin. In 1870, if figures from the principal boardinghouse serving the Gwin can be trusted, Irish made up about a quarter of the Gwin workforce.[35]

McSorley struck out on his own in 1878, joining a partnership to operate the Green Mountain drift mine. By that time six children had been added to the household, with more on the way. Their school was in Mokelumne Hill, a good walk uphill from their first home in Chili Gulch. One son, Tom, raised in mining, left for a time after graduating from high school and became a machinist and electrician. But after his father died he returned to the district, eventually managing several hardrock and placer mines in the area and abroad.[36]

Though Calaveras had a smaller share of foreign immigrants than the two counties farther south, many languages could be heard in the diggings. Germans represented more than 14 percent of the 1850 Calaveras census, and as late as 1858 Mokelumne Hill had a German-language newspaper. San Andreas also had a German contingent. One Gold Rush arrival was Joseph Zwinge.

His first commercial building in town was a tent-house restaurant that served food and drinks until fire destroyed it in 1853. A wood-frame replacement suffered the same fate in the "great fire" of 1857. Zwinge built his third structure well. A massive stone building, the oldest in San Andreas, it has stood the test of time. Remodeled in 1937, it served until 1996 as the Calaveras County Library. Now it houses the Calaveras County Archives.[37]

Many of the Germans were Jewish, both miners and merchants. In the thirty years after 1850, Jews held 123 mining claims recorded in Calaveras. Among the Jewish immigrants of 1850, Morris Cohen was one of the more prominent. Known as the "Merchant Prince of Calaveras," he held large mining as well as business interests, opened stores in Angels Camp and San Andreas, and eventually relocated in San Francisco after business in the mining counties declined. Less well known in Calaveras during the Gold Rush but ultimately a household name in men's western wear was Levi Strauss, who evidently visited the area before heading for better prospects in San Francisco.[38]

The French were also numerous in 1850, with slightly over 7 percent of the Calaveras population. As in Germany, France at mid-century was in turmoil. Economic instability, social disruptions, and political repression motivated thousands to seek better opportunities. For French capitalists and entrepreneurs, as well as for urban workers and peasants, California beckoned. French immigrant companies helped 23,000 French reach California by 1853. At Mokelumne Hill in 1851, J. D. Borthwick found a French physician at work, providing various medical services and dentistry, and selling rat poison on the side! Among the French who reached California before 1852 were two brothers, André and Louis Chavanne, and a friend, Jules Fricot. Eventually, they arrived in Calaveras after working for more than twenty years in the quartz mines of Nevada County. Their experience was put to good use in opening the Chavanne mine at Sheep Ranch.[39]

Like other early foreign immigrants, most French in the Southern Mines initially kept to themselves, partly for cultural reinforcement, partly for protection. Borthwick found French miners on Coyote Creek near Columbia living together in cozy wooden cabins enjoying the conviviality of close association and common language. But their isolation made them suspect, for Americans were wary of strangers who spoke little or no English. Besides, they seemed "more congenial" to Latin culture, and were often suspected of conspiring with Mexicans and Chileans.[40] After 1849 the anxiety level in the Southern Mines increased with the deluge of newcomers.

French success in the placer fields around Mokelumne Hill precipitated a clash of arms in 1851. Americans had been incensed by the arrival months earlier of 140 aggressive Frenchmen, part of a contingent of Mobile Guards the French government, hoping to calm troubled waters at home in the wake of the 1848 uprising, had dispatched to California in the fall of 1850.[41] Armed "to their teeth with Double barreled guns and bowie knives," as one angry American reported, French squads marched through Mokelumne Hill like a conquering army. With bugles and drums sounding, they set up camp near town and hoisted the tricolor, but lowered it under threat of court action by Americans. When the French struck rich pay dirt after prospecting the Tertiary gravels on a nearby hill, Americans rushed in to stake claims, but were rebuffed by French miners who "claimed the whole Hill and was Determined to keep of[f] every American if they could," said John Hovey, one of those chased away. The news roused Americans to a spate of jingoism. At San Andreas a few miles away, Lucius Fairchild blamed government officials for allowing "themselves to be run over by the off scourings of all Gods creation who are taking the bread out of the American miners mouths, or the Gold which is the same." Closer to the scene, Hovey spoiled for a fight. A Missourian who had been mining in Calaveras since the fall of 1849, he fumed at the thought of being "Driven of[f] our own Soil by a set of Foreners." With the backing of their American friends from surrounding camps, he and his partner decided to "go back again and *Dig at all hazards and find* out our rights."[42]

Hovey's journal is an eyewitness account of what was later labeled the French "war." Filled with self-justification and national pride, it predictably blames the French for causing the conflict and praises the Americans for their "victory" over the "enemys to this Country." The incident began with a claims dispute, but the details are unclear. Hovey said he and his partner, with the backing of other Americans, tested the French strength first by jumping a claim. Other accounts attribute the conflict to a group of Americans who confronted two French miners trying to hold a claim they had staked but not yet worked. At any rate, both sides quickly gathered reinforcements, and the quarrel erupted into a brief but deadly gunfight. One eyewitness, fifty feet from the action, said the "pistol balls flew thick and fast." When the smoke cleared an American, Leroy Jones, lay dead and two French were wounded. Both sides then reorganized. While the French prepared their defensive lines, the Americans held a rally that was attended by hundreds of men from surrounding camps, plus a few Indians from nearby rancherias. An old army officer, Colonel Meads of New York, was elected brigadier general in command of all the American forces. To Hovey, "It sounded like an old fashion muster in the States to see squads of miners comeing in over the mountains and hills with drums and fife with the Stars and Strip[e]s a flying to give Battle for the wrongs Done by these invaders." In military fashion they appointed a delegation to negotiate for the surrender of the "man who shot Jones." They also demanded the French "surrender their arms and leave diggings in 12 hours." The French, seriously outmanned, withdrew without firing a shot. Whether they gave up their arms and their protagonist is unclear. The Americans evidently did not pursue them but burned the French camp, letting the Natives reap the spoils. Hovey said it was "laughable to see the Indians . . . put on three to four wooling Shirts one over the other, and pile the blankets on the squaws back, to carry them to their Ranchere." [43]

The Americans swarmed over French Hill the next day, but rumors that the French were gathering reinforcements at San Andreas for a counterattack prompted another mass rally at Mokelumne Hill. Men were dispatched on fast horses to spread the alarm to camps up and down the Mokelumne River. Even the Indians were called in; under Chiefs Kossouth and Captain Charley, some eighty natives appeared "with their faces all painted and ready for battle," as Hovey recorded in his journal. For a day the Americans continued to gather in the village, accumulating a massive defensive force but waiting anxiously for word of the French. At last the "enemy" appeared, their arrival announced by Indian scouts running down from the hills shouting "mucha merlo, mucha merlo." Immediately, the Americans took up defensive positions and prepared to fight, but the French wanted to parley. Negotiations began as delegations from both sides met under flags of truce. Prompted by his men, Colonel Meads delivered an ultimatum: surrender all arms to the county authorities and leave the diggings in thirty minutes or face the consequences of withering fire from hundreds of impatient "yankees [who] . . . wanted to give them battle." The French talked it over while the Americans waited, growing more restless with each passing minute. Colonel Meads, seeing that his "men were getting passionate," formed them into battle and told the French they had five minutes to surrender to the county sheriff. This threat did the trick: the French gave up, and the Americans celebrated. As Hovey recalled, they "blazed away for about fifteen or twenty minutes wich made our frenchmen stare." The French later tried to enlist their government in protesting the American action, but the French consul in San Francisco turned against his own countrymen and sided with the Americans. "Thus endeth," wrote Hovey with unabashed pride, "one of the greatest battles on record, not a gun fired not a man killed or wounded." [44]

Troubles with the French stirred tensions and for months afterward made life uneasy for all foreigners. Any rumor could be a cause of alarm, such as the charge in 1852 that French and German mining companies were jumping claims on the Mokelumne.[45] But Europeans rallied when necessary to a common defense of their interests, and American antipathy gradually faded.

Barely visible among the Gold Rush throngs in Calaveras, and less subject to overt discrimination, were Italians. They were probably less than 2 percent of the 1850 population, although their numbers increased to 16 percent of the foreigners in the county by 1870. Most of the early Italians in Calaveras had lived in the East or in Argentina or Peru before coming to California. They did not congregate in ethnic enclaves, but dispersed throughout the county. Most tried mining at first, but as the Gold Rush faded only a few Italians remained directly in that line of work. Yet many of their other jobs—in agriculture, merchandising, saloon keeping, freighting, hostelry, and other forms of service—depended on the mining industry for business.[46]

The Lagomarsino family story illustrates the diversity and enterprise of Calaveras Italians. Andrea Lagomarsino arrived in San Francisco in 1856 after borrowing two hundred dollars from family members to escape the Genovese draft. Luck followed him to Calaveras, where he and a partner struck a rich pay streak on the Calaveras River. With his share he bought land, but before switching occupations he tried his mining luck once more in the Fraser River rush. After it "busted," he returned to Calaveras late in 1858 and teamed up with two other Italian farmers to raise fresh vegetables. By 1870 he had earned enough to move to town, buy a house, win the hand of a young Italian girl, and start a family. His wife, Angela, took over her husband's farming interests after he died accidentally in 1897, gradually bought out the partners, and ran the business herself before selling out to a man from Sardinia.[47]

Equally enterprising was another Italian lady, known as "Grandma" to hundreds of Calaveras miners and their families. Olivia Rolleri arrived in California with her new husband, Geronimo, in 1861, surprising her father, Giovanni Antonini, who had preceded the family to Sonora in 1849. He had arranged for his daughter to marry a Sonoran gentleman, but Olivia had refused to follow Old World customs. Except for a brief residence at Reynold's Ferry on the Stanislaus River, the Rolleris lived in Tuolumne County for more than twenty years. Geronimo died in 1888, leaving Olivia with ten children to raise out of thirteen born to the family. At the urging of friends, she then bought a boardinghouse in Angels Camp and started the Calaveras Hotel. It was an opportune time, for the Angels Camp district soon blossomed with bonanza ore from the Utica, Lightner, Angels, and Sultana mines. Her business grew rapidly, and she added other properties, including a couple of ranches to raise beef and a butcher shop to supply the hotel. With the help of her children at first, then adding staff as the business grew, she prospered.[48]

Like other European immigrants whose mother tongue was not English, Italians had some initial difficulties of adjustment, but they were minor compared to those who looked different from the white majority. Although *eth-*

"Grandma" Olivia Ellen Rolleri.
(Courtesy Calaveras County
Historical Society)

*nicity* was not a term used in the Gold Rush, it was clear that skin color made
a difference on the scale of values in California. The delegates at the Constitutional Convention in the late summer of 1849 banned slavery, but not for
humanitarian reasons. The motives were racist and economic; keeping slaves
out would send a signal that blacks were not wanted in California, either slave
or free. It would also prevent the spectacle of free and unfree labor competing
side by side for the same limited resources. The convention delegates considered banning free Negroes as well, but rejected the idea, fearing Congress
might reject the Constitution and delay statehood, just as the double ban
by Oregon's provisional government in 1843 had delayed Oregon's territorial
organization.[49]

Despite these unwelcome signs African Americans did come to California
during the Gold Rush, both as free persons and as slaves. The numbers were
not large, but their presence was felt in the mining counties. Most of the slaves
arrived before the constitutional ban and may have amounted to only a handful. After ratification some southerners left California and took their slaves

with them; others remained, allowing their slaves to work out their freedom in the diggings. The institution lingered into the 1850s in the more remote areas. Slaves either labored alongside their masters in the mines and ranches or were hired out to work for others, as was the fate of Reuben, a faithful but disconsolate slave from Kentucky who died of cholera in 1851. Reactionary southerners kept a few unfortunates who challenged the legality of slavery in bondage, either by force or by manipulating the legal system. A state fugitive slave law was passed in 1852, and in 1858 the infamous Archy decision by the California State Supreme Court returned an escaped slave to his master. Ramon Gil Navarro noted the irony of black slaves in a free state but saw nothing wrong with using Chilean peons to dig gold for him on the Calaveras and the Mokelumne. "My God!" he wrote. "What an impression it makes on me to see slaves in California, slaves chained by Americans, who, more than any other nation in the world, stand for freedom!! These poor Negroes can hardly hold their picks they are so cold, and yet they do not move from the place were their masters told them to stay."[50]

In Calaveras black men and women appear in every census roster from 1850 to 1880. The state census of 1852, for instance, reported a total of 169 blacks and mulattoes in Calaveras, or 0.8 percent of the county population. At San Antonio Bar in the spring of 1850, Leonard Noyes found slaves working under contracts that granted them freedom in two years. They worked during the week for their masters and for themselves on weekends. Not surprisingly, reported Noyes, they found little gold between Monday and Saturday, but "all their Sunday claims were good."[51]

Despite discrimination some free African Americans did very well indeed. Two black miners coyoting near Mokelumne Hill recovered some eighty thousand dollars from a single hill claim in a four-month period. It was said one of them had been told as a joke to dig on a spot considered barren by local whites; the unexpected riches triggered a rush of miners to "Nigger Hill."[52]

Not all persons classified by degrees of color were African American, however. In these years some Hawaiians and Mexicans, for example, were listed as "black or mulatto," while others passed as "white." Almost all free black miners were single males, living alone, earning a modest living from placer claims or as hired laborers. Some were reclusive enough to be legendary. Wade Johnston remembered a singular old bachelor, Ben Buster, who lived in an old cabin at Red Gulch and ran off attempted visitors with a gun. Rumored to have buried thousands of dollars in gold nearby that he had gleaned from the streams and gulches near Mokelumne Hill and in the San Antone district, he died alone, emaciated, with only a few dollars in his pocket. The gold he allegedly buried was never found—or at least never reported.[53]

Black women were scarce in California, and scarcer still in Calaveras. Of the 169 blacks counted in the 1852 county census, only 15, or less than 9 percent, were women. One of the most noteworthy was Margaret Binum, born a slave in 1826. She came west to join her husband, Edmington Binum, who was brought to California in the late 1850s by two Mississippians, Robert Newton Cloyd and Judge William B. Norman. The two white men formed a partnership, established a ranch on the Calaveritas near Yaqui Camp, and let Binum work for wages. He accumulated enough to send for his wife and three children, and after they arrived they all worked on the ranch to buy their freedom. A stone corral they built still stands on the Cloyd and Norman ranch site.

After liberating themselves, the Binums moved to a small farm and raised fruit and vegetables for the local retail market. Margaret also found a ready market for her services as nurse and midwife. She covered the county, responding to the needs of others, regardless of race or circumstances. As Wade Johnston recalled in the 1920s, "She was a famous nurse, and no doubt, many living here today owe their lives to the tender care of 'Aunt Margaret,' who never failed them, and they are sad over her passing."[54]

Although not the county's largest foreign element, the Chinese in Calaveras had a significant impact on the regional mining industry. Like African Americans they were segregated because of skin color, but language and cultural differences added to their problems. Lillian Gorham Murphy, as a young girl in her father's store at O'Byrnes Ferry remembered being frightened by a passing Chinese who pulled her long hair braids and exclaimed: "All same as Chinyman." Her mother thereafter kept her hair short. At Murphys, a Mexican War veteran, outraged at seeing a Chinese flag flying over a whorehouse, impetuously burned it to the ground. White housewives at Robinson's Ferry eagerly bought huge heads of Chinese lettuce sold door-to-door by peddlers carrying baskets balanced on long poles across their backs. But as Archie Stevenot recalled, the women lost their appetite for lettuce after they learned that the Chinese "saved their human refuse in a large vessel and then used this to fertilize their lettuce plants."[55]

Most California Chinese were part of the influx of working-class young men from Kwangtung Province. Beginning in 1849 with only a few hundred, their numbers rose by the early 1850s to several thousand, and by 1863 peaked at thirty-five thousand. Many of these sojourners headed for the Southern Mines through Stockton, the southern gateway, or the "Pekin of California," as one editor dubbed it in 1851. In Calaveras nearly one-fourth of the immigrant population in 1860 was Chinese.[56]

Chinese living conditions reflected the meager resources as well as the communal lifestyles of these Asian argonauts. Though some Chinese set up temporary quarters alongside their claims, most were more comfortable—and safer—in the "Chinatowns" that could be found in almost every mining camp or supply center in the Mother Lode. In Calaveras sizable Chinese districts sprouted in Angels Camp, Mokelumne Hill, San Andreas, and Jenny Lind, with lesser camps dotting the tributaries of the Stanislaus, Calaveras, and Mokelumne Rivers. Angels Camp had one of the largest Chinese quarters with about five hundred residents in the 1850s. At first it comprised a cluster of tents along both sides of Angels Creek—not much different from the rest of town. Gradually, canvas was replaced by wood-frame structures, many with common walls, and a few brick buildings, built by more affluent Chinese merchants and professionals. J. D. Borthwick visited this district in a "gulch near the village" in 1851, finding the inhabitants friendly and hospitable. He declined their invitation to taste some "dubious-looking articles," but accepted a "pannikin full of brandy" and a few "cigaritas."[57]

Borthwick reflected the initial attitude of Americans for these Asian immigrants, "whose strange costumes and uncouth language have excited the curiosity of the natives," as one journalist put it in 1851. But curiosity soon turned to contempt. The rising Chinese presence precipitated a backlash among nativist Americans who resented the additional competition. When a group of '49ers found a Chinese man and a "Malay boy" laboring with an English company, they declared that "coloured men were not privileged to work in a country intended only for American citizens." After learning the "slaves" were working for wages and free to leave on their own volition, the Americans tried to hire them away, but, as one of the Englishmen claimed, "nothing could shake their allegiance to us."[58]

Pressured by violent street demonstrations in San Francisco and other major communities, and by discriminatory state and local taxes, the Chinese statewide growth rate slowed temporarily after 1852. But the decline of high-grade surface placers by the mid-1850s discouraged individual white claim owners and opened new opportunities for these sojourners from the Far East.[59]

In the early 1850s Chinese Camp in Tuolumne County became the principal distributing center for Asian labor in the Southern Mines. From there Chinese labor gangs, some under white management but by the 1860s increasingly isolated from other miners, disbursed in groups of as few as four and as many as four hundred men along the streambeds to rework recent river gravels that American miners considered played out or too low-grade for profitable operation.[60]

Despite the ever present resentment among white miners in the most

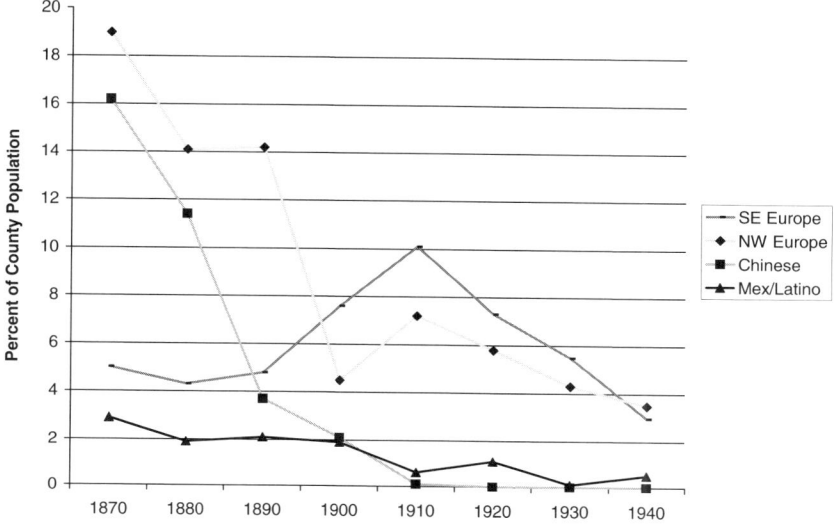

Foreign-Born in Calaveras, 1870–1940

popular camps, along the Stanislaus and Calaveras Rivers and their tributaries the Asian presence caused less of a stir. Americans did not feel as threatened by foreign operations in low-grade deposits requiring prodigious amounts of work to mine successfully. But the hardworking Chinese were adept at river mining and ground sluicing. Combining intensive labor organization with ancient placer mining and construction technology, they were able to work surface gravels more systematically and productively than their American counterparts. Where Americans usually rejected placer ground that paid less than five to twenty dollars per cubic yard, Chinese companies managed to make money on claims yielding as little as two dollars per yard.[61]

The work of the Chinese was not exclusively linked to river gravels, however. In some California mining camps before the 1860s, as one- or two-man diggings gave way to large-scale hydraulic operations, Chinese also found work as wage earners in white-owned mining corporations, yet not without renewed agitation from the white rank and file. And at least until the anti-Chinese agitation reached dangerous levels, white lode mine operators hired Chinese both as common laborers and for underground drilling crews. Wages for these Chinese underground "strikers" as well as for Asian surface workers averaged 50 to 60 percent less than their white counterparts.[62]

Anti-Chinese discrimination coincided with the declining quality of placer ground after the Gold Rush peaked in the early 1850s. In addition to the revived state-imposed foreign miners' tax of three dollars per month that fell almost exclusively on the Chinese, Asian miners after 1852 suffered under a series of local ordinances and resolutions intended to drive the Chinese from the mines. Some prohibited whites from employing or selling claims to Chinese, others ordered Chinese to leave within six months, and still others called for the government to ban further Asian immigration and to send home those that were already here. White miners at Douglas Flat expressed the common rationale for these actions, arguing in the preamble to their banishment call that unless steps were taken to protect "the present hardy, intelligent, independent miners and useful citizens," mining, in their opinion, would soon be reduced to a monopoly of bourgeois capitalist management and coolie labor.[63]

Forced out of all but the least desirable districts, Chinese miners in Calaveras and elsewhere confined themselves in large part after 1860 to low-grade placer gravels other argonauts had previously worked or passed over. Most of those were exhausted within a decade, forcing thousands of Chinese miners to find work elsewhere. By 1870 half the Chinese mining population had left the interior for other parts of California. Those that remained had to work harder for less. Mining continued along the Calaveras and Stanislaus streambeds into the 1890s and beyond, but with ever diminishing returns and with the number of placer miners dwindling to a couple of hundred by the turn of the century.[64]

## Hispanic Confrontation and Conflict

If race and language differences from the dominant Euro-American majority made life harder for the Chinese in California, those same distinctions also hurt Hispanic immigrants, but conditions back home were worse. Mexico and Chile contributed almost as many miners to California as Britain in the early years of the Gold Rush. The Mexicans were largely Sonorans who had suffered miserably from decades of Indian raids and reprisals, economic instability, declining resources, and government neglect. They came north by the thousands, primarily in the fall and winter of 1848–1849, traveling by caravan from Hermosillo to the Southern Mines. An estimated five to ten thousand reached California by the summer of 1849. The Chileans were mostly landless farmers and tenants. Some were heavily in debt and bound by labor contracts to the landed European elite. Between September 1849 and May of the next year, between two and five thousand Chilean farmers, merchants, and peons reached California by boat from Valparaiso.[65]

A small party of Mexicans entered San Andreas as early as 1848, mining a gulch just above the present town. Another Hispanic party in August 1848 arrived from Los Angeles. Headed by Don Antonia Coronel and his partner, Augustin Janssen, a Frenchman, the group located placer ground along the Stanislaus by following Indians they had traded with in the Central Valley. Leading a mixed party of thirty southern Californians and Sonorans, Coronel was camped on the lower San Joaquin one night when four Natives "came into camp carrying bags full of gold." After an evening of extraordinary deals, during which one Indian "poured a tin plate with gold" to buy Coronel's saddle blanket, another offered nine ounces for a second blanket, while a third acquired a serape for more than two pounds of the yellow metal, Coronel and his men were consumed by avarice. When the Natives left the Angelinos secretly trailed them to their diggings and spied on them as the Indians "took flat wooden stakes and began to dig gold out of the creek bed." Satisfied they had found "the richest part," and heedless of Native feelings, the Coronel party jumped the claim and "in no time" were filling leather pouches with pay dirt. The next day Coronel and two servants "picked up forty-five ounces of big nuggets, not counting finer particles to be washed out later." Other members of his party did even better. Unabashedly, he reported that "we . . . were well content with the results."[66]

Exploring upslope from the mouth of what became Carson Creek, Coronel and his party uncovered a rich placer deposit filled with small nuggets resembling melon seeds. A camp sprung up nearby that Hispanics called Melones (not to be confused with the later town of the same name located on the Stanislaus), although that label never caught on with Carson's men or other Americans who arrived about the same time.[67] For a brief time Mexican Melones flourished with an estimated population between three and five thousand. It was a rough-and-tumble community of few amenities and fewer constraints. Heavily armed Mexican miners slept in shelters made of brush and mud. Saloons were the feature attraction, where crude liquor and card games ran around the clock. Money flowed freely in the form of gold nuggets and dust.[68]

The Hispanic population in the Calaveras mines continued to mushroom until the summer of 1850. During the 1849 mining season Mexicans and Chileans began working the "dry diggings" at Campo Seco. Others explored the streambeds along the Stanislaus and its tributaries. By 1850 an estimated two to four thousand Hispanic miners were at work in the Calaveras County area.[69]

In the hardrock districts, as we have seen in chapter 2, experienced Hispanic miners at first found ready employment. On Carson Hill, for example, American mine owners hired Mexicans to manage and work their claims. In river, drift, and hydraulic mining, where intensive labor was required, white operators also employed Mexicans and Chinese.[70] But in the surface placer diggings, the Hispanic influx crowded the camps and heightened racial tensions. Americans quickly grew belligerent, a reflection of the national mood at mid-century, when the Stars and Stripes swept across the continent under the fervent tide of Manifest Destiny. In July 1849 near Mokelumne Hill, Americans chased out a Mexican party and jumped their claims. "There is no other tyranny or arbitrariness as great as that carried out by this nation of free and republican people," wrote a Chilean, Ramon Gil Navarro, whose diggings were nearby on the Calaveras River. "There were groups of men who were told to leave within fifteen minutes or else their lives would be at risk. They have gotten rid of anyone who was in their way." Navarro himself escaped the fate of his compatriots. Most Americans thought he was French.[71]

Anti-Mexican sentiment ran particularly high among California veterans of the late war with Mexico, whether in the mines or the cities. On Stockton streets in 1849 local authorities could do little to stop white gangs from indiscriminately attacking Mexicans and Chileans passing through town. One veteran summed up his feelings in a letter to a Stockton paper: "Mexicans have no business in this country. I don't believe in them. The men were made to be shot at, and the women were made for *our* purposes. I'm a white man — I am! A Mexican is pretty near black. I hate all Mexicans."[72]

Contributing to American attitudes were the racial and nationalist dogmas of the day. The debate over slavery and "free labor" had heightened awareness of labor servitude in other parts of the world.[73] Antislavery crusaders saw little distinction between slavery in the American South and debt peonage in Mexico, Chile, and China. In 1848 the Free-Soil Party had rallied its minions on the platform, "free speech, free soil, free labor, and free men." In California the tainted brush of slavery was inferred when Hispanic "peons" and Chinese "coolies" were mentioned, although neither term was accurate to describe immigrants to the United States from Latin America or China. True "peons" and "coolies" were legally bound to work off their debts, but U.S. law prohibited the immigration of any bound debtors. Of course, informal contracts that lacked the force of law could still be enforced on the streets, as Chinese or Hispanic debtors, beholden to labor contractors or family associations or tongs, well knew. The use of these degrading appellations contributed to the dehumanization of both Hispanic and Chinese laborers in the mines and fields of California, and provided additional grounds for discrimination and violence.[74]

The first serious clash over peonage arose on the North Fork of the Calaveras River near San Andreas in December 1849, prior to the organization of Calaveras County. Encamped at San Antonio diggings about three miles from a group of Americans were a large number of Chileans and Sonorans, driven from the Northern Mines a few weeks before. The two camps avoided each other at first, but on December 6 trouble arose when a party of Americans, prospecting near San Antonio, took control of several mining claims Chileans had worked earlier as dry diggings. Chased off by a larger Hispanic party who destroyed the rockers the gringos had left behind, they retreated to the American camp and raised the cry: "Drive the foreigners off!" The Americans convened a mass meeting three days later and passed resolutions organizing a new mining district, excluding all foreigners, and limiting claims to "bonafide citizens of the United States."[75]

To enforce the codes the Americans established an ad hoc judicial system and elected a military captain, L. J. Wood, and an alcalde, L. A. Collier. Some modern critics, sensitive to minority issues, have charged that these Calaveras miners, using the peonage issue as an excuse to oust the Chileans, empowered a "pseudo alcalde" with no legal standing, assuming there were regular officials in the area who should have been called upon to resolve the dispute. But the circumstances were far more complex. Government organization in the winter of 1849–1850 lagged far behind the need for local law enforcement in the mining districts. Congress was still deadlocked over the issue of slavery expansion and would not resolve its differences for nearly another year. In the interim California's government was still in limbo, and American residents were divided over what laws were in effect. The interim governor, Bennett Riley, an American army general, reiterated the position made clear by his predecessor, Colonel Richard B. Mason: that Mexican laws and customs not incompatible with the United States Constitution remained in force until changed by a legitimate civilian government. Some Americans in California, however, chafed under the old system that had never functioned well in the remote Mexican borderlands. They argued that Mexican rule ended with the American conquest in 1847, and that Americans should write their own laws and institutions. In the coastal cities the need for change was acute. Three alcaldes had been authorized for each city, along with other officials, but these functionaries had little or no formal training. Most were arbitrary, and some were incompetent or corrupt. The result, in the words of a fulminating American journalist, was "anarchy." In San Francisco, flooded with Americans after 1848, the streets were dangerous, with land titles insecure, growth chaotic, drinking and gambling unrestrained, rowdiness rampant, crime rising, and public needs unmet. Americans there had tossed out the last vestiges of Mexican officialdom in 1848, when they established a city council and replaced alcaldes with justices of the peace.[76]

In the remote mining camps, however, the office of alcalde lingered in lieu of any good alternative. Under Spanish and Mexican laws, each community of fifty or more residents could select an alcalde, an officer with combined executive, judicial, and legislative functions. His decisions could theoretically be appealed through the Mexican judicial system via courts of first, second, and third instance, all the way to the Supreme Court at Mexico City. But in practice, as Charles H. Shinn wrote in 1884, the only real chain of command was from alcalde to governor, who between 1837 and 1850 "held the only true appellate court in California."[77]

Lacking a state legislature or a county government, or even a coherent set of laws or statutes to follow, and facing the need for practical solutions to problems related to crime and property rights, Gold Rush miners in 1848 and 1849 resorted to the common American practice of appointing or electing their own courts and officers until regular officials took over. Alcaldes were absorbed into the U.S. system as an elected judicial and administrative officer. On his tour of the mining camps in the summer of 1849, Governor Riley found them "to be generally peaceful, every little settlement and tented town having elected its own alcalde and constable."[78]

Miners on the North Fork of the Calaveras that winter thus followed an established Gold Rush precedent, sanctioned both by traditional Spanish and Mexican laws and by the interim governor. Even though a new constitution, written and ratified in September 1849, established a different system of local government, the first state legislature—itself an extralegal body prior to California's admission to the Union in 1850—would not implement the new system for another two months. The "pseudo alcalde" at the Iowa Log Cabins was as legitimate as any other elected official in the waning days of military rule following the U.S. conquest of California.[79]

At first Hispanic miners took no notice of the American action on the Calaveras. The Chileans offered little resistance when a squad of armed men under Captain Wood arrived at the San Antonio camp and arrested forty men for continuing to work in the diggings, but they burned in resentment. As Ramon Gil Navarro earlier told his diary: "Bandits and murderers! It is almost impossible to hide from the men all the hatred I have built up for Americans. I wish I could find a worthy way to avenge all those they have repressed. Then they would see what it is like to abuse the power that the struggle between peoples and nations has placed in their hands."[80]

The Chileans were marched back to Judge Collier who fined them one ounce per man, but upon learning they were peons, Collier told them to collect the fines from their masters and return with the money. When they did not show up the next day, Collier sent another squad to the Chilean camp.

They arrested fifty Hispanics, sorted out the peons, and fined their masters. Dr. Julian Concha, with thirteen peons, was fined four ounces of gold dust for himself and one ounce each "for his slaves which he forked over," in the indignant words of John Hovey, one of the Americans who participated in this "Chilean war."[81] To the Chileans it was nothing more than extortion, a view Hovey unintentionally confirmed: "This morning," he wrote, "all those Concern'd went before the Alcalde and received our share of the spoils which we have obtain'd from the mexicans in the shape of fines."[82]

The Hispanics retaliated with violence. During the night of December 27, 1849, they raided the Iowa camp of Americans at San Andreas, killing three men, taking twenty-one prisoners, and leaving at least one Chilean corpse behind. Shocked Americans the next day found him shot to death next to the body of a dead miner from New York. John Hovey, whose Missourians were camped next door, walked the battleground in horror, shaken by the bloodshed and lamenting the loss of good American family men with widows and children left behind. He joined others in calling for vengeance, although Judge Collier told them to wait until he heard from the Stockton alcalde, the appellate judge of first instance under Mexican law. The answer he received jolted the Americans, for they learned that Dr. Concha, prior to the raid, had filed a complaint in Stockton against the Americans, charging them with armed robbery and obtaining a warrant for the arrest of their local alcalde, along with, in Hovey's words, "such others as a certain Chilian should recognize, as haveing been in the affair." Without bothering to investigate, the Stockton magistrate, Judge Reynolds, had issued the arrest warrant and had evidently given it to the local sheriff to serve. That worthy, lacking travel funds, was easily persuaded to empower Concha to organize his own posse to arrest the offenders. With a force of some fifty or sixty Chileans, Concha had responded with the surprise raid.[83]

While the Americans buried their dead and sent an armed force in pursuit of the Chileans, Concha and his men headed for Stockton with their prisoners. They were in hostile territory, unsure of themselves, and as they marched both prisoners and captors began to slip away. Near Double Springs they had tried to see another alcalde, presumably to sanction their actions, but he had refused to see them, and they pressed on toward Stockton.[84] On December 28 the diminished Chilean force were themselves surprised by the remaining prisoners one night. The captives overpowered their captors and held them until Calaveras vigilantes arrived to reinforce the American presence. The new prisoners were then marched back to Calaveras, exhausted and badly abused, although rumors that some had been hanged on the road proved false. In the meantime the Americans who had been released marched on to Stockton to tell their story to an indignant crowd, who passed resolutions absolving the miners of any wrongdoing and praising them for acting "with a high regard to the good order and welfare of this District, and a laudable respect for the rights of themselves and their countrymen." When they learned of this turn of events, the Stockton judge and sheriff, according to the reminiscences of an eyewitness, "took hurried departure for San Francisco in a small boat," fearing a lynch mob.[85]

Back at the Chilean camp near San Andreas, the remaining Hispanics on December 31 were surrounded and captured by an American force after a brief skirmish in which, as Hovey recalled, "nobody [was] kill'd except a dog who was more spunkey than his masters." Sixteen captives and one teenage boy were brought back to the American camp, joining those captured by the vigilantes on the march to Stockton. Whether Concha himself escaped or was among the captives is unclear. Hovey noted dryly that some Americans wanted "to take all the prisoners out and shoot them, but this was overruled." Instead they were tried by a miners' court, and all found "guilty of murder in the first Degree." But since most of the prisoners were "Peones or Slaves, and had to do just as ther Masters said," they were "sentenced to have heads shaved & given 100 lashes on bare backs, & one had ears cut off." Two were given "honerable Discharge," one a man who proved he had saved the lives of some American prisoners after the Chilean raid, and the second a boy who had turned state's evidence and testified against the others. The remaining three leaders were sentenced to death.[86]

On January 3, 1850, the prisoners were marched to a clearing by a twenty-man firing squad. Hovey, convinced the American action was justified, pondered the human consequences as he watched the victims prepare to die. One was leaving behind a young boy, alone and sick with scurvy, lying in a dirty tent in the Chilean camp. "Think of that little boys feelings," wrote a contrite Hovey, "when his father enters and tells him he's come to see him for the last time." In homespun English, Hovey's journal records the final scenes of this tragedy:

[T]hen—reading prays in spanish and counting their beads as is coustume with the Catholics, they had their arms tied behind them, and their eyes blind folded, and was asked if they wanted to say anything, and one of them made a confession of his guilt and said he murdered three be sides these, and then puting his arms around the Alcaldes neck, and kissed him and shook hands here the alcalde weeped like a child. They then kneeld and the hankercheif droped, and then the word was given, and then they was sent in to eternity.[87]

With the Chileans no longer an immediate threat, the Americans on the North Fork of the Calaveras turned from revenge to charity. They gathered all the belongings of the deceased Hispanics, sold them at auction, and made up provisions for six months to help sustain those in the Chilean camp who were unable to help themselves. One American volunteered "to take the little boy that was left by his father, and bring him up and take care of him." The remaining Chileans, and their Mexican friends from Sonora, abandoned their claims and headed south, resentful but hopelessly outgunned and out-manned. Their bitterness was expressed in the words of a condemned col-league, recorded moments before he was shot: "I only regret not being able to kill two or three more of these bandits before dying." In the words of Susan Johnson, the incident demonstrated "the power of the state to back up the Anglo will to rule."[88]

The Chilean incident on the Calaveras added to the rising demand by Americans throughout the Southern Mines for legal action against all for-eigners. Early in 1850 the California state legislature, dominated by represen-tatives from the mining counties, passed the foreign miners' tax, imposing a confiscatory levee of twenty dollars per month on all foreign mining claims.[89] The victims of this fiscal form of ethnic cleansing protested, but they were confronted in turn by gangs of Americans who used tax resistance as a pre-text to drive the foreigners out. The tax applied to all foreigners, but was only selectively applied. An unwritten, unspoken rule seemed to prevail: the darker the skin, the more rigorous the enforcement. Some foreign miners refused to pay or simply moved to more remote diggings, less likely to be bothered by official tax collectors.[90] Others ignored the tax apparently with little fear, especially if they were white and spoke English. Chinese and Mexicans were the primary targets, although Indians were also taxed if not under hire or not from the area. Wade Johnston recalled Ben Thorn's spirited story in 1856 about confronting a Yaqui Indian working a claim at Yaqui Gulch, a gathering spot for Mexican and Indian miners from Sonora. Thorn had just taken office as collector of foreign licenses, a job that preceded his long tenure as Cala-veras County sheriff. Johnston had a gold scale at his cabin when in walked Thorn with a big Indian in tow, nursing a head wound. Asked what hap-pened, Thorn replied: "Here's a man that owes me two licenses. I asked him for it and he asked me if I wanted to fight. So he showed fight, and I hit him over the head two or three times with the six-shooter." Thorn then turned over a confiscated goose quill full of gold dust, which Johnston weighed and found to be worth $5.75. Thorn pocketed the gold and gave the Indian credit for one license. "When he pays the rest I'll give credit for the other licences," he said.[91]

In Sonora a protest rally in 1850 led by French and Mexicans turned out thousands of foreigners. They were challenged by a company of Americans, all Mexican War veterans, but fortunately the confrontation ended without much bloodshed.[92] Yet the message was clear: non-American miners, particularly those from Mexico, were under attack from all sides by an aggressive protagonist who controlled the courts as well as the streets. Under these circumstances many foreigners abandoned their claims. Between May and August 1850 some five to fifteen thousand people, mostly Hispanics, left the Southern Mines. Among those who left was Don Antonia Coronel, ironically chased from his claims by Americans after he had jumped the claims of Native Americans. He "rationalized the treatment he received," in the words of one scholar, blaming gold fever and American jealousy for his fate.[93]

The Hispanic exodus had unanticipated economic effects. American merchants, suffering a sudden loss of business, pressured the lawmakers to reconsider. The next year the tax was repealed, only to be restored in 1852 at a more modest three dollars per month. In the meantime some of the Latinos returned to the Southern Mines, but their numbers were about half what they were in earlier years.[94]

Even though greatly reduced, the Hispanic population continued to represent a competitive threat to American miners during the early 1850s. If the hardrock camps first accepted the Hispanic miners as a necessary evil, the placer camps were even less hospitable. Instead of recognizing the value of Mexican placer experience for the region's development, Americans saw dire consequences. At a rally in Columbia in 1852, white miners said at least the Chinese were content to work old diggings, but Mexicans were a serious threat because they prospected the creeks and canyons and found the richest diggings.[95] Furthermore, said one nativist editor, Mexicans were "perfect Vandals in a placer," high-grading the best ore and leaving the mine a shambles. Blind to similar actions by Euro-American miners, he concluded that the mines needed "a steadier and more permanent population" of Anglo-Saxons. His words were echoed a few years later in a nativist editorial against Chileans working on Jesus Maria Creek, a tributary of the Calaveras. These miners, said the indignant editor, were "merely coyoting here and there where they can find a rich crevice. This class of foreigners are very expert in finding rich deposits, and work without regularity, thus, in most cases, spoiling a claim which otherwise would have afforded regular and profitable employment to other miners."[96]

Harvey Wood, a New Yorker who later ran the ferry at Robinson's Bar on the Stanislaus, described the problems inexperienced Americans had working beside Mexican placer miners. At Columbia in 1850 he and his partners reopened an old prospect hole on a small claim and lowered it twelve feet to bedrock. They struck a pay streak and began following it, but "soon a sharp crowbar came through from the other side and a Mexican sang out: 'caramba!' Then the strife was who could get the most of the rich dirt; I am sorry to admit the Mexican beat the Yankee, for he had seven ounces in his pan and we only had $70." Mexican mining skills also enabled them to use "rat-hole" methods in deep placer ground, sinking narrow shafts and drifting underground without a visible presence on the surface. Wood described a "sadly disappointed" American who spent several days' hard work sinking a shaft, only to suddenly fall through the bottom into "a drift occupied by a Mexican busily engaged in taking out rich pay. The air was blue with curses for awhile, but as the best of his claim was worked out by the Mexican before the American struck bedrock, he had to make the best of it."[97]

For mutual protection Hispanics stuck together in the diggings and in camp, but their concentrations in the southern camps, even after the 1850 exodus, alarmed whites. Byron McKinstry early in 1852 found a small group of Chileans working his claim on Nigger Hill. Unable to oust them himself, he hurried away to gather reinforcements. By the time he returned with an armed force of friends, some thirty Chileans had "collected together in a few minutes," and some of them "showed fight." The Americans, he reported smugly, "after some difficulty[,] succeeded in driving them off." J. D. Borthwick, no friend to Mexicans, in 1851 described an incident in San Andreas that was indicative of the xenophobia. He found the town crowded with Mexicans, on the streets, in the saloons, and "loafing about in their blankets doing nothing." Spanish music came from every gambling establishment, for each had a "Mexican band playing guitars, harps and flutes." A canvas building with a small wooden cross over the door served as the Catholic church; inside was an alter adorned with candlesticks, some made of "old claret and champagne bottles, arranged with due regard to the numbers and grouping of those bearing the different ornamental labels of St. Julien, Medoc, and other favorite brands." During the Sunday morning mass, filled with worshipers, two "swaggering" Americans entered and "jostled" their way forward, showing "supreme contempt for the congregation, and for the whole proceedings." The tense moment ended in comic relief when the faithful suddenly dropped to their knees as the mass proceeded, leaving the Americans "as sheepish-looking a pair of asses as one could wish to see."[98]

Borthwick made light of the incident, but it illustrates the racial disharmony that continued unabated through the early 1850s. Growing economic pressure was behind much of this tension. Mining yields were declining after 1850, and miners had to work harder to earn less and less. To strike out at strangers was one way frustrated miners tried to cope with their disappointments. Few Americans openly opposed this majority tyranny, but in private there were some protests, as reflected in this diary entry by a semiliterate miner at Sonora in 1852: "Nov. 1. Working. a great fuss About the foreigners with the Native born americans because the Foreigners had Privilege to dig the gold Equally so with themselves wich they thought was Not justice due to themselves. And so they went to driving all foreigners off without Law or Authority but did not succeed."[99] Yet with the lawmakers, the courts, the press, and the public all lining up against the foreign element, minority voices were drowned out.

Racial tensions in southern Calaveras County nearly exploded in 1852 as the result of a rumor that Mexican miners from Melones had organized a plot to kill all Americans in the vicinity. A group of white miners on the Carson side of the hill organized a relief party and rode toward Melones, but stopped when they learned that the sources of the rumor were American gamblers who had been fighting their Mexican counterparts. On the heels of the American "relief" expedition came a party of fifty armed militia from Angels Camp, led by Captain Thomas Matteson. When they found out who the fighters were, they turned back as well. If gamblers were fighting gamblers, it did not matter who whipped whom.[100]

In September of the same year, near Campo Seco on the Mokelumne River, an armed clash between two river mining companies was averted with the help of another local militia unit. An international group that included French, Italian, Hispanic, and Hungarian miners built a wing dam that inadvertently backed water over an American claim upstream. Failing to negotiate a settlement, which meant tearing down the dam, the Americans brought the case to a miners' court that ruled in the Americans' favor and empowered a committee to "have the obstruction removed." Seeing the foreigners preparing to fight, the committee called for reinforcements. The next day a militia company arrived from a nearby mining camp, giving the Americans a decided advantage. Without firing a shot the "insurgents" gave up, surrendering their inferior weapons, "which consisted of one hundred stand, mostly double barreled fowling pieces and six shooters." The Americans then completed a dam of their own, and reaped rich rewards. The first washings from bedrock gravels yielded fifty dollars a pan.[101]

While economic competition over declining resources accounts for much of the ethnic turmoil in the southern camps after 1850, adding to the fear and hatred of foreigners was the rising level of criminal violence after 1850. Crime has been a favorite theme of pulp literature from the dime novel to the Hollywood western, but the imaginary West is more distorted than false. The question is not whether the West was really "wild," but whether it was wilder than anywhere else. Historians have long debated this issue, but recent empirical studies are beginning to lend weight to the argument that murder rates, at least, were higher in the frontier West than in more settled communities then or now.[102]

What is less clear is the correlation between race or nationality and violent crime rates after 1850 in western mining camps like Sonora or Mokelumne Hill. Did the number of murders, armed robberies, and aggravated assaults increase because of declining economic opportunities, rising ethnic tensions, a combination of both, or for other reasons entirely? These issues are as debatable today as they were a century ago, for they raise questions of motivation that remain problematic no matter how much information is available. Adding to the difficulty is the lack of accurate data on nineteenth-century criminal activity. For California in the 1850s, most of the evidence is fragmentary and anecdotal—newspaper accounts, letters, personal reminiscences, and so on. Historians must therefore proceed with caution, and regard dogmatic assertions on the subject with a healthy degree of skepticism.[103]

Unlike modern scholars, contemporary Americans in the Southern Mines had few qualms about identifying the causes of crime. To Joseph Pownall, a physician-miner at Hawkin's Bar in Mariposa County, organized gangs were the culprits. In August 1850 he predicted trouble in a letter to a friend:

> A great many murders and robberies have been committed within 50 miles of this place during the past 6 or 8 weeks—chiefly in the diggings on and near the Stanislaus. I believe that gangs of desperadoes both foreign and American have banded themselves together for this illicit purpose and that the past is but the foreshadowing of that which will be for months to come. The victim is murdered while asleep in his tent at night or traveling upon the public road in the day. I anticipate a horrid state of affairs for the next 12 months or more to come.[104]

Pownall's forecast was unfortunately all too accurate. Criminal violence ran high for the next several years. Reports of claim jumping, robbery, murder, and mayhem filled the regional press, with Hispanics most often taking the blame. Typical was a newspaper account in the summer of 1850 reporting the murder of a man near Carson's Creek "supposedly . . . killed by Mexicans. The *Sonora Herald* that same summer loaded its columns with Mexican horror stories: a lethal encounter between two Mexicans, a robbery of Chinese by two or three Mexicans, a threat by five Mexicans against an American who ran them off with a gun, the murder of one American, and the fatal wounding of another by five Mexicans at Mormon Gulch on the Stanislaus. Near

Mokelumne Hill early in 1851, one miner concluded his letter: "I can think of nothing more to write of importance, except a man was shot yesterday a short distance from here for murder. Such cases are very frequent."[105]

Coming on the heels of the discriminatory foreign miners' tax, this intermittent violence may have represented acts of retaliation by retreating Hispanic miners. On the other hand, Mexicans and Chileans made good targets for accusations later proved false. In July 1850, for instance, the *Sonora Herald* reported the near lynching of three Mexicans and an Indian accused of murdering two white miners they were caught in the act of burying. Rescued in the nick of time by civil authorities, they were eventually acquitted. In the *Calaveras Chronicle* two years later, "Uncle Ephriam," a jocular correspondent, said that "greasers," suspected of stabbing a Vallecito man, were given twenty-four hours to leave town but "on reflection" were invited back. Some returned to find their claims occupied by gringos.[106]

Reports of Mexican criminality in 1851 led to accelerated vigilante action in the southern camps. When a New York desperado named Jim Hill was captured in 1851 by vigilantes in a "Spanish house of ill-fame" at Campo Seco in Tuolumne County, he was identified as a gang leader, and the unfortunate house proprietor, one "Guadalupe," was tagged as an accomplice. The vigilantes evidently released Guadalupe, but Hill suffered a different fate. He was turned over to a jury composed of "our most respectable citizens," said the *Sonora Herald,* who gave him "a fair and impartial trial" and sentenced him to hang. At the last minute the county sheriff arrived to plead for a regular trial in district court, and the people's court reluctantly released him to the sheriff's custody. Dissatisfied, some of the vigilantes raced ahead of the sheriff's party and organized a mob in Sonora who recaptured the hapless prisoner and immediately strung him up on a tree limb. "The crowd were deeply impressed" by the sight of a man hanging, wrote an obliging journalist, "but all were satisfied of the righteousness and necessity of the punishment."[107]

That same year in neighboring Calaveras County, J. D. Borthwick was in San Andreas when two vigilantes from Mokelumne Hill rode in, seeking the priest. He was French, but had a good command of Spanish. They wanted him to "confess a Mexican whom they were going to hang that afternoon, for having cut into a tent and stolen several hundred dollars." When his fate was announced the doomed man showed "indifference," and died quietly after all of the committee members took hold of the rope to string him up, "thus sharing the responsibility of the act." At Rich Gulch that same fall, Alfred R. Doten reported two Mexicans hanged by vigilantes after one had killed an American miner in a bar. In a related incident some friends of the dead miner confronted two Mexicans in a store on the Mokelumne after one of the Hispanic men acted belligerently. A fight broke out that almost led to bloodshed after pistols were drawn, but others stepped in to stop the row. Doten blamed Mexicans for numerous thefts from winter piles of pay dirt stored for later washing, and in one incident caught two Mexicans working his claim. According to him they were all armed but dropped their buckets and left when he threatened to shoot. He minced few words in telling his diary what he thought of them: "God damn their thieving Mexican souls eternally to the hottest corner of hell and may every sack of the dirt which they have stolen from me turn into brimstone to help roast their damned infernal carcasses— Amen—."[108]

For the majority of miners and merchants in the southern camps, escalating crime meant increasing vigilance and self-protection. The halcyon days of the Gold Rush, when cabins and tents were left open and untended for days at a time, when weapons were abandoned or left in camp while miners worked their claims, gave way under the hostile climate of the early 1850s. Since federal troops were unavailable and the state militia had not yet organized, local communities took up their own defense. Most of the major camps organized raiding parties against Hispanic enclaves, sent headhunters on search-and-destroy missions, and raised money for arms and ammunition. More evident were pistols and rifles, armed escorts and local militia units, locked cabins, and camp guards. Even recreational events required reinforcements. At a dance in a Calaveras River camp, James Megeath found that "almost every man had a pistol or bowie knife in his belt and some had both." Frequent were mass meetings called to drive out all foreigners, or to seek action against "continued acts of atrocity and bloodshed perpetrated . . . by a band or bands of miscreants," as the *Sonora Herald* reported in 1850, or—more racially specific by 1853—to "exterminate the Mexican race from the country."[109]

The tense atmosphere had a chilling effect on travel, as suggested in this journal entry of a miner approaching Sonora: "[D]id Not sleep much for there were a Number of Murders Committeed arround there. So we were Afraid to Close our Eyes into Sleep for fear of the greasers so Called or Mexicans." Most travelers took sensible precautions, but William Brown, a '49er who found modest prosperity as an express agent in Stockton, made light of the travel danger in letters to his parents. Despite frequent trips to and from the Southern Mines carrying mail and gold dust, Brown told his worried mother: "[T]he fact is that I never have seen any danger in my road yet nor do I think to stop to look for it." He should have been more cautious: Early in 1853 he was murdered.[110]

Elias S. Ketchum, another frequent traveler on Calaveras roads in 1853, was more circumspect than the ill-fated express agent. A Michigan artisan and devout Christian who had sought divine guidance before deciding on a trip to the California goldfields in 1851, he worked several placer claims in Calaveras and neighboring counties, but—like so many other argonauts—could not make a living as a miner. By early 1853 he and his partners had opened a general store in Murphys to supplement their earnings. They also put their carpentry skills to good use, manufacturing and marketing wooden utensils and tools on the side.

On January 24, Ketchum was jolted by "shocking news . . . of the murder of six persons by a party of Mexicans, or greasers, as they are termed." A citizens' posse had caught and hanged one of the alleged perpetrators; the rest had escaped. Cognizant of the moral and legal ambiguities, in his journal he questioned whether justice was served by extralegal action: "The pursuers are reported to have destroyed all the Mexican tents or dwellings that came in their way, which [I] believe to be cruel & unjust, the innocent must suffer for the guilty in that case—But many persons who are prejudiced say they are all alike, 'a set of cut throats & should be exterminated or drove out of the country.' "[111]

Alfred Doten, at Paloma, made note of the same event two days later, but had no qualms about the subsequent vigilante action. With obvious relish he described the attack on the Mexican camp above San Andreas in retaliation for Mexican atrocities on the upper Calaveras, "robbing, murdering, stealing horses, etc." He did not question the lynchings or the orders to drive all Mexicans "away from the mines and set fire to all their houses," nor did he express any regrets that refugee families were headed for Stockton with "what they could pack on their backs." More important to him was the news that more than two hundred Americans had gathered at San Andreas "to rout the greasers from there."[112]

Doten's view was far more common, and more widely expressed, than the moral perplexities recorded in the private diaries of conscientious moralists such as Elias Ketchum or John Wallis. At Double Springs on January 26 a mass meeting endorsed the vigilante rout and passed resolutions "making it the duty of every American citizen at all events to exterminate the Mexican race from the country. The foreigners should first receive notice to leave, and if they refused they were to be shot down and their property confiscated." Not every community called for such draconian measures, but the stated intent, even if not implemented, was enough to drive some Hispanics from the district, and to send others into hiding. Yet even the threat of indiscriminate reprisals was not enough to reduce the predatory attacks on miners and merchants. As a correspondent from Mokelumne Hill wrote on February 14, "The necessity for immediate and strenuous measures for the protection of the life and property of the citizens of this county is evident, from the fact that not less than twenty innocent persons have been murdered in this vicinity within a month, and that robbery is an every-day occurence."[113]

But what more could be done? Rather than wasting energy and resources in a futile and unjust attack on all Mexicans, sounder heads called for concerted action against those directly responsible for criminal behavior. By the spring of 1853 this line of thinking had centered on a Mexican gang led by a bandit named Joaquin. The name had surfaced for the first time early that year in Calaveras newspapers.[114]

This narrative is not the appropriate forum to follow the lives and legends of the various Joaquins that kept the Southern Mines in turmoil over the next five months. A dedicated amateur historian, Frank Latta, did the best work on the subject. After spending some sixty years ferreting out the details of Mexican gang activity in the Southern Mines and tracing the origins of gang members and relatives in southern Sonora, he concluded that there were at least four different but coordinated Mexican gangs operating in Gold Rush California. Their principal business was capturing wild horses and driving them to Mexico, but by the early 1850s they had taken to raiding ranches and remote camps in northern California. Sometimes they sold stolen stock not far from its owners, as John Wallis found out in 1853 when he decided to leave mining and begin ranching. He bought a cattle herd from "an American a Chillanian and some Greesers" for more than six thousand dollars in gold. But the sellers "scarcely got out . . . from . . . sight when some men came and claimed the . . . [herd] as their property [and] identified the cattle by the brand on them."[115]

Family ties related gang members and leaders, although they were not always in agreement. The main leader was a married Sonoran former miner with a burning grudge against gringos after he was run off his claim on the Stanislaus and his wife abused. The details of his mining troubles and retaliatory attacks are sketchy and have given rise to fanciful tales of blood and thunder, first by the contemporary journalist John Rollin Ridge and later by a host of poets, playwrights, and pulp writers. Despite the bewilderment evident during the crescendo of violence in the spring of 1853, when newspapers published contradictory reports of Joaquin's whereabouts and confused the principal character with several related gang leaders with the same given name, it is a mistake to dismiss the Joaquin story as the work of fevered fictionalists and mythmakers. Joaquin was no figment of the imagination to fearful miners and merchants in the Southern Mines. Even Chinese miners, normally ignored or abused by Euro-Americans, won sympathy in Calaveras after a Joaquin-led gang terrorized their placer diggings at Fourth Crossing.[116] As the toll rose the victims clamored for protection, and the state finally responded, first by placing a price on Joaquin's head, then by financing the organization of a company of rangers "for the arrest or killing of the robber Joaquin." The California Rangers under Captain Harry Love acted quickly once authorized; in two months they swept south through the southern camps and caught up with an assortment of banditti near the San Joaquin River in eastern Mariposa County. In the ensuing battle Love's men dispatched three men; one they were convinced was the main leader, Joaquin. They sent his head and the hand of a three-fingered lieutenant back to their headquarters, where the specimens were pickled in alcohol and later placed on display, a grisly reminder of the wages of sin.[117]

Whether the jar held the head of the "real" Joaquin is incidental to the main objective of the Rangers's brief career. Their two-month adventure in the Southern Mines calmed troubled waters, eased public anxieties, and chased out at least some of the more notorious desperadoes. That may have translated into fewer crimes in Calaveras and other southern counties, at least for a time, but more important was the easing of pressure on the Hispanic population. With the notorious Joaquin now eliminated, public attention shifted away from the remaining Mexican and Chilean sojourners, letting them quietly resume their quest for riches along with everyone else. As the Gold Rush placer era gave way to the era of industrial mining, some Hispanic men found employment in hardrock mines, others took up farms and ranches, still others gravitated to service industries such as blacksmithing, drayage, and carpentry. Through the tragic times of the early 1850s, and despite sporadic patterns of discrimination, Hispanics learned to cope in Calaveras and remained a visible, and vital, thread in the social fabric.

## Miners at Work and Play

Social disorders did not alter the daily rituals of a miner's life. Racial or ethnic distinctions were set aside in the ordinary routines of work and play. Miners and their families, regardless of nationality or skin color, were socially leveled by the common struggle for life's necessities in an unfamiliar environment.

For the majority of miners, the venturesome boys and men who left mothers, wives, and family back home, adjusting to mining camp life meant a series of personal compromises. They either had to do or had to do without things that ordinarily were done by others. Preparing meals, washing clothes, nursing the sick, and tending the fire were the most essential tasks, with mending, cleaning, decorating, gardening, and other chores relegated to a lower order of priority. How well men adjusted to "women's work" of course depended on individual backgrounds and circumstances. Some men made good surrogate wives and mothers; others failed miserably or did not bother even to try. Riley Senter, writing to his cousin from Angels Camp in 1850, was proud of his newfound cooking ability. "I have just served out my week," he announced, "and have beat the whole mess at making bread, there being some strife lately as to who showed up the best." Though most miners in the area lived on "fried pork and pancakes," Senter and his company, determined "to live somewhere near well if we can earn enough to live upon," supplemented the standard fare with potatoes, dried apples, onions, and fresh beef.[118]

In contrast to Senter and his happy crew, a lonely Angels miner described a bachelor routine that kept him so busy washing, mending, cooking, and packing in provisions that he did not have time for personal care. "I haint shaved since I left Mass," he explained to his sister in New England, "nor a had my hair cut. I should be all right if it warnt for the sand fleas. They are thicker than the flies are in the States dog days."[119] Poor furnishings was the main complaint of a Campo Seco miner who took over an old cabin at Campo Seco:

> My bed consists of two pairs of blankets and an inverness cape, a cotton tick without straw, spread in a rough bunk with canvass stretched across—the balance of the furniture consists of three chairs, something of the bar room style, two of them home made—and all three have been through the wars. The table is of pine, tolerably well covered with grease. . . . [Dishes include three teacups], one with the handle broken off, and two that never had any . . . , [and] three forks with here and there a tang missing.[120]

The results of hygienic neglect or privation in a primitive environment were predictable, with high rates of dysentery, scurvy, and other diseases attributable to deficiencies in diet or sanitation. "A man who is usually successful, and there are not so many," wrote John Woodhouse Audubon from the Tuolumne River in 1850, "may have acquired five or six thousand dollars, but he has usually aged ten years." The miserable conditions help explain why the white mortality rate in the West was higher than the rate for Chinese miners. Despite their minority status and distance from home, Chinese ate better and had superior health care. The ironic contradiction between white miners losing their health in the West and promotional images of the West as a "place where ailing people went to recover and regain their health" has not escaped modern historians.[121]

Even conscientious white men had difficulties keeping up with contemporary standards of diet and cleanliness on the mining frontier. Matthew Scott's

1849 letter from Mormon Diggings on the American River to his sister in Massachusetts provides a glimpse into the culinary limitations of mining camp fare:

> I must give you a little discription of the way we live—in the first place it is very simple—our pork we buy & bread we make[ . Y]ou can, no you cannot immagine what pallateable stuff it is, however we make more pancakes than bread[,] these we wet up at every meal with a little flour water & saleratus [baking soda], for a change[ . W]e make our bread in the same way only make it a little thicker[ . C]akes & a slice of fried pork constitutes a meal.[122]

Month after month on such a diet made some men desperate if not ill, providing a golden opportunity for single women willing to risk their reputations, if not their lives, on a trip to the mines where they found more work than they could handle as cooks, seamstresses, laundresses, and housekeepers. Jane Steele, a young and ambitious English girl, earned more than ten dollars a week as cook and seamstress at a hotel near Folsom, while her mother pondered the American law that said "her parents cannot receive her wages without her leave." Lucy Stoddard Wakefield, a hardworking divorcée, made a good living baking pies and pastry in Placerville.[123]

Married women also found good opportunities if they were willing to endure the physical and psychological strain. One young man near Mokelumne Hill considered asking his mother to come from Massachusetts to run a boardinghouse and bar where she could "make money fast," but decided she had "to[o] much temperance" to consider it. Harriet Behrins, sitting in the doorway of her husband's cabin at Quartzburg, saw some rocks shining in the distance. Upon closer inspection she discovered a shallow pocket that yielded an ounce of gold. Three wives walking on a trail near Railroad Flat found a quartz lead by accident. Stepping into the bushes to relieve themselves, they "suddenly noticed gold glistening on the wet rocks where they had lifted their petticoats." Their husbands named it the Petticoat mine.[124]

Near the North Fork of the Calaveras River "at the foot of the mountains" on the Mokelumne Hill road, twenty-four miles from Stockton, a New Zealand couple in 1849 built a roadhouse they called "Oak Ranch." For two years or more Charlotte Harrold and her husband fed eager travelers at one dollar per meal, importing food from Stockton supplemented with meat and vegetables grown locally. They sold "a great deal of pastry," but liquor and gambling added to the popularity of the place. Charlotte wrote her family that there was always "plenty to do, though, at the same [time], we are making money." Cash bolstered her courage, for she could write nonchalantly of shooting wolves from her backyard, or grizzly bears in the mountains, the latter rendered into hair grease that commanded a ready market. There were also human dangers in operating a roadhouse in the early 1850s, but they were left unspoken in her letters that have survived. Despite the anti-Mexican sentiment in Calaveras, Hispanic customers were numerous and welcome at the Harrolds.' "I am trying to learn the Spanish language," Charlotte wrote, "for we have so many in for drink and meals, I have some trouble to understand them. A female here is treated with the greatest respect, there are so few of them here."[125]

How women were treated in the mining camps depended to some extent on the color of their skin. The news spread quickly when a visitor to Winters Bar on the Mokelumne brought his wife along in the summer of 1850. In surrounding camps "there was a general cleaning and sprucing up," wrote one participant, "and a run to Winters, to see the face of a white woman once more." Women in Hispanic mining parties were in great demand as cooks and servants by male miners of all races. Some Mexican women made money selling tortillas on the streets of Sonora, while others earned extra income as laundresses in remote Mother Lode camps. But Mexican women earned less than white women for comparable work, and they earned even less respect. At his store in Volcano, for example, Alfred Doten politely courted white women, but was lusty and wanton around Indian, Mexican, or other women of color. Even census takers used color as a criterion for job descriptions. White prostitutes were rarely identified as such, while the same scrupulous regard did not apply to their Chinese and Mexican counterparts. San Andreas in 1870, for instance, had mixed brothels, but the only named prostitutes in the census are Chinese.[126]

Regardless of color or class, among the argonauts gambling and drinking were ubiquitous vices. The Gold Rush tested conventional morality, allowing young middle-class males a chance to experiment out west without forsaking traditional values. Unrestrained and unregulated, liquor dealers set up shop wherever miners congregated. Supply towns offered the widest variety of bottled goods. Calaveras miners coming to Stockton, for instance, could sample the wares of dealers on nearly every street corner, much of it manufactured on the spot by adding various flavors to raw alcohol. William Ryan, an English refugee from Colonel Jonathan D. Stevenson's regiment, with a friend entered a Stockton "groggery of the lowest description," bought a glass of liquor "for which we paid very liberally in gold-dust," but found that "[i]t was of execrable quality, and comparable only to vitriol in its effects on the stomach."[127]

Outlying camps offered similar diversions. Lucius Fairchild, later the governor of Wisconsin, explained immorality in the diggings as a product of indolence and isolation. He and his mining partners wintered on the Cosumnes River in 1849, holing up until the rains ceased and they could return to work. "We lived, Eight of us in a cabin all alone away from every body, being five miles from any house and I tell you we had fine times doing nothing, eating, drinking, Playing cards & Smoking." A young river miner on the upper Mokelumne in 1852 admitted to his father that he was broke and out of work with no prospects in sight, but that his complaints were common and due to hard luck, for "the river was a total failure." A later admission was more revealing: "Now about disipation I am trying to get ahead as fast as I can."[128]

Charlotte Harrold, at Stockton before taking up her roadhouse life, said the gambling houses ran twenty-four hours a day, seven days a week, pulling in miners "every day till they get into such a loose habit. . . . Some men will come from the mines, and put a pound of gold on the table at a time, and in less than an hour lose a fortune." At Weber's diggings in 1850, Thomas Wylly watched miners work hard all day and gamble hard all night, placing even sucker bets with professional con artists and losing heavily. At Lower Bar on the Mokelumne in 1852, Sunday fun-seekers crowded into Alfred Doten's general store. "Today I was as full of business as a dog is full of fleas," he wrote. "All day the store was full of drunken Chilenos, French &c and the day passed off finely with plenty of jabbering and quarreling and several fights in which some eyes were blackened and noses bled—but no one was hurt very bad."[129]

Early river camps on the lower Mokelumne lacked local gambling houses, but two ranches near Ione on the road to Sacramento provided a tolerable substitute for miners looking for fun on Sunday. Visitors could place bets on cockfights, or watch skilled Mexican vaqueros break and race wild horses, fight bulls in a makeshift ring, or pull the heads off roosters buried to their necks in dirt while galloping by at top speed. For upper-river entertainment, however, the bull-and-bear baiting at Mokelumne Hill drew enthusiastic crowds. One young miner early in 1852 described the spectacle in a flood of words to his mother: "This is sunday night i went down to the hill to day they had a second bear and bull fight they intended to have one chrismas but it rained so that it did not come off but it come of[f] on new years day and i went to see it they got a young bull with his hornes sawed of[f] about two inches and chained by the one fore foot and the bear to the bull had no show at all the best fun was to see them get him into the cage again."[130]

To nineteenth-century moralists mining camp vices grew inversely with the scarcity of "respectable" women. Modern scholars have debunked most of these gendered frontier stereotypes, but contemporary culture mavens believed that the lack of women and children loosened the fetters of masculine self-restraint. As Louis B. Wright observed long ago, western leaders who sought stability and "progress" tried to imitate the social structure they were most familiar with. Bringing families, churches, and schools to the mining camps was thus a high priority.[131]

Even without many women masculine indulgence in the remote camps was tempered by anxious letters from relatives and sweethearts back home, and to a lesser extent by the restraining hand of religion. Correspondence to and from the mining camps, as recent scholars have demonstrated, reveals the conflicting dimensions of pleasure and pain that "seeing the Elephant" evoked in family relationships during the Gold Rush. "You know not my feelings," wrote a frantic wife to her husband en route to the goldfields. "I cannot live if you go any further. Oh return home, sell your things & return to me." A miner in Sutter Creek, responding to his sister's chiding letters, wrote defensively: "I once more take my pen to assure you again that I have not forgotten you[ . W]hy do you need the assurance[? D]o you think I have become a heathen living in a heathenish land[?]"[132]

Despite such promises Gold Rush widows had reason to worry. The mining excitement drew both sinners and saints like a moral vacuum. Saloons competed with teetotaling reform groups; preachers vied with prostitutes for the miner's eye and purse. The moral dilemmas of mining camp life are visibly captured by the paintings of Charles Nahl, a contemporary artist. They also surface in miners' journals. Peter Cool, an Amador quartz miner in the 1850s, struggled daily with trying to reconcile old values and new opportunities. His journal is filled with self-righteous disdain for the human blight he witnessed around him: crude men and degraded women, drunks and addicts, the helpless and the hopeless. Trusting his salvation to religion and temperance, he mined during regular working hours and lectured and preached at night and on weekends. In the words of a biographer he "lusted after gold and repenting the lust still sought the gold."[133] After three or more futile, frustrating years in the diggings, he finally conquered the sins of mammon and became a full-time preacher.

As the camps died or matured, and as wives and children arrived to reduce the demographic imbalance, the type and location of entertainment changed along with the rising desire for respectability. Sonora took the lead in the Southern Mines, seeking Sabbatarian legislation as early as 1852. One pious but semiliterate miner canvased the town with an omnibus petition "to Suppress bull Bear & Cock fighting Auction Theat[r]ical and other places of Amusements gamboling Saloons Stores Shops Brokerage houses Exchanges & Banking houses Likewise Mechanicks farmers And trademens shops etc."

Popular Sunday afternoon amusements in Mokelumne Hill gradually shifted from bull baiting to family-oriented picnics and church socials. San Andreas promoted musical entertainment and fraternal organizations as alternatives to hurdy-gurdy dances and other barroom amusements. The saloons themselves improved in style and appearance, replacing tents with wood and brick; upgrading the quality of food and drink; and adding more attractive entertainment in the form of traveling troubadours, mesmerizing magicians, theatrical stars, and occasional divas and basso profundos from the operatic world. During the boom years of lode mining, the Altaville Iron Works added a comfortable and heated reading room for its employees "so as to compete with the saloons" that were "the only comfortable places in the mining camps for winter assemblages of men." Under the banner of social reform and civic improvement, change also came to the brothels and their occupants. They were banned outright or relegated to the minority enclaves. In the 1890s, Angels Camp officials, to clean up the town, exiled white prostitutes to Chinatown.[134]

More worrisome than moral issues were the economic uncertainties associated with seeing the elephant. Building a grubstake by mortgaging the family farm and borrowing from friends and relatives seemed worth the risk to novice fortune hunters in the early years of the Gold Rush. But gold fever wore off quickly, and the realities of trying to earn a living in California ended many golden dreams. John Fletcher, a Boston '49er, considered himself lucky after a year in the diggings to have six ounces of gold left and nothing worse than a poison oak infection. "It is a hard case to get a fortune out of California," he wrote, "for everyone who goes home with his pile there are six who find their graves here." A cynical William Brown, writing from Stockton, told his mother, "This is the greatest country in the world for a crowd from the fact that there are so many more lofers in this country who have been sent out by their friends at home to get rid of them knowing at the time that the young sprout never could get money enough together again to get back. Now this must have been my fix." Out of luck and burdened with debt in Murphys, Francis Xavier Hall pleaded for understanding from a male creditor. He also sought advice on how to explain things to a woman who had loaned him money: "[S]he is a verry [sic] sensitive person," he wrote. "I hardly dare to let her know the facts as they are." Asked by a friend when to expect him home, William Henry Newell, a placer miner on the Stanislaus, admitted: "I have seen three men who have returned here who have been home and they advise me to stay untill I make somthing worth carrying home." William Hanson in Hangtown began "to feel homesick" after failing at both mining and butchering. He promised to book passage home in steerage as soon as he could collect on some unpaid accounts receivable. But promises faded with empty pocket-books, as a youthful miner near San Andreas recalled: "It is hard for a man to leave here, where ther is so much money made, with nothing, still clinging to the hope that he will strike it soon he hangs on untill he spends what little he has and is then forced to stop."[135]

For Gold Rush widows financial responsibilities increased the burdens of homemaking and added a plaintive note to domestic relations. Those left behind assumed more authority over their own lives, but many were ill-prepared to be the major family breadwinner. Bad news from the goldfields increased anxieties at both ends, but some families drew closer in adversity. As Malcolm Rohrbough has written, the Gold Rush, despite its individualistic implications, "became, at its heart, an intensely cooperative venture," both in California and back home. A letter Jonathan Frost Locke wrote from Mokelumne Hill in the fall of 1852, just before turning homeward, is indicative. "Dear Wife," it begins, "by what you write I fear you have felt oblidged to part with those specimens I sent you[.] I hope not[;] you are as unfortunate in respect to getting boarders and keeping them as I am in getting a pile. [B]oth seem a thing that is not to be."[136]

## Calaveras Literature and Lore

By the 1860s, Calaveras mining towns that had survived the early booms began to lose their rough edges. The frantic quest for riches gave way to more orderly growth as industrial mining slowly emerged. Of the Gold Rush miners and merchants remaining from the days of 1849, hundreds left for the Comstock or for new booms farther inland; others gave up scrambling after the quick dollar and moved to the coastal cities or took up farming or ranching. For a time Civil War politics and passions revived Mother Lode energies and generated a mining boom in the copper districts of Calaveras, but at war's end the county economy sagged, and so did the population. A small but steady flow of European immigrants continued to flow into Mother Lode counties late in the nineteenth century, but most newcomers bypassed the foothills, heading for farms in the Central Valley or jobs in the cities.

The remaining Gold Rush pioneers and their families, firmly rooted to the land after just a decade of settlement, grew more reflective over the passing years. Without education or experience they had confronted and conquered the challenges of a new frontier, pushed back the Indians, solved the geologic mysteries, and generated a vast new wealth for the nation. They were proud of their legacy, celebrating it in annual speeches on the Fourth of July, cherishing it in pioneer reunions and picnics, and preserving it in statewide historical organizations such as the Native Sons and the Society of California Pioneers, or in county groups such as the Calaveras Reunion, which held its first grand picnic in 1878 at Woodward's Garden in San Francisco.[137]

The heroic spirit of these proud rural folk was not universally admired, however. In the urban centers of the United States and Europe, mining culture had a troubling legacy. Cultural critics belittled the manners, methods, and bucolic ways of these Gold Rush pioneers. Professional engineers scoffed at their makeshift machines and mining methods. Reformers mourned the human victims of material greed.

Contrasting images thus confronted Americans eager to learn more about Gold Rush people and places. To address regional themes without alienating readers was a considerable challenge. The center of West Coast literature was San Francisco, a brash young city with a dozen newspapers that vied for public attention along with literary magazines such as the *Golden Era* and the *Overland Monthly*. In the 1850s aspiring literati danced around Gold Rush issues using a jocular style of satirical journalism that titillated but did not offend. Bret Harte's "Heathen Chinee," for example, mildly rebuked Americans for cheating Chinese but at the same time reinforced white fears of an influx of cheap Asian labor. Harte's model for "Truthful James" allegedly was "Lying Jim" Townsend, a colorful and peripatetic journalist, briefly on the staff of the *Golden Era* before taking his talent over the mountains to Virginia City and the *Territorial Enterprise*. In 1850, as editor for the *Sonora Herald,* Townsend spooked gullible Mother Lode residents with tales of Horrendus Min, the "Calaveras Constrictor," a midnight monster with huge claws and an appetite to match.[138]

By the 1860s satire had given way to romantic sentiment in portraying the mining heritage of the Sierra foothills. In the words of historian Kevin Starr, regional authors relied on the "evanescent gifts of sentiment, charm and humor" to entertain audiences. Writing about the picturesque past was fashionable in the Civil War era, but as Franklin Walker wrote in his pioneering study of literary San Francisco, local colorists "nearly missed the deeper significance of the scene in picturing its external trappings."[139]

For Francis "Bret" Harte, the first California writer to attract a national audience, the Gold Rush days had ended when he first visited the mines in 1854 as an unemployed teenager looking for work. For a few months he taught school, possibly at La Grange in Stanislaus County, but when school ended he toured the diggings on foot and wound up on Jackass Hill on the south side of the Stanislaus River, broke, nearly shoeless, and discouraged. A kindly pocket miner named Jim Gillis gave him twenty dollars to get back to Oakland. Harte took other brief trips to the Mother Lode over the next three years, once as an express rider on a stage line, collecting stories and ideas as he traveled, and fixing locations he would later describe in popular yarns dripping with romantic sentiment.[140]

Harte's excursions into Calaveras County are not well documented, yet the region had a powerful grip on Harte's literary imagination. Although scholars still debate the location of particular stories, the camps along the Stanislaus fit the descriptions, if not the names, in some of his most famous yarns. Yet Harte's sentimental tales, for all their popularity, conveyed a false image, a stereotypical view of Gold Rush life that unfortunately still shapes popular culture today. As Mark Twain later wrote, if Harte "were to write about an Orphan Princess who lost a Peanut he would feel obliged to try to make somebody snuffle over it."[141]

Better known, and better documented, is the one-month Calaveras visit of Samuel L. Clemens during the winter of 1864–1865. The fledgling humorist, still not published nationally, was cooling his heels at Jim Gillis's cabin on Jackass Hill after fleeing San Francisco, where both Twain and Jim's brother Steve had incurred the wrath of the San Francisco police, Steve for a bar fight and Twain for his critical remarks while a reporter for the *San Francisco Morning Call*. As a pocket miner, Jim Gillis worked both sides of the Stanislaus, and Twain accompanied him while he prospected. In this period Carson Hill was infested with claim jumpers and pocket miners, and the notebook Twain kept during this period contains a few cryptic notes on the Morgan dispute and the rich "nuggets" discovered. He found nuggets of another kind in Angels Camp, where he and Jim Gillis spent several weeks in late January and February waiting for the rains to stop. Most of the time Twain sat around the stove in the Angels Hotel saloon, complaining of the food and trading tall tales with patrons. Some scholars have attributed the frog story Twain made famous to Ben Coon, an old river pilot from Illinois, but Twain's notebook makes no "explicit connection" between the two, although his notes outline the plot and the punch line as well as other yarns he later used.[142]

Twain left the Mother Lode on February 23, 1865, riding by horse to Copperopolis to catch the stage to Stockton. He had been persuaded that the Calaveras route was faster than walking to Sonora from Jackass Hill, but when he reached Copperopolis he was chagrined to learn that the stage did not arrive until the following morning. Even though the copper boom was still on Twain was anxious to leave—he had lost his pipe and could not buy another in "this hellfired town," as he described it. To kill time he rode a cage down the Union shaft and walked through "all the ramifications of its six galleries

& numerous drifts," seeing the "very rich" ore that had earned "Hardy" more than half a million greenbacks when he sold his half interest a few days before. The next day he climbed aboard the stage and left Calaveras for good.[143]

Twain's brief stay in the southern foothills left a lasting legacy that not only propelled him to international fame when his celebrated frog story was published in December 1865, but also gave Calaveras an invaluable publicity asset. But Calaveras residents were slow to grasp the significance of that humorous little yarn. For more than a half century it lay dormant in the minds of Angels Camp residents, resurfacing only in the late 1920s when an Oakland preacher, invited to speak at a Booster's Club luncheon in Angels Camp, suggested the county colors should be gold for mining and green for Jim Smiley's frog. That was enough to launch the "Jumping Frog Jubilee," still a major attraction that each spring brings thousands of visitors to Calaveras.[144]

Mark Twain never returned to Angels Camp, but his visit may have inspired local residents to perpetrate a hoax that brought Calaveras international notoriety. Just a year after Twain was exchanging tall tales around the stove in an Angels Camp saloon, on Bald Hill in nearby Altaville James Mattison's crew found a "curious object" while cleaning out cemented gravel at the bottom of a Tertiary mine shaft 130 feet below the surface. Thinking it was petrified wood he took it to R. C. Scribner, a merchant in Angels Camp, who bought it and gave it to a clerk to clean. When it turned out to be a skull Scribner sent it to a physician and amateur naturalist, Dr. William Jones of Murphys. He recognized the enormous implications, and wrote a detailed letter to Josiah Whitney, head of the state geological survey.

A scientist of national prominence, Whitney had a reputation to protect and enhance. He arrived in Angels Camp personally to study the skull and investigate the circumstances. The Mattison shaft was temporarily filled with water, preventing direct examination of the site, but Whitney interviewed the discoverers and became convinced that the artifact was an authentic fossil of revolutionary significance. Though the scientific community at the time generally acknowledged a much older timetable for the origin of man than the 4004 B.C. date endorsed by Christian theologians, Whitney's revised view pushed back the antiquity of Homo sapiens by millions of years. At a sensational meeting of the California Academy of Sciences in 1866 he announced his findings, leading to an international debate that continued for more than a half century. A second skull, somehow mixed up with the first, added confusion to the lengthy dispute. Ultimately, radiocarbon dating proved both skulls were of relatively recent origin. The truth that so eluded prominent nineteenth-century scientists was widely known locally, however. Even Bret Harte could not resist poking fun at the great minds so troubled. In Angels

Camp, Scribner had a reputation as a practical joker. He had acquired the skull to fool Dr. Jones, but when the story got out of hand he was afraid to tell the truth.[145]

* * *

In Calaveras and other southern mining counties in the waning years of the Gold Rush, physical violence and social reform were two sides of the same coin. Striking out against foreigners, either in the diggings or on the streets, was one way white Americans in the early 1850s reacted in the face of declining economic opportunities. The bonanza years of placer mining were coming to an end. By 1852 nearly all surface placer ground along the main streams, tributaries, and creeks in the Mother Lode had been prospected. Newcomers from the "states" still poured in, but throughout California the placer camps had "lost much of their attractiveness" to old and new alike because of the diminished returns per man. At Angels Camp, for example, where in 1849 Thomas Jefferson Matteson had recovered nineteen ounces of placer gold in just four hours, by 1855 quartz mining was the big attraction. Placer ground at San Andreas and Campo Seco was worked out by early 1857.[146] Surface works at Murphys held out until the early 1860s, but by that time Chinese had taken over most of the diggings. In the Calaveras region there were still remote districts such as Mountain Ranch and Railroad Flat in the East Belt as well as West Belt deposits yet to be thoroughly explored, but the best surface ground was already discovered.

Despite the diminishing chances to "strike it rich" as an individual entrepreneur, there were still vast quantities of gold to be mined. Gold production in California actually reached its peak in 1852, even though per capita returns were declining. The discrepancy between individual and aggregate production data reflects the changing nature of mining. Already the goldfields were experiencing a perceptible shift in patterns of organization and operation. Industrial mining had its foot in the diggings.

Josiah Royce, a California-born philosopher, recognized the importance of economic growth as an antidote to social disorder. In his classic account of the Gold Rush era, published in 1886, he offered two explanations for the unrest during the early 1850s. He blamed the violence on the excessive nationalism of the period and on the "irresponsible freedom" of a transient, masculine society "who sought wealth and not a social order." In his view order was restored only with the arrival of women and families, the emerging leadership of "sensible men" who found minority oppression and mob rule abhorrent, and the rise of industrial mining.[147]

Even though Royce and other prominent observers condemned the social turbulence of the fifties, the dark clouds cast a thin shadow in the eyes of American Gold Rush pioneers. They aged amicably, proud of their heritage, remembering the good times, forgetting or romanticizing the rest, emphasizing the material wealth, and taking due credit for lasting accomplishments. This buoyant mood was manifest in the 1871 report of the Calaveras County assessor. Looking both to past glories and to future prospects, he wrote:

> The placer mines of this county are "things of the past." True, now and then, in one's journeyings, you will startle some old "forty-niner," in his secluded ravine, with pick, pan, and shovel, mayhap a rocker; now and then a squad of Celestials working, for the twentieth time, old tailings. But if the bright yield of placer mines has passed, we are content with the more resplendent glories of cement [that is, compacted Tertiary gravels, a reference to drift and hydraulic mining] and quartz; of the first, we are but in our infancy.[148]

What he lacked in foresight he made up in hindsight. Though wrong about hydraulic mining, which essentially ended within fifteen years of his report, the assessor correctly marked the cusp of two different mining eras. The Gold Rush era of entrepreneurial mining ended in the mid-1850s with the decline of surface placers. The era of industrial mining began with more complex technology, corporate organization and financing, improved management, and a wage-labor system. Though hydraulic operations in the 1860s and 1870s laid the groundwork in the Far West, nationally the industrial age of mining came to fruition underground, in the hardrock mines of Michigan, Pennsylvania, Nevada, Colorado, and, to a lesser extent, California.

# 4: Ancillary Industries

The transition from pioneer to industrial mining affected both the scope and the direction of economic development in the Mother Lode. Mining remained the primary engine of growth for more than a half century after the Gold Rush, but intensive lode mining and milling needed a more complex and more integrated economic system than was available in an earlier era of extensive surface digging. Some of the structural changes in corporate finance and management have already been explored. Fundamental improvements in mining and milling technology and efforts to consolidate key properties and develop a reliable labor supply will be discussed in later chapters. Here we are concerned with important subsidiary activities that arose in conjunction with mining and that continued to grow after mining declined. They contributed to the diversification essential to sustained regional economic growth and development.

In the new industrial age after the Civil War, prospecting and entrepreneurial mining dwindled as the number of individual working claims declined. Some of the most promising early claims were absorbed into larger, better-capitalized and -managed industrial operations. Others were abandoned and later restaked, while still others were held for years as unpatented prospects or patented under the liberal 1872 mining law, eventually becoming more valuable for real estate than for mineral content. Ownership in some cases stayed in family hands for two or more generations. Of the many hundreds of individual claims in Calaveras, only a very few became significant mines. The rest failed not because of lack of promotion, but because they had no real potential.

While major mines accounted for most of the production, smaller mining operations made significant contributions to regional economic development. They required the same services and supplies as their bigger neighbors, though on a much smaller scale. The combined economic weight of big and small mines was felt throughout the Sierra foothills right up to World War II. In Calaveras and other Mother Lode counties, as the mining industry expanded or contracted, so did the regional economy.

Mines of every size relied on local merchants for the ordinary supplies and foodstuffs needed to sustain an active operation. Some merchants also distributed mining machinery, blasting powder, and other products, in some cases holding exclusive distribution rights from major mining machinery and supply companies. Any mine manager in northwest Calaveras needing dynamite, for example, went to Treat's or Tiscornia's General Stores in San Andreas. More specialized parts and equipment that local stores and foundries did not carry came from dealers and distributors in the larger supply centers such as Sacramento, Stockton, and San Francisco. Mine operators ordered merchandise from catalogs provided by the major suppliers, or gave orders directly to "drummers," the ubiquitous traveling salesmen of the trade.

Handling mining goods had both rewards and risks for the local merchant. So long as the mines were booming and the bills paid on time, the increased business raised the profit margin and helped stimulate the local economy. But whenever mining slowed or shut down for a variety of reasons, the creditors suffered along with the miners. Some merchants resorted to the courts for redress, suing failed companies or managers, sometimes taking over the remaining assets of a defunct operation. Others waited it out, or accepted mining stock or options in lieu of cash, or simply wrote off the loss as part of the cost of doing business in a small town.

The mining industry also employed an extensive nonmining surface crew of blacksmiths, carpenters, mechanics, timber framers, woodcutters, and assayers. Most active mines had their own specialized crews, although small operators usually did the work themselves or used temporary help to reduce labor costs. From small beginnings some services grew into specialized industries.

Mother Lode mining also affected statewide economic conditions, though to a lesser degree than the more important centers of mining farther north. Major mines in the southern Sierra foothills stimulated linkages to grain and cattle producers in the Central Valley as well as to bankers, brokers, express company operators, suppliers of mining equipment, and other service providers in Stockton, Sacramento, and San Francisco. Industrial mining helped diversify California's economy and generated new lines of commerce and communication between the coastal supply centers and the hinterlands.

## The Rise of Agriculture

Providing food to remote mining camps was a problem of production as well as logistics. California was not attractive to small farmers prior to the Gold Rush: it was too remote, too expensive, too dry for traditional small farming methods and crops, and too uncertain. Most of the best land was already claimed by land barons such as John Sutter in the lower Sacramento Valley and Luis Maria Peralta in the East Bay. Only a few hundred American farmers had arrived before 1848, and those held either clouded land titles, or no titles at all but simply squatted and took their chances. Commercial farming could not flourish under these circumstances. A few orchards and some grain could be found under cultivation on former mission lands in the Bay Area and inland on Sutter's domain, but for Californios and *estranjeros* alike the primary staples were beef and mutton. With a population of some fifteen thousand, not counting Native Americans, northern California in the mid-1840s lived simply in the pastoral era.[1]

The Gold Rush radically transformed California agriculture. Unlike many mining regions that failed to diversify and died quickly after the mines declined, California offered a diversity of mining opportunities over a long time period. Its mineral wealth was broadly disbursed in deposits that varied widely in minerals, size, complexity, and value. Although individual mining districts often boomed overnight and died almost as quickly, the aggregate impact of sporadic but continual developments over a long period provided the impetus for an expanding population and a new and better food supply.

Rising demand and skyrocketing prices after 1848 soon exhausted local supplies of cattle and sheep, sending drovers far to the south in search of bigger herds. Southern California rancheros temporarily prospered but ultimately declined due to increased competition from Texan and midwestern cattlemen, rising indebtedness, and a devastating drought. By the mid-1860s, California's open-range livestock industry had nearly disappeared. In the meantime wheat production mushroomed in the warm valleys of the Bay Area and inland where rangelands gave way to bonanza wheat farms. For the first few years the mining camps and supply towns in northern California consumed all of the grain grown locally, but production soon outpaced regional demand. From the late 1850s to the early 1890s the state's primary export was golden grains of wheat.[2]

Fresh fruit and vegetables were in high demand in the early mining camps but nearly impossible to find. Scarcity and inflation combined to turn the heads of many immigrants who found farming more profitable, and more predictable, than mining. Some traded claims for farm acreage; others used the last of their grubstakes from home to buy land; still others simply found an empty plot and staked a preemption claim, hoping the government would sustain them later. Despite the rich soil and the vegetation that "grows luxurian and wanton," as one traveler reported while following the Cosumnes River toward Calaveras, squatters and homesteaders had difficulty finding and holding suitable land to cultivate. From the 1850s to the 1870s many of the best tracts were purchased or otherwise acquired by speculators such as William S. Chapman, Henry Miller, and James Ben Ali Haggin.[3]

Because of transportation problems, most of the early perishable crops

Delivering vegetables in San Andreas from the Italian gardens, ca. 1895. (Courtesy Calaveras County Historical Society)

grown along the watercourses of the Central Valley went downstream to San Francisco rather than upstream and overland to the mining communities. Early Mother Lode miners and farmers managed to produce a limited supply of grain and a few vegetables in backyard gardens, but orchard crops took longer to grow. The Gold Rush was over by the time California fruit and vegetable production reached significant levels.[4] Nevertheless, the mining excitement brought a permanent population to northern California, which in turn provided the stimulus for sustained agricultural production and diversification.

Commercial agriculture in the Mother Lode counties began near the mining camps where conditions were most suitable for ranching and crop cultivation. One of the earliest farm sites in Calaveras was Salt Spring Valley, just over the Bear Mountain Range, ten miles southwest of Angels Camp. Situated on a direct route between Stockton and the Southern Mines, its fertile and well-watered alluvial soil attracted farmers and orchardists as early as 1850. That year Jacob Tower, a New England argonaut, went to work at the Peach Orchard farm, and within two years had started his own ranch nearby. Tower and a partner, Wilson Bisbee, raised and sold vegetables as well as livestock, hay, and grain. Their ranch became an important rest stop for teamsters en route to and from the mines. Another Peach Orchard employee, Thomas McCarty, started his Log Cabin ranch and way station south of Tower's on another well-traveled route. By the end of the decade more than fifty farms had been established in Salt Spring Valley.[5]

As placer mining slackened, water from old ditch systems provided farmers with a vital resource that was widely distributed through extensions and laterals. Where feasible, farmers also built terraced streamside gardens near the principal routes to town. After 1862 new land laws encouraged land-hungry immigrants from Europe and the eastern states to take up foothill farms and ranches on land now considered marginal. This rural population increase made up in part for the loss of miners after the Gold Rush.[6]

Many of these newcomers earned a decent living by specializing in potatoes, nuts, apples, dairy products, and other cash crops that could be readily marketed in nearby towns, or by raising draft animals and fodder crops. Adolphe Zwinge, for example, a Prussian immigrant in the 1850s, supplied regional miners with fruits and vegetables from his Murray Creek homestead for a decade before moving to San Andreas and entering the freight business. A Scottish immigrant, Alexander Love, ran a dairy farm near Avery after trying his luck in the diggings, and then operated a sawmill before entering politics as the county assessor. Lucca Canepa, arriving in Calaveras from Italy in 1860, mined for gold at first, and then sent for his family. After acquiring bottomland near Vallecito in partnership with his brother-in-law, he planted fruit trees and grapes, grew and sold vegetables while the orchard matured, and prospered. On the Stanislaus, Melones was entirely dependent on local farms for produce during its active mining days.[7]

By the end of the nineteenth century, improved transportation and increasing competition from Central Valley farmers darkened the futures of most of these small foothill producers. Intensive agriculture in the foothills gave way to stock raising and cereal production, primarily wheat and barley. By 1884 cereal crops covered nearly 41,000 county acres. Calaveras had 575 farms and ranches in 1900 with an average acreage of 370. Many rural families remained in the county on rented or leased land, but often had to find other lines of work to supplement their income. By 1940 the number of farms had shrunk to 468, though they averaged 807 acres per unit.[8]

## Developing a Water Supply

Water was as vital as lumber in the early mining camps. Waterwheels, steam boilers, sluices, Long Toms, hydraulic monitors—the primary machinery of an active mining camp in the pioneer period—all depended on adequate water supplies. Unless the values were high, early-day American placer miners had little interest in such dry diggings as Campo Seco, where Mexicans and

Chilenos eked out a living with blanket sorters and rockers. Even rockers needed some water, which soon became thick and opaque with constant reuse. The extensive Campo Seco placers, many of which were exposed remnants of the lower course of the Tertiary Calaveras River, became much more attractive when the Mokelumne Hill Canal reached the area in the mid-1850s.

In the Southern Mines water shortages occurred much more frequently than in the northern Sierra foothill camps, which had higher rainfall, more extensive watersheds, and bigger water companies. All along the Mother Lode in the dry season, shortage of water was a perennial problem.[9] Mining camp population declined drastically in the scarce water months, then rose again when the rainy season returned. In the fall of 1850, for instance, the placer camps along Carson Creek were practically abandoned as miners went prospecting or headed for San Francisco to see the sights, but newspapers predicted a population of eight to ten thousand during the winter months. At Mokelumne Hill in 1851 the rains were so late that miners were still waiting in December for enough water to work their claims. A year later rising water had driven river miners to higher ground, yet the ravines and gulches were still too dry to work.[10]

Too much rain was as serious a problem for placer miners as too little. Swollen streams crumbled the sides of unsupported coyote holes and pocket diggings; wrecked wing dams, waterwheels, sluice boxes, and other equipment; and forced miners from their claims for months at a time. On the Stanislaus River in March 1850, for example, three men took five pounds of gold each from a placer claim in two days, but then had to abandon their mine because of rising water.[11]

River mining was particularly difficult because of the amount of work involved in building wing dams and other means of draining the water away from the river bars. By 1852 miners on the Mokelumne were forming companies to "turn the river at nearly every eligible point," wrote a newspaper editor. Miners on the Calaveras also organized construction gangs to divert the water. A French company in the fall of 1851, following the undulating contours of a bench nearly parallel to the riverbed, spent six months constructing dams and pumps to drain excess water. By July of the following year they were working the richest part of exposed gravels that yielded up to fifty dollars per pan.[12]

Occasionally, heavy rains can expose gold deposits in gulches and ravines overlooked by earlier prospectors. Modern recreational miners still turn out in droves after an extremely wet year and a prodigious runoff, sometimes with interesting results in the bottom of gold pans or in the riffles of a suction dredge. Even in the Gold Rush era, latecomers sometimes benefited from soggy weather. In January 1853 a small ravine running into Calaveras Rich

River mining on the American River using water- and steam-powered pumps. (Courtesy Holt-Atherton Library, University of the Pacific)

Gulch produced a thirty-ounce lump of gold that had lain hidden until recent rains washed off the overburden. Miners reported that the exposed gravel in the ravine "was literally pregnant with gold." A nostalgic correspondent covering the story wrote that "it makes one feel as if it were again '49 to see so many well filled buckskin bags hauled from the bottom of men's pockets."[13]

Where water shortage was the major problem, the miner's answer was water diversion, a practice in the Far West that dates back to the first Native American squash and maize cultivators along the lower Colorado. Franciscan and Jesuit missions in the Southwest had also diverted water for domestic use and for irrigation, using open ditches and clay pipes built by neophytes. It has been argued that Brigham Young's Mormon farmers near Salt Lake City also resorted to water diversion prior to its utilization by California miners. The extent and the actual date of these earlier precedents are difficult to determine, and probably not very significant anyway, for clearly the California miners were the first to practice American water diversion on a large scale, and out of their precedent developed a new form of water law in the West.

Diverting water from a streambed without regard to downstream users was prohibited under the doctrine of riparian rights, rules inherited from England and practiced in most U.S. states and territories prior to the Gold Rush. The riparian doctrine worked well enough where water was plentiful. For nearly

250 years after Jamestown, English precedent served as the primary model for American water law. But beyond the hundredth meridian, positive law came face-to-face with the realities of nature. In the water-scarce provinces of the North American continent there was simply not enough surface water to support a doctrine that gave everyone along a watercourse the same usufructuary rights. Upstream farmers or miners in a semiarid environment could not use water without reducing its volume or degrading its quality for the downstream user.

The legal and environmental consequences of water diversion were not immediately apparent. Fortunately for California miners, there were few downstream competitors for the same water in the early years of the Gold Rush, aside from riverboat owners and townsmen in the Central Valley and San Francisco Bay. Large-scale downstream agriculture came after, not before, large-scale upstream water diversion. By the time farmers and merchants realized the adverse consequences of diversion, a precedent had been set that could not be undone despite years of protest and litigation. The result was a new water doctrine, "prior appropriation," a principle based on a familiar American maxim, frequently invoked whenever eager competitors confront scarce resources: first come, first served. In many western states prior appropriation replaced the old riparian doctrine and opened a new era of water law in the West. In California, where riparian rights had been adopted and implemented before prior appropriation became a recognized legal doctrine, a complex double standard arose that was called the "California system." It recognized the existence of both doctrines and applied each according to the pragmatic circumstances.[14]

In Calaveras County miners along the Mokelumne River resorted to a combination of ingenuity and hard work to divert water to their placer claims. At Big Bar in 1851 five miners kept two Long Toms in continuous operation by using a hundred-foot canvas hose fed by a horizontal pump rigged up to a waterwheel that turned in the river current. Later that year, a particularly dry one in contrast to the preceding season, a company of fourteen miners at Murphys dug a 6-mile canal to divert water from an upstream source to their claims in Murphys Flat.[15]

These early efforts demonstrated the feasibility of water diversion in Calaveras as well as the necessity of working together, both lessons learned well by hydraulic miners. For many years after the Gold Rush they were the largest water users in the California mines. Their huge monitors could shoot a three hundred–foot stream with killing force (and sometimes did if man or beast got in the way), and their long sluices daily washed thousands of tons of debris to the valleys below, consuming enormous amounts of water. To provide constant water supply for all kinds of mining and for other industrial as well as domestic uses was the primary task of early water companies.

In Calaveras County two important water companies appeared in the early 1850s, one on the Stanislaus and the other on the Mokelumne River. The Union Water Company grew out of early efforts to solve the perennial water problems of Murphys Flat. Late in January 1851, William H. Hanford and several others organized a company to bring water from an upper tributary of the Stanislaus River to the mines around Murphys. Apparently, this was entirely separate from the 6-mile canal mentioned above. The next year, before the Hanford enterprise had begun construction, a party led by J. Curtis began digging a ditch to transport water from Angels Creek 2 miles above Murphys to hillside placer operations in the area. Sometime in 1852 these two companies consolidated into the Union Water Company. Raising two hundred thousand dollars for construction, the company hired T. J. Matteson to survey a 15-mile ditch from the upper Stanislaus. With picks and shovels, an occasional mule-drawn plow, and an ox-powered railroad to haul logs to a mill where they were sawn into flume boards, the Union Company completed the work by January 1853.[16]

The Union Company later absorbed another company that had ambitious plans it could not complete. In 1856 the Calaveras Water Company was incorporated for the purpose of building a reservoir and ditch system to serve the southern part of the county. It built the Salt Spring Valley reservoir in 1858, but its ditches never extended beyond Altaville. By 1863 it was bankrupt and taken over by the mortgage holders, Morris Cohen and Isaac Levi. Five years later the Union Water Company bought the Calaveras ditches, and the reservoir in Salt Spring Valley went to the Quail Hill Mining and Water Company.[17]

Meanwhile, in the fall of 1852, the Mokelumne Hill Canal and Mining Company filed articles of incorporation and almost immediately set to work on a canal from the South Fork of the Mokelumne River to the Mokelumne Hill diggings, a total of some 18 miles. Within a month 6 miles of flume and canal had been constructed, and by June of the next year water reached the first mines. Soon laterals covered the hillsides, and eager miners zealously worked the piles of ore they had stockpiled in anticipation of the achievement. Visitors were impressed by the topographical difficulties overcome by innovative engineering. J. D. Borthwick marveled at the sight of water flowing across rugged canyons, spanning deep ravines and broad valleys, carried by flumes high above the trees or sometimes attached to them, supported on "graceful scaffoldings of pine logs." The coming of water stimulated the political organization of the district, for claim disputes increased as mining property rose in value. To reduce confusion and settle conflicts, a miners' court was established and a local civil government elected to maintain order.[18]

By the fall of 1853 the Mokelumne Hill Canal was supplying all the water needs of the district. Small placer and hydraulic companies gladly paid about seven dollars per day for enough water to work a claim.[19] With business booming the company made plans to extend the system to the dry diggings around Campo Seco. Eventually, the sinuous canal system spread some 60 miles of ditches, flumes, tunnels, and laterals thorough the mining districts of northwestern Calaveras County.

Many smaller companies dug ditches on a lesser scale at the same time. By the beginning of 1854, Calaveras County had seventeen ditch systems and 325 miles of artificial waterways. Conflict inevitably arose between competing interests for scarce water resources. As Douglas Littlefield has pointed out, miners demanded "free access" to water as a "fundamental right" whether they held a riparian claim or not, whereas water companies treated water "as a commodity to be bought and sold." In the evolution of California water law, the latter view proved more compatible with the doctrine of prior appropriation.[20]

Water companies expanded so rapidly that in the low-water years of 1857–1858 riparian users faced severe shortages because of upstream diversions. In the summer of 1857 one ditch company, learning that downstream users had called a meeting to seek redress, turned the water back into the natural channel to "satisfy the miners." Other companies were not so accommodating. Wade Johnston recalled that Irish miners along San Antone Creek destroyed some of the local ditch company's diverting flumes in protest. By 1858, however, the rains returned, cooling tempers and ending the drought.[21]

The Mokelumne Hill Canal and Mining Company was by far the biggest supplier in the 1850s and 1860s, but financial troubles led to foreclosure in 1859 and major reorganization in the 1860s. Under Samuel Linus Prindle, the renamed Mokelumne Hill and Campo Seco Canal and Mining Company completed extensions to Chili Camp and Spring Valley (near Valley Springs and Burson today). For many years the most important customer was the Gwin mine, which used large amounts of water to run its hoists, mills, and air compressors.[22]

To the southeast, along Angels Creek and the Stanislaus River, the Union Water Company bought out the Calaveras Water Company and gradually absorbed smaller suppliers. By the 1860s it was the primary source of water for Murphys, Douglas Flat, Vallecito, Carson Hill, Albany Flat, Altaville, Angels Camp, and other mining areas. Over the next decade, using Chinese ditch diggers, the Union Company expanded services to domestic users and farmers, but bigger was not better, for rising operating costs and declining placer operations undercut its financial position. In the 1880s the company changed ownership and direction under the leadership of George W. Grayson and Archibald Borland of San Francisco, the principal stockholders. With the emergence of consolidated lode mines later in that decade, the Union Company began to shift its customer base. It also modernized, anticipating the advent of electricity and building a power plant that initially used steam to turn its dynamos. The Utica in Angels Camp was its first customer, but the mine soon reconverted to water-powered hoists and compressors after the Union Company raised the price of electricity to cover the higher price of cordwood charged by woodcutters supplying the steam plant.[23]

Expansion of the Utica mine in the booming nineties led to a management change in the Union Water Company. To secure water and power for its growing mining and milling operations, the Utica bought out Grayson and Borland and took control of the water rights on the North Fork of the Stanislaus River and Angels Creek. Essentially a Utica subsidiary after 1890, the Union Company by 1903 no longer operated as an independent entity but retained its name until 1914. It also retained its larger customer base, but its main interest was serving the Utica's power and water needs. Yet despite these corporate changes, Union facilities continued to grow and modernize. During the 1890s the company constructed new dams and reservoirs to store more water upstream for its power plants, and in January 1895 began generating the first hydroelectric power in Calaveras County. By 1903 it had sufficient capacity to expand electric service to the big mines at Sheep Ranch, Hodson, and Carson Hill.[24]

For smaller water and power consumers in southeastern Calaveras, however, the Utica takeover proved unwise and unwelcome in the Populist-Progressive Era. Farmers and urban users dependent on a limited water supply now found themselves at odds with an aggressive mining company that diverted most of the flow to its mills and hoist works. In dry years peripheral users were cut off entirely. Cornelius Demarest, brother of the founder of the Altaville Iron Works, left a bitter account in his memoirs, written after his last visit to the Angels area in 1903: "Although more gold is extracted than ever before, nothing whatever has been done to improving or beautifying the town that has given the owners of the Utica all their wealth and whose beauty they have destroyed, together with a large region around it. The principal 'improvement' has been to increase the population by about four thousand foreigners, and to support some forty saloons."[25]

The Utica held on to its water and hydroelectric facilities in the Angels area for fifty-nine years despite such complaints, serving water and power customers in southeastern Calaveras until the 1940s even though the Utica Water Company closed down all its mining operations by 1921. When gold mining at Carson Hill ended in 1942, the Utica Company decided to end water delivery to the area, but landowners fought back, physically by reopening the head gates above Murphys that the company had closed and legally by taking their case to the courts. After a long period of litigation between the Hobart estate, representing the Utica heirs, and affected landowners, now organized as the Calaveras Water Users Association, a settlement was reached that restored the landowners' water rights in the Carson Hill area. The Utica Company sold out to Pacific Gas and Electric (PG&E) in 1946, and in the 1950s the Calaveras Water Users Association reorganized as the Union Public Utilities District (PUD).[26]

For the next half century PG&E and the Union PUD served the southern portion of the county. But the era of deregulation opened new avenues for electric-power generation and distribution in the Mother Lode. In 1995–1996 a consortium of three local public agencies purchased the facilities from PG&E and now operates the system under a familiar name, the Utica Power Authority.[27] The old Union and Utica ditches and flumes today still supply industrial and domestic water to the southern part of the county.

## The Evolution of Regional Transportation

Cheap, reliable, and efficient transportation was as important to the economic prosperity of Calaveras County as the wealth contained in its fields, forests, foothills, and streambeds. The pack mules of the early argonauts gave way to freight wagons pulled by horses, oxen, or mules. Trails widened into wagon roads and county arteries that connected the mining camps to each other and with their supply centers in the valley. Federal subsidies materially aided overland stage companies that carried mail as well as passengers and package express, but the federal largess at first did not extend to the mining camps. Private companies, many of them one-man operations, arose to fill the regional need. By the 1850s nearly sixty small express companies served the remote hinterlands.[28] The attempt to expand transportation and communication connections with the outside world was a major theme in early Calaveras history, a theme that can be followed right up to modern times.

The city of Stockton was, and still is, a major supply and financial center for Calaveras and the southern Mother Lode.[29] Situated within thirty miles of the Calaveras boundary and within fifty miles of the county seat, it was known in the early days as the "Gateway to the Southern Mines." Much closer for miners from Calaveras and Tuolumne Counties than Sacramento, Stock-

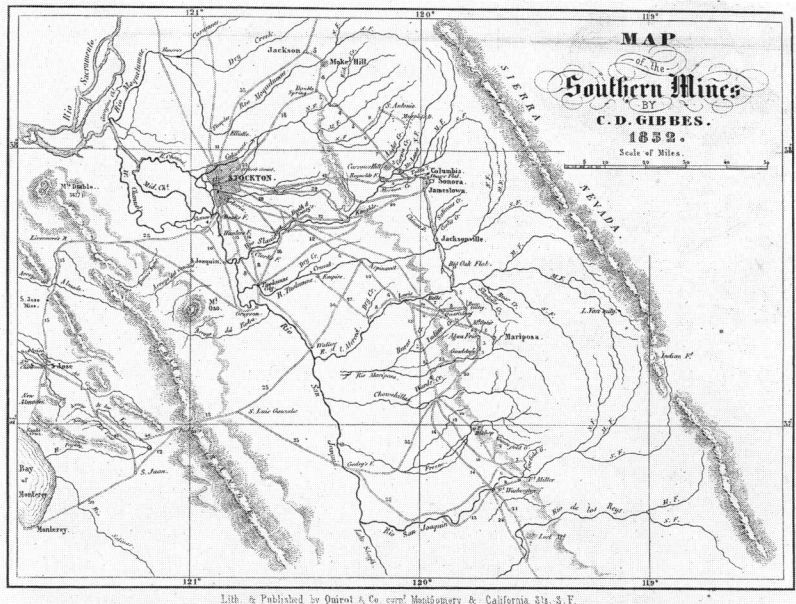

Map of Southern Mines. (Courtesy the Huntington Library and Art Gallery)

ton was a geographic funnel through which ores, grain, and timber products flowed on their way to distant markets, and through which supplies from the outside world reached foothill miners, farmers, loggers, and merchants. Up to 75 percent of the tonnage shipped inland from San Francisco was destined for the mines.[30] In 1851, for example, river steamers from San Francisco arrived daily with nearly two hundred tons of freight that was unloaded on Stockton wharves, most bound for the mining camps. About five hundred horse or mule teams hauled freight between Stockton and the Southern Mines, pulling heavy wagons in long chain lines of up to sixteen well-trained animals driven by expert teamsters, many of them Mexican. Ten or twelve daily stages also carried light freight, mail, and passengers out of Stockton toward Sonora and other Mother Lode camps.[31] Augmenting the drayage were some fifteen hundred pack mules.

As surface mining played out in the early 1850s, some diggers turned to freighting. John Woodhouse Audubon, working along the Tuolumne in January 1850, discovered that packing to remote camps was much more lucrative than mining. In Calaveras, George Underhill began hauling freight full-time late in 1849 after his brother, using the same wagon and team that had carried him across the plains, sold a load of goods at Lower Bar on the Mokelumne for a handsome profit. Purchasing two more wagons and seven yoke of oxen, the Underhill brothers spent a profitable but hectic four months in the freight business. Winter rains soon turned mountain roads into impassible bogs. One of their wagons, mired in mud at Double Springs, resisted the efforts of three yoke of oxen to pull it out and had to be left "stuck fast until Spring." An assistant, hired at seven dollars a day plus board, quit after an eighteen-day trip in inclement weather, "refusing to work longer at such low wages," Underhill reported sarcastically. To avoid the Double Springs bog, the brothers took a pack train on a detour through the hills but missed a turn and courted disaster trying to follow a soggy mountain trail. One heavy-laden mule wrenched its knee trying to free itself from the mud and had to be left, but its load was added "to that of the others." Another mule frequently mired as they slogged along. Each time the sweating teamsters had to unload it, roll it on its side, pull one leg out at a time, "and then apply the whip." On a ridgeline still another animal slipped, rolled downhill head over heals but landed on his feet, miraculously unhurt, "with his load hanging under him." To get back to the main road, the men led the train down a slope so steep they had to secure the packs with a rope around each animal's rump to keep them from slipping forward. But one contingency they did not anticipate, as Underhill's journal recounts:

> Before starting the Mules down this path I sent Kelson to the bottom of it to receive them on arrival, it was well that I took this precaution as we had packed upon one of the Mules a sheet iron stove with the cooking utensils packed inside of it, on the other side of the Mule the pipe was slung, the stove pipe became loose and commenced rattling, that started the Mule upon the run, which commenced rattling the cooking utensils inside of the Stove, which made so much noise that the whole train became frightened and ran to the bottom, where Kelson stopped them, there was a general scattering down the mountain.

Fed up with mountain freighting, the brothers sold all their teams and equipment that spring for up to three times the price they had paid, and went into the mercantile business in Stockton.[32]

Passenger and freight rates varied with the season and circumstances. In summer and fall stage passengers paid fifteen dollars for a sixty-five-mile trip to Sonora, three times the cost but only half the distance for a boat ride between Stockton and San Francisco. In winter prices were higher, and during periods of heavy rain travel was nearly impossible no matter how much money the traffic would bear. At those times visitors tried to avoid the place

Sixteen-mule team with freight from Milton. (Courtesy Calaveras County Historical Society)

altogether. One man heading for Murphys was unimpressed by the "gateway city" when he passed through it in November 1852: "I found it a little, miserable inland town surrounded by mud and water," he wrote, "and as soon as possible I left it behind me." The papers estimated that daily about one hundred people left Stockton for the mines on various conveyances or on foot.[33]

Stockton grew rapidly as a result of this booming commerce. In the early 1850s it was the state's third-largest city, after San Francisco and Sacramento. Local boosters estimated its growth rate at nearly 100 percent per year, primarily because of its connection with the mines of Calaveras and Tuolumne Counties.[34] But both regions were mutually interdependent. The economic rise or fall of one had an immediate and direct impact on the other. Much of the investment capital in Calaveras enterprises, for example, came from Stockton, and many of the leading merchants and entrepreneurs in Calaveras had Stockton backgrounds, addresses, or connections. This naturally made Stockton the chief advocate of improved transportation facilities to the Southern Mines as well as the principal protagonist of Sacramento, whose boosters also wanted to tap the economic markets and resources of the southern Mother Lode.

While the story of Theodore Judah and the Big Four is well known to

California schoolchildren, much less known are Stockton's efforts to secure
a transcontinental railroad either by way of Calaveras and Carson Pass or up
the San Joaquin Valley to connect with a thirty-fifth-parallel route. Only a few
years after Asa Whitney voiced his dream of a great transcontinental route
linking the two edges of the continent, local visionaries imagined such a railroad coming through Stockton en route to San Francisco. To encourage that
eventuality, wrote a Stockton newsman in 1851, a regional railroad should be
chartered to lay tracks south and east, tying the San Joaquin Valley and foothill commerce to the head of navigation on the San Joaquin River. If Stockton were already on the map with a regional line, he argued, transcontinental
promoters would naturally want to connect the great national line with the
"nearest railroad that might be in existence east of San Francisco."[35]

Before San Joaquin Valley boosters could act on this sage advice, Sacramento Valley boosters fifty miles to the north came up with a similar idea,
which they implemented much faster than their potential southern rivals.
The result was the first railroad in California, the Sacramento Valley Railroad, built by an eastern engineer who became Sacramento's biggest railroad
promoter, Theodore P. Judah. Stockton thus lost the battle of the continental
visionaries nearly twenty years before the last spike at Promontory Point. But
it still had regional pretensions as an economic hub, and Calaveras County
played a key role in Stockton's economic vision.

The California Gold Rush was over by the time a transcontinental railroad
became technically and politically feasible, but the Calaveras copper boom
directed attention once again to the crucial problems of cheap transportation
between Stockton and the mines. In order to transport bulky, heavy copper
ore to distant smelters, Calaveras miners had to pay winter freight-wagon
rates of eight dollars per ton from Copperopolis to the head of navigation
on Stockton Channel. Railroad promoters felt they could reduce that figure
by nearly two-thirds. Another factor was weather; tule fog kept the humidity
high in winter, and rain turned clayey dirt roads into impassable mud bogs
that isolated both Stockton and Sacramento from regional markets and in
turn hemmed in residents in the interior for months at a time. In the business of getting products to market, mud was the great equalizer in the horse
and wagon trade. Even though Stockton's phenomenal growth impressed one
winter traveler, he underscored its limitations by referring to the town as "a
very flourishing mud hole."[36] But the prospect of a railroad changed the picture entirely. Stockton railroad boosters preached the wonders of the iron
horse as a year-round transportation miracle that would, incidentally, leave
rival teamsters literally stuck in the mud.

Stockton waterfront, 1852. (Courtesy Holt-Atherton Library, University of the Pacific)

The leading advocate of Stockton railroads was Erastus S. Holden, a Maine native whose pharmacy training provided the foundation for a successful drugstore business in Stockton after coming to California in 1849. A transportation pioneer as well as a druggist, in 1850 he organized the first passenger stagecoach service between Stockton and Sonora.[37] As transcontinental railroad speculation increased, so did Holden's efforts to place Stockton on the main line. Through his active campaigning the West Coast Railroad Convention in 1859 proposed a transcontinental line that terminated in San Francisco by way of Stockton and San Jose. But Holden's local lobbying was no match for the brilliant performances of Theodore Judah, who managed to locate a feasible route over the Sierra, to incorporate the Central Pacific Railroad and, most important, to secure congressional aid in the Pacific Railroad Bill of 1862. Temporarily, Judah's coup damaged Holden's transcontinental enthusiasm, although not his hope of improving Stockton's regional importance as a transportation hub.

A few months after Judah's triumph in Washington, Holden and a group of Stockton and Copperopolis promoters, including Thomas Hardy and William K. Reed, organized the Stockton & Copperopolis Railroad. They counted on widespread local support, but immediately ran into opposition from teamsters who did not want to lose their lucrative freighting business hauling Copperopolis ore to Stockton. The opponents raised enough doubts in the public's eye to demolish Holden's fund-raising efforts.[38]

He tried again in 1865 after hearing reports that the Sacramento Valley Railroad, soon to be absorbed by the Central Pacific, had made plans to extend a branch line from Latrobe to Copperopolis via Ione and Campo Seco. Scared by warnings that Stockton was about to lose a large share of Mother Lode commerce and would soon become the "Sleepy Hollow of the State," local merchants and officials this time joined hands to encourage the reorganized Stockton & Copperopolis Railroad. By February 1866 the route had been surveyed, rails ordered, and contracts signed for grading the roadbed. To Holden it seemed a good time to revive Stockton's transcontinental ambitions.

While Central Pacific crews battled the snow and rocks of Donner Pass, Holden asked Congress for a land grant to extend the Stockton & Copperopolis Railroad south over Walker Pass where it could move either directly east into Nevada or south to meet a southern transcontinental line coming from the Mississippi Valley. A bold proposal, perhaps, but one of many brought before Congress in this speculative era. Unfortunately for Stockton, Holden's effort was too little and too late. Congress in 1866 awarded the Southern Pacific Railroad, a San Francisco firm not yet in Central Pacific hands, a land grant down the California coast to connect with the southern transcontinental system. Even those profligate lawmakers could not be persuaded to favor a second southern transcontinental connection. However, with the help of California senator John Conness, a zealous advocate of western railroads, Holden's fledgling line did get a modest grant of ten miles of land on each side of the track all the way to Copperopolis, provided the rails were laid in four years.[39]

The land grant was insufficient to speed construction, for railroads, as local financiers argued, were built with cash, not land grants that could not easily be liquidated.[40] But raising cash proved increasingly difficult for Holden and his associates. By the late 1860s the copper boom was fading rapidly. C. T. Meader's bankruptcy in 1867 ended both his mining career and his role in Copperopolis railroad promotion. Another railroad director, T. R. Anthony, in the meantime suffered the indignity of an indictment for embezzling money from Wells Fargo, and even though a hung jury kept him out of jail, he lost his seat on the railroad board. On top of all this, the grading contractors sued for back pay and damages, and Holden had his hands full trying to settle out of court. Work on the line stopped altogether after twelve miles of track bed had been graded.[41]

Despite these troubles Holden did not give up. To attract a larger market he proposed an extension from Copperopolis to the Nevada silver mines at Austin by way of Murphys and Carson or Ebbetts Pass. Notwithstanding the technical difficulties, at least one hopeful but mistaken engineer claimed that he had located a Carson Pass crossing of the Sierra that would be no higher than five thousand feet.[42] Other dreamers sought extensions to Sonora and Lake Tahoe, but while these schemes aroused some interest, they failed to attract much-needed capital. In December 1869 the Stockton City Council for the second time rejected Holden's appeal for a city subsidy. He had reached the end of the line, and a major corporate change soon after shook up the Stockton & Copperopolis directorship. Although Holden still touted the railroad, after 1869 he was no longer the controlling voice.[43]

While the Stockton & Copperopolis Railroad struggled for survival, the Central Pacific forged ahead. Completion of the transcontinental link in May 1869 was followed three months later by the arrival in Stockton of the Central Pacific's branch line, the Western Pacific (not to be confused with a later railroad of the same name that merged with the Union Pacific in the 1980s). By September the first transcontinental train passed through Stockton on its way to Oakland. By that time, however, the political manipulations and the ambitions of the Central Pacific's "Big Four" were beginning to arouse widespread opposition, for it became clear that they wanted to monopolize California rail and water traffic. Stockton therefore welcomed the appearance of a major competitor to the Central Pacific. The contender was the California Pacific, a San Francisco–based firm that proposed to build a line from Stockton to Visalia. This route would rival the Central Pacific's proposed Western Pacific extension south from Stockton.

The California Pacific plans revitalized the lagging Stockton & Copperopolis Railroad, although the details of the corporate changes did not become public until the summer of 1870. In the meantime J. C. Sullivan, a California Pacific official, took over management of the Stockton & Copperopolis and rolled up his sleeves. In the spring of 1870 he hired several hundred Chinese and put them to work regrading the roadbed that had been abandoned three years before. By December his crews began laying track, and just before Christmas the new railroad company launched its first locomotive. E. S. Holden led the van to the cheers of an enthusiastic audience that acknowledged him to be the "father of the Stockton & Copperopolis Railroad," although by then he was more of a figurehead. By February 1871 the line had been completed as far as Peters so that regular passenger and freight service could commence. There a branch line was started toward Visalia, but after reaching Oakdale in October it stalled. In the meantime the main line moved east, crossing into Calaveras County at last and arriving in Milton by June. There it ended, far from the mines and forests its original promoters had hoped to reach. The loss of federal subsidies (since the congressional land grant deadline passed in April), the high engineering and construction costs facing the road along the Rock Creek grade to Copperopolis, and the decline of copper mining all contributed to the termination of construction. The line was left with grain shipments and limited freight and passenger traffic as the primary revenue, and those were insufficient to cover construction costs.[44]

The economic realities were more apparent than the corporate changes in 1871. In July rumors surfaced that the California Pacific—the "great iron hope"—had secretly merged with its rival, the Central Pacific, but it was not clear whether the Stockton & Copperopolis line was included in the bargain. The takeover was made public in September, although not until October did Stocktonians learn that the Central Pacific indeed controlled both the Stockton & Copperopolis and the Visalia line as well. Possibly the Big Four had a controlling interest as early as the spring of 1871, but lack of corporate records (which C. P. Huntington destroyed) makes it impossible to trace the corporate changes precisely. At any rate Stockton had lost its long battle to become the "Chicago of the West," and the Central Pacific emerged with a monopoly of local lines.[45]

Big Four control of the Stockton & Copperopolis turned out to be more expensive than profitable. Without long-haul connections or major industrial traffic, it lost money from the start. In 1877 the Central Pacific directors reorganized the line, merged the Oakdale subsidiary with the Milton main line, and leased the whole thing to themselves for twenty-five thousand dollars a year, a figure representing the average cost of running the road.[46] Grain revenue continued to provide the basic income during the 1870s, but the books never showed a profit throughout the period.

But even if not moneymaking, the little line was politically useful to the Big Four. During the 1870s, while the state legislature debated proposals to curb Central Pacific's power, Leland Stanford threatened to shut down the Stockton & Copperopolis if antirailroad legislation passed. Senator Rienzi Hopkins, a Calaveras resident, obligingly opposed any unfavorable railroad bills.[47] That was a pattern repeated time after time. Through an artful but ruthless combination of extortion and bribery, the Central Pacific–Southern Pacific worked its will in late-nineteenth-century California.

Grain farmers' complaints against high freight rates, plus renewed interest in reaching the Sierra forests, led to Calaveras County's second railroad. Falling back on a standard pattern established in the 1860s, the directors of the reorganized Southern Pacific Company charged grain shippers on their

Stockton-Oakdale division as much as the traffic could bear. By 1880 grumblings of discontent were loud enough to attract the attention of a San Joaquin farmer-entrepreneur, Jacob Brack, who opened a river landing on his huge grain acreage along the South Fork of the Mokelumne a few miles west of Lodi. Even though the site was located on a shallow slough, Brack dredged a six-and-a-half-foot canal to the main channel of the river, opening an eighty-seven-mile all-season water route that was thirty-five miles closer to San Francisco than the rival Stockton route.[48] Some northern valley farmers began using the alternate waterway even before the rail line was built.

In the meantime James L. Sperry, proprietor of the Big Trees Hotel and owner of large timber acreage along the Stanislaus above Murphys, became interested in a railroad to Big Trees in order to attract both tourists and timbermen. Finding their interests coinciding, Sperry and Brack, together with a committee of railroad promoters from the northern San Joaquin Valley, joined forces to raise capital. Their appeals fell upon the ears of Frederick W. Birdsall, a New York forty-niner who had earned his fortune shipping and milling Comstock ores in the 1860s. After the Comstock decline he invested in California real estate, and by the 1880s was a Sacramento bank director and prominent Placer County landlord.[49] With the Birdsall family's financial backing, supplemented by a subsequent bond issue, the Lodi and Calaveras promoters incorporated the San Joaquin & Sierra Nevada Railroad on March 28, 1882. Work began immediately on a narrow-gauge line whose eastern terminus was initially planned for Big Trees even though a few visionaries wanted it extended to the silver mines at Bodie.[50] The firm ran out of money long before reaching either goal. By the fall of 1882 tracks had been laid to Wallace, just across the Calaveras County line at the fringe of the foothills. It took three more years to reach Valley Springs, but by that time the Birdsall family had withdrawn its support, leaving creditors clamoring for their money. The long arm of the Big Four reached down to pluck this little line virtually out of the jaws of bankruptcy, thus reasserting the Southern Pacific's monopoly on routes to the central Sierra.

Lacking cheap transportation to the coast, Calaveras County's timber industry remained small and regional. Grain growers in the northern San Joaquin now had a slightly different route to water, but by the late 1880s grain production was dropping in the Central Valley, and shipping by way of Southern Pacific's Lodi line rather than over its Stockton route could hardly have made much difference. Besides, the original promoters had been too optimistic. They had hoped that narrow-gauge construction would serve both to lower capital investment and to forestall absorption by the standard-gauge Southern Pacific. Furthermore, by providing a direct route to water, they expected to avoid having to transship on standard-gauge track. But they did not anticipate the financial difficulties encountered in the mid-eighties, which gave the hated Big Four a chance to move in. After taking over, the Southern Pacific began transshipping most of the narrow-gauge traffic onto standard track at Lodi, adding to shipping costs and leaving Brack's Landing high and dry.[51] Thus, the second railroad to serve Calaveras was almost as disappointing as the first. Its terminus at Valley Springs provided an alternative for northern county residents who found the stage and wagon run to the railhead at Milton inconvenient, but most Calaveras traffic continued to flow over the Stockton-Copperopolis route.

## The Lumber Industry

Calaveras timber resources have played a central role in the county's economy since Gold Rush days. The county has a mixed conifer forest, part of the southern section of the Westside Sierra Pine Region. Mixed-conifer stands average twenty-five thousand board feet per acre, with some of the richest areas yielding up to one hundred thousand board feet. A forest of this type between two and five thousand feet in elevation produces a variety of marketable species, including ponderosa (yellow) pine, white fir, sugar pine, white pine, and incense cedar. Calaveras County is also the northern limit of giant sequoia, growing at Big Trees State Park at an elevation of about five thousand feet. The most valuable Calaveras commercial tree is sugar pine. Tall and straight, with few knots or blemishes, it is the largest pine species in the world. Some trees reach ten feet in diameter and more than two hundred feet in height.[52]

In the early mining era, timber supplies were crucial for all types of fuel and construction needs, including steam boilers, flumes, sluices, water pipes, and buildings. Before the advent of the California oil industry, charcoal was the primary fuel for blacksmith forges, kilns, and other high-temperature needs, although some lignite from the Ione Basin was used locally and exported to the valley. Calaveras woodcutters and ranchers produced charcoal in "coal pits" that consisted of trenches loaded with oak. The wood was then set afire, covered with dirt, and allowed to smolder for several weeks. After it cooled, hardworking pit operators recovered the sooty product and sold it for twenty-five cents a sack. Coal pits were scattered throughout the county, though concentrated in the Glencoe area in the midst of lush, mixed pine-oak woodland.[53]

After lode mining became important, mine timbering demands consumed large quantities of Calaveras lumber. While sugar pine was the prize of build-

A jerk-line team delivering peeled logs to the Utica mine, Angels Camp, 1890s. (Courtesy David Clarence Demarest Collection, Holt-Atherton Library, University of the Pacific)

ers, miners in the late-nineteenth-century West preferred spruce and Douglas fir for structure and mine support. At the Oneida mine in Amador County, for example, 50 percent of the mine timbers were Sabiniana (digger) pine, although the superintendent, according to a report in 1908, preferred spruce, yellow pine, digger, sugar pine, fir, and cedar in that order.[54]

So important were good timber supplies that in some major mining communities like the Comstock in Nevada or Grass Valley in the Northern Mines of California, large mining companies purchased sawmills and prime timberlands as part of their mining operations. Calaveras County mines relied primarily on private contractors, but their lumber needs nevertheless placed heavy demands on the county supply. The Calaveras pattern doubtless followed that of Tuolumne, where as early as 1890 two-thirds of the local timber production went into mining. The lumber demand was so great in the golden years of lode mining that county and town roads were clogged with freight wagons. Growing up in Angels Camp, Ruth Lemue remembered watching jerk-line teams of twelve or more horses and mules rumble down the dusty roads, with teamsters tossing rocks at the lead animals to keep them on course.[55]

Mining needs continued to consume the bulk of Calaveras County lumber until the end of World War I, although Douglas fir was by that time the timber of choice for most mine managers. Not native to southern Sierra forests, "Oregon Pine"—the old name for Douglas fir—became popular to mine managers after 1895, when lower shipping costs made it feasible to import bulk quantities by rail from forests farther north. In Calaveras the Gwin mine had been the first major importer of Douglas fir, and by the turn of the century most Calaveras mines had followed suit.[56]

Gold Rush miners usually served as their own loggers, denuding the immediate vicinity of anything that could be used for fuel or building materials. Most of the lumber came from adjacent public-domain timberlands, which Congress obligingly allowed miners to appropriate in unlimited quantities and for any purpose.[57] The barren countryside portrayed in early mining photographs attests to the clear-cutting practices of early miners and woodcutters who had no time for aesthetics or conservation.

By the early 1850s more systematic logging began in Calaveras County as full-time timbermen built mills and organized companies to handle the increasing demand. To provide lumber for their flumes, the Mokelumne Hill Canal and Mining Company in November 1852 erected a sawmill at Glencoe near some of the county's best pine forests. By the summer of 1853 the Glencoe mill was producing fifteen thousand board feet per day, a figure that rose to twenty thousand feet within three years.[58] Pine logs, floated down flumes to the mining camps where they were cut into mine timbers or boards, were in such great demand that the company found it profitable to increase production in order to sell excess lumber to eager miners and merchants.[59]

James Madison Grover worked for the Mokelumne Hill Canal and Mining Company in the early 1850s, and his memoirs provide a fascinating glimpse of the early Calaveras canal and lumber trade. An immigrant from Maine, he knew the lumber business firsthand and was put in charge of the company's lumberyard at Mokelumne Hill. There a circular saw was erected, and logs were cut into twelve-foot planks for the retail market. He personally delivered orders downstream by building a raft of the boards and riding them down the three-foot-wide main line, carrying a pike to lean on and a pistol on his hip. "It was one sport where everything went well," he wrote, "to glide away in the early morning before the miners had opened their gauges which reduced the quantity of water in the flume, to look across the broad valies [sic] and see the curling smoke from the miners cabins, to startle the coyote and jack rabbit and perhaps send a bullet after them as they scamper away." The trip went smoothly until he came to a "shoot" (chute) of rapids or a tunnel. Then he jumped off, scrambled around or over the obstacle, and jumped back aboard the raft when it caught up. "When the order was delivered the romance ended," he wrote, for there was no way back home but to walk.[60]

The county lumber business mushroomed in the 1850s, a formative decade for industrial mining and the service industries that supported it. Water diversion for sluices and hydraulic operators absorbed thousands of board feet. Lode mining was developing an appetite for mine timber and cordwood. Mining camps rapidly transformed from tent towns to wood-frame dwellings and stores. Major fires in Mokelumne Hill (August 1854), Campo Seco (1854), Angels Camp (1855, 1856), and San Andreas (1856, 1858, 1863) also increased the demand for building material.[61]

On both sides of the county the lumber trade blossomed for both big and small dealers. While the Mokelumne Hill Canal and Mining Company handled much of the business on the Mokelumne watershed, the Union Water Company expanded on the Stanislaus side. It sold excess lumber from its water-powered sawmill, erected in 1852 eight miles above Murphys on present-day Love Creek.[62] Several independent operators entered the field by the mid-1850s, starting with Kimball and Cutting, two partners who constructed a mill on the Big Trees Road above Murphys in 1855. The next year William H. Hanford erected a mill on Angels Creek that had a capacity of twenty thousand feet per day. In 1860 Dr. Fisher established a mill at West Point about three miles east of town. By that time Calaveras lumber production of sawn boards had risen to at least nine million feet annually. High-quality timber from West Point, Independence, and the Big Trees region sold for fifty-five dollars per thousand at Mokelumne Hill, and at Stockton, Calaveras lumber was reported to outsell redwood supplies from the north coast.[63] Lack of good transportation retarded growth of distant markets, however.

During the 1860s the lumber industry continued to expand. Even though surface placer mining was in decline, the demands of hydraulic and drift mine operators increased significantly. Disaster also made a contribution to the industry. Severe winter storms and major flooding in the 1861–1862 seasons wiped out bridges, flumes, head works, and shafts throughout the county. In addition, the copper boom in the decade consumed large amounts of mine timber as well as thousands of cords of wood fuel for boilers, kilns, and furnaces at Copperopolis and Campo Seco. By 1868 the county had in operation five steam-powered and six water-powered sawmills. Because of its close proximity to commercial supplies, Murphys became the central manufacturing and distribution point for lumber and mine timbers in Calaveras.[64]

Two decades of steady growth could not be sustained in the 1870s, however—a period of consolidation, retrenchment, and economic decline both nationally and regionally. In Calaveras copper mining had ended, drift mining was falling off, few lode mines were operating, and the mining camps themselves had dwindled in population. Only hydraulic mining continued to increase, but most of the major flumes supplying water to hydraulic pits had been constructed ten or twenty years before. Good transportation facilities to the major population centers of the state had still not appeared, and as a result Calaveras by 1875 had only three operating sawmills with an annual production of two million board feet. Indeed, despite wasteful practices between 1848 and 1870 less than 1 percent of the available state lumber supply of pine and fir had been logged, according to a modern estimate. The bulk of industrial production had been confined to the most accessible areas where transportation to market was less of a problem.[65]

Just as fir and pine lumber production was tapering off, commercial loggers began to take interest in another Sierra resource, the giant sequoia. Though the largest groves are located in the southern Sierra, Calaveras first opened the world to this magnificent tree, one of two redwood species endemic to California. John Bidwell claimed to have discovered the Calaveras Grove of big trees during his epic trek across the Sierra with the Bartleson-Bidwell party in 1841, but his account was not published until after the grove was internationally famous. More important was the rediscovery by Augustus T. Dowd, a Union Water Company employee who in 1852 was led to the trees by a wounded grizzly bear he was tracking. Dowd's published account first called public attention to these Calaveras giants, and soon the most accessible part of the grove was attracting a steady stream of tourists.[66]

Initial fascination gave way to timber speculation in the minds of some early visitors to Calaveras Grove. Euro-Americans were caught between the logic of material progress and the mystical reverence for what were then considered the world's oldest living things. Ironically, the big trees in Calaveras were saved from the saw blades probably more because of lack of good transportation than desire for preservation. Although the Calaveras North Grove is confined to a few acres of magnificent trees, loggers had their eyes on them after the first tree was felled in 1853 and its bark stripped for exhibition in New York. European exhibitors the next year paid one thousand dollars for the bark of this or another tree—an indication of its novelty as well as its commercial value.[67] James L. Sperry, who with a partner took over the Big Trees Hotel from its builder, A. Smith Haynes, had logging interests in mind when he helped organize the San Joaquin & Sierra Nevada Railroad in 1882. This is not to say he would have felled the big trees in the North Grove, but his logging plans probably included the South Grove in what is now the state park, as well as his extensive holdings nearby. The failure of Sperry's plans saved at least some of the Calaveras sequoias, but elsewhere they were not so

Manuel Company's logging team in the woods near Big Trees, ca. 1890. (Courtesy Calaveras County Historical Society)

lucky. Indeed, the indiscriminate destruction of giant sequoias in Tuolumne and Fresno Counties was so alarming by the early 1870s that the state legislature in 1873 passed a bill to discourage the practice.[68] Unfortunately, the bill was never enforced, and probably could not have been implemented anyway, since the groves were on federal land. Commercial logging of big trees began soon after implementation of the federal Timber and Stone Act of 1878, which opened the door to corporate exploitation of huge timber tracts in California. In the 1880s, for instance, the Kings River Lumber Company acquired nearly thirty thousand acres of prime timberlands in the southern Sierra, and in the following decade cutting began in Converse Basin, the largest giant sequoia grove on record.[69] Sierra redwoods continued to fall before the ax and saw until World War I. By that time more enlightened loggers decided giant sequoias were too brittle to make good lumber. Now they are protected, but many scars remain.

Outside the mining camps and the big-tree groves, upland timber in Calaveras County remained relatively untouched for decades except for the shake makers. Their camp near John Gardner's trading post, later renamed Dorrington, was by all contemporary accounts a veritable wasteland of butchered sugar pine. D. C. Demarest remembered touring the area and seeing magnificent sugar pine trees "lying prone, abandoned and completely wasted, because the first log cut from its trunk had revealed the grain of the wood to be unsatisfactory for proper splitting." In the midst of the rubble was a row of twelve cabins occupied by single or widowed men ranging in ages from seventeen to sixty-eight, all listed as shake makers in the 1880 census.[70] Using only a small fraction of the logs they felled, they worked unhindered among some of the best timber stands in the country. John Muir encountered shake makers throughout the Sierra, and his critical appraisal of their freebooting lifestyle provided impetus to the conservation movement at the turn of the century: "Happy robbers!" he wrote in *Our National Parks*:

> [D]welling in the most beautiful woods, in the most salubrious climate, breathing delightful odors both day and night, drinking cool living water. . . . There is none to say them nay. They buy no land, pay no taxes, dwell in a paradise with no forbidding angel either from Washington or from heaven. Every one of the frail shake shanties is a centre of destruction, and the extent of the ravages wrought in this quiet way is in the aggregate enormous.[71]

But Muir was not the only voice calling for protection of the forests. As early as 1881, California senator Miller introduced a bill in Congress that would have placed practically the entire southern Sierra timber belt in a national park. That was too drastic for most Americans, but the tocsin of public opinion rang in 1889, when the secretary of the interior received the first known petition from American citizens urging forest conservation in the southern Sierra. The seeds of Sequoia National Park and of Big Trees State Park as well were germinated by these early appeals.[72]

Notwithstanding the demands of shake makers, timber speculators, miners, and town builders, the lumber industry in Calaveras County remained in infancy until the 1880s. By that time the county's most prominent lumberman, John Manuel, had purchased the Kimball and Cutting mill (1878) and had constructed a new plant near Arnold (1881).[73] The Manuel family eventually became the county's most important lumber manufacturer, but they were small-time operators compared to modern counterparts. Not until major lode mines in the Angels and Copperopolis districts reopened in the late 1880s did the demand for lumber increase. More than anything else, lode mining triggered the modernization and expansion of the county's lumber industry.

## Mining Machines and Regional Suppliers

Modern mining is a high-tech business that requires elaborate corporate organization, sophisticated management and engineering, state-of-the-art tools and techniques, specialized labor, and economies of scale to effectively mine and mill low-grade ore deposits. These are essentially twentieth-century characteristics, although part of the foundations of modern mining was laid as early as the 1850s in California.

The first gold seekers at Sutter's mill used ancient tools and techniques, but technology made rapid advances in the early 1850s as successive waves of placer miners began to explore the deeper gravels and the lower-grade deposits left behind earlier in the mad dash for nuggets. Lode mining after the 1850s also required increasingly sophisticated tools and techniques. The result was a rising demand for bigger and better mining machinery—hydraulic monitors, iron pipe, amalgamation pans, camshafts and cast-iron mortars for stamp mills, and thousands of specialized hand tools and replacement parts. Eastern and European suppliers for a time tried to meet this demand, but the products were extraordinarily expensive and often unworkable or at least unadaptable to particular needs. Out of this need arose regional equipment manufacturers and a skilled labor pool of draftsmen, machinists, and millwrights familiar with regional conditions.

On the West Coast the mining equipment industry gained its first and most important foothold in San Francisco. Beginning with a single foundry in 1849, within a decade the city's industrial district south of Market Street had sprouted fourteen machine shops and foundries. During the Comstock era the industry mushroomed both in size and in sophistication, for Comstock ores required much more complex mining and milling machinery than the relatively simple equipment used in California gold mines and mills. Not until after the Comstock faded and the expanding transcontinental rail system lowered freight rates for eastern competitors did San Francisco's mining equipment and supply business begin to diminish.[74]

While most of the major equipment needs were supplied by San Francisco shops, in both the regional trade centers and the local mining camps an assorted collection of assayers, mechanics, millwrights, blacksmiths, surveyors, and consultants could be found to supply the regional mining industry with specialized services not provided by the mining companies themselves. William A. Hallidie, for example, was a blacksmith in Calaveras and other Mother Lode counties before opening a machine shop in San Francisco. He gained international recognition for his innovative hoisting and surface tramming equipment, including the flat wire rope and the famed cable car.[75]

Hallidie was a practical engineer, not a school-taught specialist, but the lack of formal training was not a serious handicap in the early years of California mining. Most technical specialists in California before the 1880s learned their profession through practical experience and applied research. Typical was the local inventor at Mokelumne Hill, William Higley, who in 1860 built a homemade pump for lifting water thirty feet above a flume, using two water-wheels each twenty feet in diameter placed so that the water turned both wheels. The mining industry at first welcomed—indeed, demanded—engineers trained in the field, not in the classroom. Many early lode mining ventures failed because of managerial incompetence or errors of judgment, but often those who took the blame were professional engineers. To the mining public in this era, experience counted more than background, education, or social status. College-trained engineers, fresh out of school, were thought impractical, too filled with book learning and lofty theories to have any common sense. American nationalism and resentment had a lot to do with this attitude, for westerners retained a lingering suspicion of European-trained professionals. European engineers and metallurgists had played prominent roles in several early western hardrock mining and milling ventures, often with very poor results. Others had been openly critical of American mining methods and technology, arrogantly rejecting American advice, only to be upstaged in the 1860s and 1870s by pragmatic Americans who adapted European technology to meet American needs.[76]

This popular emphasis on proven ability rather than book learning helps explain the success of Thomas Fullen, another practical engineer with Calaveras experience. Born near Boston in 1849, he came west with his family in 1863 and learned the machinist trade as an apprentice under S. N. Knight, founder of the Sutter Creek Iron Works, more familiarly known today as the Knight Foundry. After working as machinist and master mechanic for several regional mines, he came to Angels Camp in the mid-1880s, taking a job at the Angels mine. D. C. Demarest, son of the proprietor of the Altaville Iron Works and a trained engineer, met him there in 1886 and was taken by his practical skills and youthful energy. Demarest claimed that Fullen as a young man was a problem drinker but was cured of booze after too many inebriated escapades racing buggies through the streets of Angels Camp "performing dare-devil and hair-raising stunts." As a practical engineer Fullen was sober and innovative. He joined Demarest's firm as a full partner and for a decade employed his experience in devising new safety devices, improving old tools, and building a popular line of machinery for small mining companies.[77]

The Altaville Iron Works, founded in 1854 by J. M. Wooster, was the oldest and longest-running foundry, pattern shop, and machine shop in Calaveras. Its success stemmed from an ability to find and employ good men regardless of training. In the late 1850s, Wooster sold out to David Durie Demarest, a practical engineer who learned the foundry trade as a young apprentice in New York. Demarest joined the Gold Rush in 1849, starting in the North-

Inside the Altaville Iron Works, ca. 1890. (Courtesy David Clarence Demarest Collection, Holt-Atherton Library, University of the Pacific)

Employees of the Altaville Iron Works pose with some of the firm's products. (Courtesy Calaveras County Historical Society)

ern Mines and working his way south. Shifting from mining to engineering in the 1850s, he ran the Union Water Company as a hydraulic engineer and superintendent before turning to the technical needs of the local mining industry. The shop grew slowly at first, but with the lode mining boom in the late 1880s the business expanded rapidly. High-quality patterns, tools, and precision parts at lower prices than distant competitors could afford made the Altaville Iron Works a popular shop in the southern Mother Lode. D. D. Demarest retired in 1891, turning over the business to his son, David Clarence, a graduate of the University of California's College of Mechanics, the first engineering school on the West Coast.[78]

As suggested by his relationship with Thomas Fullen, D. C. Demarest was a good judge of human nature. Over the years he employed or worked with men of ability regardless of their background or training. Some of his closest associates were untrained engineers with commanding managerial skills, like Charles D. Lane of the Utica and William J. Loring of Carson Hill. Others were highly trained professionals such as W. Spencer Hutchinson, a graduate of the Massachusetts Institute of Technology in 1893, who was a resident engineer at the Utica mine for a decade or more before it closed.[79] The quality of its products and the close business relationships it developed over the years made the Altaville Iron Works an important contributor to the mining life of Calaveras.

* * *

Cheap and abundant water and timber, improved transportation, and high-quality service and supply companies — these were important elements of the Mother Lode mining industry's wish list in the formative years following the Gold Rush. No less important were cheap labor, better technology, and sound financing — topics we have addressed elsewhere. Like so many other aspects of nineteenth-century industrial development, the reality of mining conditions in Calaveras fell far short of ideal, even by contemporary standards. The only dependable and reasonably cheap necessity was food, produced regionally in abundant quantities after the 1850s by California farmers, ranchers, orchardists, and gardeners — some of them former or part-time miners. Water was always a significant expense, and water delivery systems were often inadequate to meet the demand. Poor transportation was a perennial frustration and a major stumbling block to industrial modernization. The absence of bulk transportation facilities limited the production of both base metals and complex gold ores. It also confined the timber industry to local markets for most of the nineteenth century. Despite these shortcomings, ancillary industries and institutions emerged and slowly expanded. They depended on mining for their start, but developed a life of their own after the Gold Rush as the regional economy diversified.

# 5: Preparing for Modern Mining

Almost every fourth grader in California schools today knows that the Gold Rush opened with Marshall's discovery at Sutter's Mill on January 24, 1848. But when did it end? Even the "experts" cannot agree, and statistics are not much help, since the first signs of diminishing immigration rates in 1851 do not match the first downturn in production rates two years later. But the Gold Rush was much more than mere population shifts and production records. Some social historians see a never-ending Gold Rush mentality that even today affects growth rates, job searches, building booms, and lifestyle patterns. Headline writers continually find ways to inject Gold Rush imagery into sports and entertainment features, real estate promotions, lottery hype, even automobile and perfume advertisements.[1]

For mid-nineteenth-century residents of Calaveras and other Mother Lode counties, the Gold Rush closed just a few years after it had begun. The beginning of the end came by 1851, when per capita placer production began to dip. The downturn accelerated after 1855, when a credit crunch closed banks, forced some businesses to foreclose, and dried up sources of capital investment. By 1860 two-thirds of the miners counted in 1850 had left Calaveras. Paralleling declines in other mining counties, Calaveras lost 19 percent of its general population between 1852 and 1860, and another 45 percent by 1870. With the decline of surface placers, mining camps slumped to the point of collapse if their economies had not diversified. Dozens of Gold Rush communities withered to skeletons or disappeared altogether. A Stockton newsman was shocked by the "miserable wrecks of mining towns" he observed during a Mother Lode visit in 1863.[2]

The county gold mining industry remained in the doldrums through the Civil War years, with some young miners leaving for eastern battlefronts while others joined the copper rush or sought other lines of work in the lowlands and cities. After the war the ratio of miners to the total county population continued to slide, dropping to nearly 25 percent by 1880, then sliding again over the next decade. By 1890 less than 10 percent of Calaveras residents were employed in mining.[3]

Economic and social consolidation and transformation were inevitable in the post–Gold Rush years. But diminishing surface placers and population shifts were not the only causes of change. The nature of mining itself shifted as the rugged individualist mentality that characterized pioneer mining in the

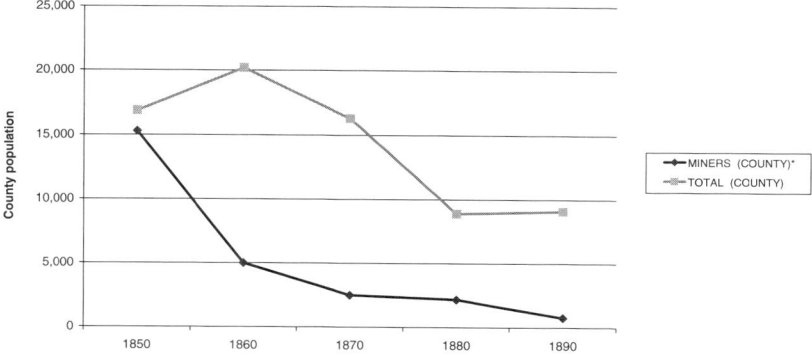

The Decline of Miners in Calaveras. (Authors' database)

Gold Rush gave way to the corporate demands of industrial development. Miners themselves faced hard choices: either work for wages or leave. Some confronted the problem head-on and signed on as hired hands; others tried to perpetuate the golden dream and pulled up stakes for new mining frontiers in Canada, Oregon, or after 1859 the Comstock and the interior West. The old ways died hard, but the reality of economic change forced itself on Calaveras and other mining counties.[4]

## The Evolution of Placer Mining

As we have seen, placer mining accounted for nearly all of California's gold production before 1860, and most of it between 1860 and 1885. Even though lode mining expanded rapidly after 1860, it was still small-scale and experimental compared to the prodigious development of Tertiary gravel deposits by drift and hydraulic mining before the 1880s. By the latter decade, however, signs of change were apparent. Drift mining fell off rapidly after 1880 because of the depletion of the best deposits, and the *Sawyer* and *Field* decisions in 1884 halted almost all hydraulic mining until the debris problem could be solved. Production figures for the period 1880–1896 do not distinguish between placer and lode mining, but it is clear that by the late 1880s placer mines no longer were the state's major gold producers.

Calaveras County generally followed these statewide trends. Among the southern counties it was the "greatest producer" of hydraulic gold after 1884, but that was nothing to brag about.[5] Hydraulic mining south of the Mokelumne never developed to the extent that it did in the north, largely because Tertiary deposits south of the American River were smaller and more scattered. In some Calaveras districts the *Sawyer* decision was inapplicable because local topographic conditions made it possible to continue operations after 1884 by containing the tailings. However, production from these few scattered surviving hydraulic pits was comparatively small. By 1896 only 5 of the statewide total of 157 hydraulic licenses issued under the Caminetti Act had been issued to Calaveras County operators.[6] Over the next ten years even these shut down as they ran out of room to store debris or because insufficient gold values were recovered. By 1907 hydraulic mining in Calaveras County and elsewhere throughout the Sierra Nevada had virtually ceased.

The demise of hydraulic mining also brought an end to the prosperity of independent water companies that had made placer mining productive for so many years. With water needs reduced, some water companies simply shut down. Others converted to domestic or agricultural uses. The Mokelumne Hill Canal and Mining Company, for example, reorganized as the Mokelumne River Water and Power Company. In 1938 the Calaveras Public Utility District acquired the water rights to the Mokelumne Hill system.[7] By that time Pacific Gas and Electric Company had picked up most of the power-generating dams and transmission lines in the county.

Placer production did not end with the close of hydraulicking and the decline of drift mining. By the turn of the century the gold dredge had made its appearance in the Oroville district.[8] When first introduced these mechanical monsters could dig 40 to 50 feet deep into the gravel of a streambed. Gradually their digging capacity was extended to depths of more than 100 feet at a rate of over 175,000 yards a month. Dredges mounted on pontoon hulls operated in ponds that constantly changed shape and size as the dredge worked. Heavy steel buckets with replaceable teeth, coupled together in an endless belt and mounted on a "ladder" or boom, projected in front of the hull that housed most of the operating machinery. A huge gantry raised or lowered the bucket line, and a "spud" kept the dredge in place while the buckets did their work. As the buckets filled with sand and gravel, they climbed up the ladder and dumped into an elaborate screening and washing plant. The fines ran through sluices where the gold was recovered as amalgam and later retorted. Cobbles and coarse material discharged from the screening plant onto a stacking belt that carried it aft and dumped it in symmetrical piles on top of fines that had been discarded from a previous position of the dredge.

Dredge machinery was designed so that any moving part or fitting could be replaced when worn or broken. This required a well-equipped blacksmith and machine shop close by. A typical bucket-type dredging operation employed a crew of a dozen men or so who worked in three shifts around the clock, all under the supervision of a dredge master or superintendent.

Good dredging ground extended not only along the stream channels but also laterally under the floodplains on each side. Many acres of prime farmland were gobbled up by these behemoths that left the soil buried under mountains of gravel. This environmental degradation occurred long before the adoption of mine reclamation laws; thus, the dredging scars on the landscape can still be seen along many of the streams flowing through the Sierra foothills. Fortunately, today many of the dredge fields have become important sources of sand and gravel aggregates, and others are being reclaimed for subdivisions or for commercial and industrial development.

Calaveras dredging began in 1904, when the Calaveras Gold Dredging Company built and put into operation a bucket-line dredge at Jenny Lind under the supervision of G. H. Shearer. Working with a crew of twelve, it operated successfully for a dozen years. Isabel Dredging Company operated the second dredge at Jenny Lind in 1908 under Fred Estep as dredge master. In 1916, after the Ivy L. Borden Company took over Isabel, the old dredge was replaced with a new one that ran until 1928. Meanwhile, in 1910 a third dredge, brought in by the Butte Dredging Company, operated for seven years under dredge master W. A. Parks and his ten-man crew.[9]

On the Mokelumne River, dredging began with David Pepper's bucket-line operation in 1904. It continued for about five years. A second company entered the Mokelumne district in 1913 when Oro Water, Light, and Power Company, managed by C. G. Leeson, installed its "Camanche No. 1." Two years later it added "Camanche No. 2," and after reorganizing as the American Gold Dredging Company installed a third unit in 1916. As the dredge ground dwindled the company shut down its dredges one by one. The first stopped in 1919, the second three years later, and the third in 1925.[10]

The impact of dredging on Calaveras can be seen in the county statistics for gold production, which after 1897 include a breakdown of placer and lode figures. In 1900, before dredges were introduced in Calaveras, lode mining contributed almost 96 percent of the county's gold production. In 1904, the first year of local dredging, the lode gold share had dropped to 86 percent. Dredging activity continued to increase up to World War I. Between 1910 and 1918, for example, gold recovered by dredges accounted for an average of 26.5 percent of the county's total gold production.[11]

With nearly three hundred thousand ounces of gold to its credit, dredging became one of Calaveras County's important mining industries, even though compared to statewide totals its production was modest. It spread rapidly because of its remarkable cost efficiency. In the early Gold Rush days the cost of working one cubic yard of gravel with a gold pan was estimated to be twenty dollars. With a rocker this could be reduced to five dollars, with a Long Tom

Butte Mining Company's bucket-line dredge working at Jenny Lind, ca. 1910. (Courtesy Calaveras County Historical Society)

to about one dollar, and with a hydraulic monitor to twenty cents. In 1910 the cost of working a California bucket-line dredge had dropped to as little as five cents per yard.[12] The difference was not in wage rates, which remained about the same from 1850 to World War I, but in the volume of ground worked. Judging from production figures, dredges mined more ground per man than any other type of placer operation, and were more efficient in washing and disposing of tailings. With monthly net profits averaging two to five thousand dollars, dredging reached a production peak in the early years of World War I, but fell off rapidly after 1918. Elsewhere, this "third placer boom" was over by 1919, but lasted into the 1920s in Calaveras County.[13] Both the state and the county would experience another dredge boom in the thirties that survived even beyond World War II.

## New Lode Mining Technology

Underground techniques changed very little during the first two decades of California lode mining. Up into the 1870s underground miners had to rely on single- and double-jack hand drilling and on black powder for breaking rock. As a result progress below the surface was slow and laborious. Few mines in the

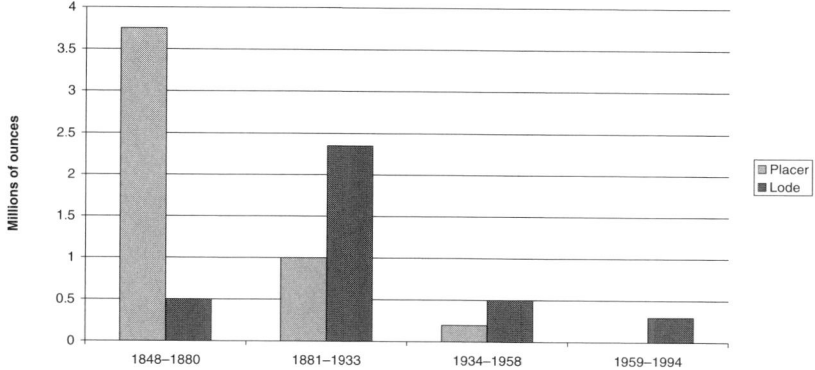

Calaveras Gold Production, 1848–1994. (Authors' database)

Mother Lode reached depths beyond five hundred feet before the mid-1870s. Alvinza Hayward's Plymouth mine, which set a record in 1871 by sinking a shaft to more than thirteen hundred feet, was a rare exception. In Calaveras County the Gwin mine reached eight hundred feet by 1874, but others were much shallower.[14]

New drilling tools and techniques made deep vein mining much less expensive and more feasible by the 1870s. Although dynamite had been invented by Alfred Nobel a decade earlier, it first came on the market as straight nitroglycerin and was thus extremely difficult and dangerous to handle. Within a few years powder companies had remedied most of the undesirable properties of nitro by mixing it with a solid base, but Euro-American miners continued to oppose its use for many years. In the West, Chinese railroad crews gained considerable experience with dynamite, and some mine owners threatened to "let the Chinese take over the entire industry" unless underground crews accepted the new technology.[15] Its adoption in California was delayed because of the underground miner's fear that "Giant Powder," as the new explosive was called, jeopardized both his health and his job, especially if he was on a double-jack crew. Dynamite fumes are dangerous, but miners learned to wait until the air had cleared before mucking out the ore. Ventilation had to be improved, particularly in mines with deep and complicated workings. More important, dynamite required smaller and fewer blast holes, hence fewer drillers, allowing mine owners to cut labor costs by eliminating double-jack crews. Miners fought these changes both in technology and in the number of underground workers. The resistance slowly receded, but not before several major strikes and a new tone of labor militancy had been injected into the California camps.[16] Labor issues will be explored more extensively later in this chapter, as well as in chapter 6.

Hand drilling in the larger mines disappeared almost entirely with the introduction of pneumatic rock drills in the latter years of the nineteenth century. One of the earliest was the Burleigh, followed shortly by the Rand, Ingersoll, and Sullivan machines. Even more revolutionary than dynamite, they drastically reduced the number of men needed underground while greatly speeding up the drilling process. In 1898, for example, using two Ingersoll Eclipse three-and-one-eighth-inch drills, four drillers and two muckers each eight-hour shift drove more than four feet in advancing the South Carolina tunnel for the Melones Mining Company.[17]

These early pneumatic drills were cumbersome and required two or three men to set up and operate. Their introduction was slower than some operators would have liked, while others resisted rapid changes in a bow to tradition, strong among Mother Lode mining men. More important, cost and miner resistance put the brakes on change. Machine drills required air compressors powered by steam or water and thousands of feet of piping in the shafts, on the levels, and in the stopes. Investors balked at the cost of adding expensive new equipment. Miners also balked, since each new mechanical drill lessened the need for workers with hand-drilling skills underground. Soon they found a better reason to complain, for the new drills generated clouds of rock dust that miners unavoidably inhaled. In mines high in quartz or other forms of free silica the incidence of silicosis rose steadily, although the connection between the dust and the disease was at first not recognized. Not until the late 1890s did the danger begin to ease, after the introduction of hollow-core drills that used water to convert the deadly dust to harmless mud. However, the change to wet drilling was agonizingly slow, especially in small mines lacking the resources to upgrade rapidly. Right up to World War II as many as 25 percent of western miners had "some form of pneumoconiosis (silicosis)," according to the U.S. Public Health Service.[18]

Calaveras County briefly went into pneumatic-drill manufacturing in the 1890s. The Demarest-Fullen Iron Works at Altaville produced the Pacific Rock drill, a solid-core piston machine that its builders hoped would find widespread use. However, J. G. Leyner's hollow-core drill, activated by the hammer principle rather than by a reciprocating piston, made this and all other piston machines obsolete.[19] By 1910 the Leyner machine had become the standard drilling tool in the industry. Unfortunately, economy-minded mine owners kept many of the older piston machines in service long after they had become outmoded by the newer, safer hollow-core hammer drills.

Driller stoping at the Carson Hill Mine, ca. 1930s.

The new double-drum hoist at the Alto mine in Scorpion Gulch, 1902. (Courtesy Calaveras County Historical Society)

As shafts deepened and lateral-level workings increased, hoisting, pumping, and other equipment had to be upgraded. Steam- or water-powered metal hoists, rope, cages, and skips replaced hand-cranked windlasses or horse-powered whims that dropped and raised ore buckets with hemp ropes. Geared hoists, wire ropes, and safer skips were common by the 1890s.[20]

For many decades the most effective dewatering system was the Cornish pump, steam-powered with a walking beam on the surface, connected to a series of pumps through a complicated wooden linkage down the shaft. The big Comstock mines used the largest Cornish pumps, but Calaveras operators installed smaller models built locally by the Altaville Iron Works.[21] Not until the 1890s did electric pumps begin to take over. Steam-powered pumps had also been tried, but the steam pipes were cumbersome and inefficient. Even after electricity arrived some Calaveras mines continued to bail water from vertical shafts where water inflow was not too large. The entrance to Utica Park at Angels Camp displays a pair of water skips once used by the famous lode mine of the same name.

Water power was used at many Calaveras mines where it was available in sufficient quantities to run huge wooden overshot wheels, sometimes forty feet in diameter, as well as inefficient undershot "hurdy gurdy" wheels. In

A Knight impulse wheel, built in Sutter Creek. (Edward B. Preston, *California Gold Mill Practices*, Bulletin 6 [San Francisco: California State Mining Bureau, 1895])

the 1870s the power demands of the industry led to a number of innovations in waterwheel design, including the development of impulse or tangential wheels powered by high-pressure jets of water controlled through riveted iron pipes or wood-stave penstocks. The wheels ran air compressors that in turn powered machine drills, hoists, pumps, and other equipment. At Sutter Creek in Amador County, Samuel N. Knight was the first successful impulse-wheel designer. In the early 1870s he devised a high-pressure, single-bucket impulse wheel and manufactured it at his foundry at Sutter Creek. Knight wheels were widely used in the industry, and the Knight Foundry continued to produce them right up to modern times. An example could still be seen in operation at the historic shop shown above as late as the 1990s.

Working independently of Knight in the same period was L. A. Pelton, a Nevada County blacksmith, who patented a split-bucket design in 1880. Pelton has received most of the recognition for developing a split-bucket wheel, although other machinists and engineers also deserve credit. Nicholas J. Coleman of Railroad Flat in Calaveras County designed and patented a split-bucket wheel in 1873, seven years before Pelton's wheel was perfected. The following year an Italian immigrant merchant at San Andreas, G. Tiscornia, working with Joseph Moore, an engineer from the Risdon Iron Works in San Francisco, fashioned a similar wheel built by the Risdon Works, but evidently never brought it to commercial production.[22]

Pelton-type wheels were more widely used and improved after the Pelton patent expired. They supplied power for a variety of mining and milling operations, and in some mines, such as the Gwin at Paloma and the Empire in Grass Valley, they lasted well into the twentieth century. Though a few operating tangential wheels can still be found in some power plants today, modern, highly efficient hydroelectric plants like those installed at Camp Nine, Melones, and Murphys in Calaveras County use high-capacity turbines rather than impulse wheels.

One of the greatest technological advances of the golden age was the introduction of electricity to the mines and mills along the Mother Lode. Hydroelectric power not only was vastly superior to wood or water, but was much less expensive as well. In 1897 the Standard Electric Company, organized by "Prince" Andre Poniatowski, erected a generating plant on the Mokelumne River and began selling power to the mines. Other power companies using Stanislaus water soon arose to serve the southern part of the county. Some mines installed their own hydroelectric-generating plants. The largest was operated by the Utica. After taking over the Union Water Company's Murphys-Altaville ditch, Utica operators built a power plant, supplied their own needs, and generated enough excess electricity to serve the town of Angels Camp as well.[23]

Many of these new techniques and types of equipment originated or were first used on the Comstock, the capstone of lode mining in the West long before California hardrock mining reached maturity. In Calaveras County one of the most important Comstock imports was square-set timbering, devised in 1860 by Philipp Deidesheimer. He developed the idea while working on the Mother Lode before he was called to Washoe. This type of timbering was essential at Angels Camp. It permitted the mines to be deepened and the wide ore bodies to be stoped in "heavy" and dangerous ground. As will be seen inadequate timbering and incomplete backfilling contributed to several disastrous cave-ins, especially at the Utica, which at one time had a reputation as one of the most dangerous mines on the Mother Lode.[24]

New tools and deeper mines made new underground techniques necessary. The most important concern of the hardrock engineer was to hoist ore as easily and efficiently as possible. By properly using gravity ore "could almost be made to fall out of the mine into the mills," as one scholar simplistically expressed it. Obviously, in shaft mining, the principal type of lode mining in Calaveras, total gravity flow was not possible. But even in shaft mines good managers sought means to maximize the use of gravity. Larger mines usually had two hoisting compartments, equipped with double-drum hoists operated in counterbalance. To mine large low-grade ore bodies, where ground conditions were favorable, Calaveras managers utilized two important innovations: shrinkage stoping and glory holing. In each case, miners drilled and blasted overhead, and trammers below filled cars with ore dropped down passes to loading chutes on the haulage level. It was then trammed in ore cars over carefully graded tracks that ran either directly to the mill via an adit or haulage tunnel or to the main shaft where the ore was hoisted, then carried by surface tram to the mill bins. The glory hole found more use in Calaveras than elsewhere on the Mother Lode due to the more abundant low-grade mineralization in the wallrocks. It was used extensively at the Gold Cliff, Morgan, Melones, Royal, and Alto mines. Diluting low-grade ore with waste, however, was a serious drawback to the glory hole technique if it was not carefully controlled. From the managers' perspective the new techniques had the added benefit of reducing labor costs by lowering the number of man-hours per ton of ore produced. Like the introduction of dynamite and pneumatic drills, new mass-production methods posed a threat to the quantity of hardrock miners. Some miners felt they also threatened quality by lowering the level of hand skills necessary to work underground.[25]

These mining improvements and new techniques made it feasible to develop and mine ore bodies that extended more than four thousand feet deep and up to forty feet wide. The new technology made Calaveras deep mining possible at the Gwin, Utica, and Carson Hill–Melones mines, but the increasing use of trained engineers by the larger mines after 1900 was perhaps the most significant factor in modernizing lode mining operations. Professional engineers enabled mine operators to make better use of the improving equipment and techniques, and to undertake more ambitious projects. Proper planning and engineering in the early phases of a project also improved the chances for success. As historians Jeremy Mouat and Hogan Lovis recently wrote, the key to understanding the impact of changing technology was "not the introduction of machinery . . . but rather the re-designing of the systems in which workers and machines operated."[26] However, it must be remembered that many gold mines on the Mother Lode and most of those on the adjacent belts were small operations too poorly financed to afford trained staff and the most recent improvements in mining machinery. In general, California mines probably modernized slower than those in other western districts, partly because of the size and cost factors, and partly because of a traditional conservatism among Mother Lode mine owners and operators that

Ralph Lemue's blacksmith shop in Angels Camp, ca. 1890. (Courtesy Calaveras County Historical Society)

persisted down nearly to the present day. As will be seen later, modern mining companies, largely operating with out-of-state capital and management, have broken away from this conservative trend and use the most up-to-date methods.

The size and scope of the workforce changed with the new technology, but the change varied from mine to mine, depending on size, financial status, nature of the ore body, and objectives of the management. Underground mechanical muckers and other laborsaving machines were not widely used until the 1930s. In the larger mines, each shift boss was responsible for a variety of men with different skills and functions. On the surface, technological changes also required adjustments in the labor force. Almost every new addition to the surface plant called for new specialized personnel. For example, when the Angels mine introduced Burleigh air drills, it hired Ralph Lemue, a trained and experienced blacksmith, to sharpen the new drill steel. In addition to these skilled employees, daily operations at the largest mines required a trained professional staff. At the management level by the 1920s professional engineers often controlled both field operations and corporate decision making.[27] Modernization resulted in increasing specialization and training at every level of the workforce in a complex mining operation.

## Milling Innovations

Throughout the mining West many mine promoters tended to overlook technical problems in their eagerness to raise capital for development. Often, expensive milling machinery was installed to impress investors even before the ore body was adequately explored or blocked out. In 1908 a mining expert estimated that 30 percent of all capital wasted in mining was due to premature construction of mills.[28]

The California stamp mill, described earlier, developed from an ancient technique for crushing and pulverizing ore, was the characteristic processing method for reducing ore throughout the West in the latter part of the nineteenth century. Cheap, simple and reliable, adaptable to almost any power source, and easily expanded or contracted by adding or reducing the number of batteries (which usually contained five stamps each), stamp mills changed little after the 1850s except for minor modifications designed to improve their efficiency and capacity. For example, while he was mill superintendent at the Utica, William J. Loring made one such modification that reportedly increased the Utica mill's capacity by four thousand tons per month. He simply reduced the space around the stamp heads, thus adding to grinding capacity without increasing power or cost.[29]

This ubiquitous stamp mill, which pulverized most of the gold ore mined in the Sierra foothills, could still be heard almost up to the middle of the twentieth century, although by that time most operators had converted to ball mills. During the 1930s, in order to increase throughput, engineers modified the stamp batteries at the Mountain King mill near Hodson by using much coarser discharge screens. This reduced the amount of stamping necessary, in effect converting the stamp mill into a preliminary grinder. A ball mill circuit was then added for final grinding. The Carson Hill mill at Melones likewise increased capacity by a similar conversion with pebble mills and, later, ball mills, using the stamps up to at least 1937.

If new mining technology made Calaveras deep mining possible, new milling technology made it potentially profitable. Before the 1870s low-grade Calaveras ores could not be successfully and profitably worked because concentrating techniques had not advanced to the stage where recoveries were consistent and sufficiently high. Free-milling ores could be handled quite well with amalgamation, although there was still room for improvement. On the other hand, millers of that day could not recover any significant amount of gold values contained in refractory ores. The unsuccessful early attempts to mill the deeper ores on Carson Hill and in the Angels Camp district have already been discussed.

Inside the Royal Company's 120-stamp mill at Hodson, ca. 1903, showing the amalgamation tables under each of the twelve 5-stamp batteries on one side. An identical arrangement was back-to-back on the opposite side of the mill. (Courtesy Calaveras County Historical Society)

Methods for concentrating both free-milling and complex ores improved tremendously after 1870. Most of the techniques had ancient origins, and were primarily based upon the separation of the heavy ore minerals and gold from the lighter gangue by the principles of gravity and inertia in a water medium.[30] One such device was the jig, which, modernized and mechanized, is still in service today in certain applications. A large variety of other concentrators was devised, especially using the revolving-bowl principle. However, an individual concentrator of this type could treat only a relatively small amount of pulp, with a low recovery of free gold and sulfides. To offset this limitation and reclean the middlings, groups of concentrators were mounted in tandem and linked with other units, resulting in a complex flow of pulp through the milling process.

One of the most widely used concentrators after the Civil War was the vanner, a device generally used for cleaning products from other concentrators. It sprayed and separated minerals while they moved down an endless shaking belt of leather. The water spray washed off the lighter particles, while the heavier ones stuck to the leather and were scraped off at the end of the

Vanners on the concentrating floor at the Gwin mine, ca. 1899. (Courtesy Calaveras County Historical Society)

belt. After William B. Frue greatly improved it, the vanner became standard equipment for collecting the sulfide content of Mother Lode ores. At the Utica and other local mills, vanners were followed by "blanket" or "canvas" milling of the vanner tailings, using the Gates canvas machine, a variation of the old blanket sluice.

Another descendant of earlier methods was the shaking table. Simpler, easier to operate, and more effective than vanners, tables were considerably improved by Wilfley and later by Deister. By World War I they had largely replaced revolving concentrators and vanners, and have continued in limited use even in recent times.

After milling the next step was treating refractory gold ores. The chlorination process offered the first real possibilities for successful treatment of the concentrates in which most of the gold was locked up in the sulfide minerals. Early attempts to roast off the sulfur, as David Strosberger tried to do near Angels in the late fifties, had been uniformly unsuccessful. Later, English and German metallurgists, working in Australia as well as in Europe and other parts of the world, developed improved methods of roasting. But until roasting and leaching were combined with the addition of chlorine to the process, millers could not recover satisfactory amounts of gold from sulfide concentrates.[31] By the mid-seventies, as we have seen, the Gwins were the first in the county to use chlorination at their mine in Paloma, to increase the recovery from the deeper ores, higher in unoxidized "sulfurets." The chlorination process, however, did not see significant use in Calaveras County until 1887. This was a complicated process, difficult and dangerous to use, and not adapted to small milling operations. But in that year, Percy S. Buckminster, who had built and operated a quartz mill for Grayson and Borland at Carson Hill for several years, built a large chlorination works for his employers to process the Angels ores. Shortly after its completion the successful chlorinator was taken over by the Utica company and expanded so as to handle not only Utica concentrates, but those from the adjoining mines. These refractory ores had frustrated Angels miners ever since they first encountered them in the late 1850s.[32] Chlorination extracted another 5 or 10 percent of the gold content in Angels Camp ores, and opened up new possibilities for development of the district.

At best the chlorination process could recover only part of the gold in the sulfides, and it was expensive as well. Direct smelting of the concentrates resulted in much higher recovery. The nearest smelter that could process iron-rich sulfide concentrates was Thomas Selby's works at San Pablo. Starting in San Francisco in the 1860s, Selby quickly gained prominence. With twenty-five furnaces by 1870 his plant handled much of the Far West's major smelting needs. From the banks of San Francisco Bay once rose a proud symbol of industrial achievement, a "huge chimney" that stood with its "lofty, smoking top from out the waters far above all the surrounding objects," wrote an awed admirer. Today the North Beach division of the complex site is buried under earthquake and fire debris from 1906, with businesses, houses, and streets layered on top. Long before the 1906 disaster, however, Selby moved most of his operation to the East Bay, directly across from Mare Island. The furnaces there kept running until after World War II, processing gold, silver, and lead concentrates from western mines, but polluting the air and water as well. Recent studies of San Pablo Bay sediments show silver and lead concentrations elevated by five to ten times above baseline levels.[33]

Only a small amount of those mineral particulates could be traced to Calaveras County, however. None of its copper rests there, since Selby's San Pablo plant was not designed to treat copper ores. Before the late 1890s most gold producers could not afford the high cost of transporting concentrates outside the district. But cheaper transport became imperative as the major mines dug deeper and exposed more iron-rich sulfides. The Gwin mine began shipping concentrates to Selby via the Valley Springs railhead after Thomas reopened the mine in 1894. When the Sierra Railway reached Angels Camp in 1902, it immediately became more economical to close the Utica chlorinator and ship the concentrates to the East Bay.

Despite its low recovery rate, the Utica chlorinator was successful because of its size and its efficient management. Thirteen years after it was shut down, the chlorinator at the Kennedy mine in Amador County closed, bringing to an end the use of chlorination in California and perhaps in the West.[34] From then on, direct smelting and in some cases cyanidation took over the difficult task of winning the gold from the obstinate ores and concentrates, and in scavenging the gold that had escaped into tailings.

Cyanidation developed over a ten-year period beginning in 1887 after its discovery in Scotland. Although not adaptable to all ores, it became a remarkably efficient process for extracting both gold and silver and was often used in conjunction with other methods to raise the recovery of gold to extremely high levels. While cyanide use expanded rapidly in most districts—even the remote mining camp of Bodie had a plant by 1893—along the Mother Lode the major mines were slow to adopt it. Almarin Paul, well known for his successful work on milling Comstock ores, attributed part of the lag to the greed of the cyanide patent holders, who wanted too much money for use of their patents. After 1896 royalties declined and cyanidation increased.[35]

On the Mother Lode cyanidation was used not only to process concentrates and slimes, but also to clean up the tailings, since one of its remarkable characteristics was an ability to recover gold from very low-grade materials. Concentrates normally were not treated without first being ground finer and, if necessary, roasted. After roasting, these finely ground concentrates were treated chemically and then bathed in a weak cyanide solution in wooden, and later in steel, tanks. The dissolved gold was then recovered by precipitation with zinc shavings in a treatment known as the Merrill-Crowe process. Because of the expense and complications of cyanidation for a relatively small volume of concentrates, most Mother Lode operators preferred to ship concentrates to the Selby smelter. Moreover, Mother Lode stamp mills did not grind the sulfides fine enough for successful cyanidation, and the arsenides and tellurides were more successfully handled by direct smelting. Finally, the graphitic content of most Mother Lode ores also prevented effective use of cyanidation.[36] Thus, it found limited application along the lode.

Use of cyanide in Calaveras County followed the pattern of other Mother Lode districts. In 1893, two years after cyanidation was first used in the United States, Dr. A. Scheidel, one of the cyanide pioneers, installed a small experimental unit at the Utica chlorinator. Mill men discovered that the existing chlorination process more effectively treated the coarse stamp mill concentrates before regrinding. On the other hand, after regrinding the fine slimes responded very well to the cyanide process, but its limited application did not appear to justify installing a full-scale plant.[37]

The Melones Mining Company had better luck. It became the principal user of cyanidation in Calaveras County after about 1903, when a plant was built at Melones to process concentrates. The Sheep Ranch mine also tried cyanidation in 1917, but without much success. A number of Mother Lode mill-tailings piles were reprocessed by cyanidation, including the large dump at the Melones mill and the tailings from the Carson Hill (Calaveras) mill. The San Francisco firm of Hamilton, Beauchamp, and Woodworth designed and operated many of these tailings plants.

In more recent years cyanidation has also been adapted to a long-known process called "heap leaching," which can recover gold from very low–grade ores suitable for cyanide treatment. Fine grinding, which is much more expensive than just crushing the ore to a one- to two-inch size, is not required for heap leaching. The crushed ore is "heaped" on pads and sprayed by a weak cyanide solution that is then drained off and washed in a process using "activated" carbon. Absorbed gold is recovered or "stripped" off by one of several techniques. In Calaveras County, Carson Hill was the principal mine to employ this procedure in the 1980s, with results not entirely satisfactory, as will be discussed later. In contrast, the much smaller Alto mine used it with good results.

On the eve of World War I a new milling process called flotation came into general use after many years of development in Europe and Australia. Similar to a laundry process that lifts dirt from clothes, this method could not only separate sulfides from gangue, but also separate one sulfide from another.[38] In a typical flotation circuit finely pulverized sulfide-ore pulp, conditioned with chemical reagents, is mixed with oils or other frothing agents and run into flotation cells. The mineral-bearing sulfides are captured by the froth generated by air pumped in from the bottom of the cells. These are then collected, dried, and shipped as concentrates to a smelter.

Like cyanidation flotation had limited application in Mother Lode mills, where gold-bearing sulfides were successfully "dressed" using traditional gravity methods such as tables and vanners. The principal use of flotation in a Mother Lode mill was to save the gold-bearing sulfides that had been overground to slimes, especially after the introduction of ball mills. It was also used to separate heavy low-gold-bearing pyrite from highly auriferous arsenopyrite.

Flotation was ideally suited to the county's copper ores, but it arrived too late to be of much help except during World War II. Although copper and gold ores generally employed similar milling processes, the copper ores contained a number of sulfide minerals, and these ores were dressed by flotation to produce concentrates that could be successfully smelted. The Penn mine at Campo Seco sent selected high-grade ore to the Eagle-Shawmut mill in Tuolumne County, and the North Keystone mine processed their hand-sorted ore at the Mountain King mill at Hodson. The concentrates from both mills were then shipped to copper smelters outside California.

The decline of the county's gold districts also made flotation less useful than in most other western mining regions. Carson Hill was the first Calaveras gold mine to employ the process. During World War I, William Loring installed a two hundred–ton flotation circuit at his new Calaveras mill at Melones, and a pilot flotation plant at the Dutch-Sweeney mine in Tuolumne County. He had been an associate of Herbert Hoover's Zinc Corporation in Australia, where flotation had its first commercial success. Convinced that flotation would be the salvation of the lode mining industry, Loring was frustrated by poor results on Mother Lode ores. The Dutch-Sweeney mine shut down because of high operating costs and poor ore grade before the pilot mill could prove itself, and a cyanide circuit soon replaced the Calaveras plant.[39] Before World War II a few small operators set up flotation cells at the Oro y Plata, the Royal, and other mills in the county, but besides Carson Hill the only other substantial Calaveras gold-mining operator to use flotation was the Sheep Ranch mine, under control of the St. Joseph Lead Company after 1937.

Despite a few setbacks, lode mining and milling technology had come a long way since the 1870s. By World War I lode miners in Calaveras County and elsewhere had all the essential technical ingredients to work successfully an ore deposit as deep as a mile below the surface. But rising costs, depletion of reserves, and declining ore grade in the face of a fixed gold price threatened the future of Calaveras mining just as it was hitting its stride.

## Expansion of the Lumber Industry

The lode-mining boom after 1885 placed new demands on the fledgling timber industry of Calaveras County. As the mines deepened, so did the necessity of more extensive timbering, especially in the "heavy ground" below the three hundred–foot level at Angels Camp. Technological innovations that required elaborate surface plants and sophisticated machinery also consumed great quantities of local timber. As in the previous period, the timber industry between 1885 and 1910 continued to be primarily an auxiliary to mining, although the volume of lumber production rose significantly.

In the golden years one of the county's largest timber producers was John

Manuel Company log pond. (Courtesy Calaveras County Historical Society)

Manuel, who began in the 1870s with a portable sawmill he moved from one location to the next as the best timber was harvested. In 1878 he purchased the old Hanford mill from Kimball and Cutting, and was thus poised to take advantage of the great demand for lumber after lode mining picked up in the 1880s.[40] Soon after he acquired the Utica in 1884, Charles Lane told Manuel the mine could use all the timber he could deliver.

Manuel was not the only timberman to flourish in the eighties. The Utica's demands also stimulated Nathan McKay, a Nova Scotian logger who had come to California in 1873. In 1885, McKay and his brother John purchased 160 acres of prime sugar-pine land on Love Creek one mile south of the Calaveras big trees. Within a year they had completed the construction of what became known as the "Clipper Mill," named in honor of Donald C. McKay, a relative and celebrated builder of the *Flying Cloud* and other famous American clipper ships. By 1888 the Manuel and McKay mills together were producing three million board feet per season; the McKay mill alone averaged this amount over its eighteen-year history. In 1904 it closed after exhausting all the stumpage in company hands. In the meantime, Manuel, with more land and mills, became the predominant lumberman in Calaveras County. After his death in 1899 his family organized the Manuel Estate Company that continued the business until the 1930s.[41]

The Manuels, the McKays, the Raggio brothers, and other loggers of the period were handicapped by the lack of good transportation facilities. In the early 1880s during dry seasons the Raggios and another independent partnership under the Jones brothers kept a continuous string of twelve- and sixteen-oxen teams on the roads between the logging camps and the mines at Angels Camp. Each team pulled two ponderous wagons with solid wooden wheels loaded with logs and secured with heavy chain.[42] To ease field conditions the McKays in 1888 built a horse-drawn railroad to haul logs from forest to mill, and in the 1890s they replaced the horse with a converted steam traction engine that can now be seen on display at the museum in Angels Camp. This short-haul line never extended beyond company boundaries, however. For long hauls to market the McKays, like most other loggers, relied on solid-wheel logging wagons pulled by combined teams of horses and oxen. The Manuel Estate Company was the first to convert to steam traction engines for long hauls. Each company tractor could pull five wagons loaded with a total of forty thousand board feet.[43]

Despite gradual improvements in local transportation, Calaveras lumber did not reach distant markets in quantity until after the construction of the Angels Camp branch of the Sierra Railway in 1902. Even then the mines continued to absorb much of Calaveras timber until the beginnings of World War I.

Corporate and financial limitations as well as transportation problems account for the local nature of Calaveras lumbering in this era. Practically isolated from major urban markets, Calaveras lumber did not attract outside investors until after World War I. Up to that time, except for the timber supply itself, the industry remained almost entirely in individual or family hands. Like its mining counterpart the wood-products industry had to await technological modernization and large infusions of outside capital before expanding to statewide or national significance.

The first efforts to consolidate Calaveras timberlands date back to the late 1880s, when land speculators began acquiring sizable tracts. One major speculator was Charles Lever Van Buskirk, a Wisconsin lumberman who, in 1893, moved to Lodi and began investing in Sierra real estate. In this era many timber entrepreneurs secured land by buying up discounted military land warrants and by paying delinquent taxes on abandoned land. In 1889, for example, Frank Solinsky picked up 160 acres of Calaveras mixed-conifer forest northeast of present-day Dorrington by "redeeming" the tract from the county treasurer for $57.83.[44]

Land speculators also capitalized on the broad loopholes in federal land legislation such as the Timber and Stone Act of 1878. In California this law provided opportunity for speculators to privatize public domain timberlands at a fraction of their value. In the southern Sierra it was possible to purchase sequoia tracts for $2.50 per acre. Three giant sequoias on one acre could produce four hundred thousand board feet of lumber valued at $4,000! Little wonder that timber barons swarmed to California. During the fifteen years the Timber and Stone Act was in force, California recorded nearly three million entries under its provisions.[45]

Van Buskirk was a timber production innovator as well as a speculator. He tried to harvest his bonanza by floating logs down the Mokelumne River and its tributaries to mills in the Central Valley. If successful, this method—widely used in midwestern and Pacific Northwest forests but rarely in California—could save transportation costs as well as capital expenses by eliminating the need for local mills. However, the experiment failed miserably. Nature failed to cooperate by providing sufficient and regular stream flow. Logjams choked local streams, while a few errant logs escaped captivity and floated all the way to San Francisco Bay! Doubtless disgusted by this fiasco, but also shifting into urban real estate development, Van Buskirk in 1902 sold three-fourths of his timberlands to Brown Brothers Lumber Company of Rhinelander, Wisconsin. Years later, by then a prominent Lodi businessman, he turned over the remaining quarter interest to Charles F. Ruggles, a Michigan timberman.[46] The Van Buskirk timber tract became a component of Calaveras County's corporate timber holdings.

The major component evolved from the holdings of Frank Solinsky and his son, Frank Jr., who organized the Big Trees Improvement Company in 1906.[47] Born in Tuolumne County, son of a Polish nobleman who came to California in 1851, the elder Solinsky was a leading attorney and banker in San Francisco during the lode mining boom era. In 1877, as a member of the first graduating class from the University of California, he received a mining engineering degree. Four years later he completed legal training at Hastings and set up law offices in San Andreas. Between 1885 and 1915, according to a contemporary source, he and his partner, Frank Wehe, helped organize practically every new mining venture in Calaveras County. Solinsky's Calaveras County Bank pioneered branch banking locally and introduced a time payment plan—later banned by the Securities and Exchange Commission—that allowed small mining investors to share in stock speculation.[48]

Solinsky's influence and financial assets proved invaluable as he moved into timber speculation. Repeal of the federal land laws in the 1890s closed the door to most public timberland purchases, but thousands of acres in private hands still awaited consolidation and development. Working independently until 1909, and thereafter as agents for Charles F. Ruggles, the Solinskys were two of Calaveras County's most prominent timber buyers. After

Ruggles folded as a result of financial difficulties in the late twenties, both the Solinsky and the Van Buskirk tracts came under control of the Calaveras Land and Timber Corporation, a newcomer that by World War II was the largest private timber company in the County.[49]

## Rails to Angels Camp

Like the lumber companies, the railroads served as auxiliaries to the mining industry in the 1885–1918 period. The lode mining boom not only revived the two existing railroads, but also attracted a third by 1900. Passenger traffic as well as mine freight increased significantly during the era. In the days before reliable, inexpensive automobiles and good roads, the rails served the essential transportation needs of miners all along the Mother Lode.

At the beginning of this era, Calaveras County had two railheads near the eastern county line: the Stockton & Copperopolis at Milton and the San Joaquin & Sierra Nevada at Valley Springs. The Southern Pacific, which by the late 1880s had absorbed both rail lines, saw no reason to extend trackage eastward despite the rising traffic. Expansion could only increase the bonded indebtedness of these two lines, and debt service still consumed much of the operating revenue. Besides, with a monopoly of local rail service, the Southern Pacific leadership had little competitive fears, at least until the late 1890s. Daily stagecoach service connected the railheads with all the mining towns, and freight companies did a land office business bringing out ore and concentrates and taking in mine supplies. The only improvement made under Southern Pacific supervision in these years was to convert the San Joaquin & Sierra Nevada to standard gauge in 1904, primarily to standardize its rail network and thus avoid the time and expense of reloading at Lodi.

Southern Pacific's rail monopoly ended in 1902 when the Sierra Railway completed its branch line from Jamestown to Angels Camp. Encouraged in part by Calaveras County lumber, tourist, and mining interests, and backed by New York, San Francisco, and European capital, the new railroad was incorporated in 1897. The fact that Crocker money went into the project led some cynics to believe it was really a Southern Pacific branch line built to forestall competition, but the new road's independent management belied that notion.[50]

Completing the main line between Oakdale and Jamestown in 1897, promoters sent feeders into the timber belt above Sonora in Tuolumne County by 1900, and at the same time they laid plans to tap the Calaveras mines and timber resources. Prince Andre Poniatowski, one of the Sierra Railway's main promoters, wanted rail lines to reach the Thorp mine at Fourth Cross-

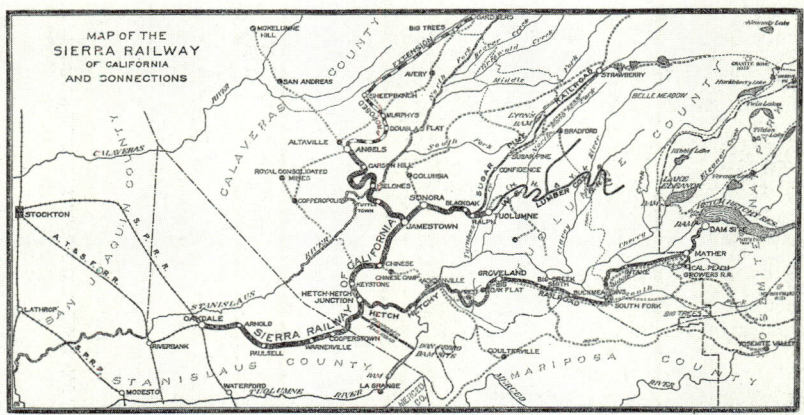

Sierra Railroad map. (Authors' collection)

ing, of which he was general manager under a syndicate that included W. H. Crocker as well.[51] James L. Sperry, related by marriage to both Crocker and Poniatowski, hoped this new railroad would reach Big Trees and provide him with the tourist and timber connections he had expected of the disappointing San Joaquin & Sierra Nevada Railroad twenty years before. But the high cost of crossing Stanislaus Canyon ended Sperry's dream, if not Poniatowski's. Sierra Railway officials terminated the branch line at Angels, thirteen miles from Sonora over a 5 percent grade that required three switchbacks and a long trestle. Although standard gauge, the route had such sharp curves that the company had to use special passenger cars with short wheel bases drawn by bevel-geared Shay or Heisler locomotives at an average speed of twelve miles per hour.[52]

Geographic circumstances thus denied Calaveras County's final try for a logging railroad. Even as a mine-freight and passenger line the new railroad, like the old ones, never quite lived up to expectations. Angels Camp travelers found it took just about as much time to reach Stockton by the Sierra Railway as it did to catch a stage to Milton and board the train there. Neither did the new line relieve merchants of high freight charges, as D. C. Demarest found out after selling all his wagons and mules in anticipation of lowered freight costs from the Altaville Iron Works to Stockton. The California Railroad Commission, instead of requiring lower rates, allowed the Sierra line to charge high fees because of the expense in bridging the Stanislaus.[53] Although the line served the mines of Carson Hill and Angels Camp, as well as the lumber companies that brought their products to the new railhead, business-

The Sierra Railroad's trestle across the Stanislaus River at Melones, ca. 1905. (Courtesy David Clarence Demarest Collection, Holt-Atherton Library, University of thePacific)

men and passengers in northern and central Calaveras continued to use the older railroads until improved auto and truck transportation made almost all Calaveras railroads obsolete.

## Calaveras Miners and Managers

Technical know-how alone was not enough to ensure success in the mining industry. Sustained success depended on combining the factors of production (land, labor, capital, and technology) with good management. In Calaveras good management led to high productivity at several major mines in the county's golden era, although it is worth repeating that financial success did not necessarily follow from large production. More often than not the amount of gold recovered was insufficient to pay the costs of mining and milling low-grade ore bodies.

At the operational level a general manager who left daily operation in the hands of a superintendent usually controlled larger mines. Before the twentieth century the quality of leadership depended not so much on training as on experience and native ability or common sense. This was due in part to a shortage of good engineers before the 1880s, and in part to a lingering bias against academically trained professionals. As we have seen the engineering profession had to struggle to gain acceptance in western mining circles. Before the 1880s trained engineers came largely from European schools and were not well liked by American miners, as John Hayes Hammond discovered as a young graduate of Freiberg, Germany's famous mining institute. He asked George Hearst, a family friend, for a job of any kind, but was dismayed by Hearst's response: "The fact of the matter is, Jack, you have been to Freiberg and have learned a lot of damn theories and big names for little rocks; that don't go in this country." Hammond got the job only after confessing he had not "learned a blessed thing at Freiberg."[54]

Calaveras had men of talent in the mining business, but it also had its share of the mediocre. Whether managers were trained in school or on the job did not seem to make much difference in the long run. If a mine had good potential, the key to success was in making the right operational decisions. In the larger mines that meant building a good management team at every level, from the general manager to the shift boss. But mining was an uncertain business. Even the most talented management was vulnerable to unpredictable technical, financial, or personnel problems that could cause a temporary or even permanent disruption of operations.

In a period when industrial productivity still depended to a large extent on hand labor, a steady supply of capable and willing workers was crucial to the success of any mining operation. Miners made up a sizable part of the labor force in the Sierra foothills, where hardrock gold mining predominated after 1885. Statewide, the number of mine workers declined 87 percent between 1880 and 1920. By the latter year only one-tenth of 1 percent of California workers were miners. In Calaveras, however, the decline of mine workers during the same period was not as steep. They dropped from 23.9 percent of the county workforce in 1880 to 13.3 percent in 1920.[55]

Industrialization altered labor-management relations in the mines, just as it had in the factories, machine shops, and mills across the United States and Europe. Hardrock mine employees were typically single young men living in boardinghouses and earning three to five dollars for a ten-hour day. Many were first-generation Europeans, though the percentage of foreign-born had declined from the Gold Rush era. In Calaveras foreign-born had been in the majority in 1852 and 1860, but by 1890 had dropped to little more than 25 percent. Their place of origin had also shifted, largely reflecting the national immigration patterns in the industrial era. In general, from the 1870s to World War I the percentage of residents in the county population from northern and western Europe declined or held steady, whereas the percentage of immigrants born in south-central or eastern Europe rose significantly. Especially noticeable in Angels Camp, the center of industrial mining during the boom years, were miners from Finland, Italy, and the Slavic portions of the Austro-Hungarian Empire.[56]

More class-conscious than Gold Rush miners, yet divided by ethnic and cultural distinctions some managers exploited to keep down the specter of unionism, mine workers struggled under adverse conditions. Their daily wage did not begin until they arrived at their work site, even if it took a half hour or more to reach from the collar by cage, skip, ladder, or tram. During the day they faced a variety of dangers, both natural and man-made. Faulty or poorly designed equipment caused some accidents; natural forces beyond human control triggered others. At Mokelumne Hill one lone miner drifting along bedrock from the bottom of his shaft broke into a "flood of water" that propelled him "whirling and bumping" back to the shaft and up sixty feet, then left him bruised and battered but still alive forty feet from the surface. He held on until morning before his cries were heard and he was rescued. A few days later he "was about, taking and giving the jokes on it." Another man in the same district, riding a bucket near the collar of his three hundred–foot shaft, suddenly slumped and fell out as his horrified windlass operator watched helplessly. He had succumbed to carbon dioxide or "choke damp," the noxious fumes that took the lives of many unsuspecting miners before modern detectors were available.[57]

Carelessness was also a frequent culprit in mine accidents. Staying alive meant staying alert. As an old-timer told Chris Porovich, a young mucker just starting to work at the Oro y Plata mine, "Son, there's only two kinds of people down here—the quick and the dead." In a perverse application of Murphy's Law, some miners who were more experienced became less alert and more careless on the job, often with tragic results. All too often during the labor-intensive boom years, miners suffered serious job injuries or deaths after mishandling blasting caps, cutting fuses too short, drilling into "missed holes," returning to check a slow fuse, or some other heedless act. Drago Metrovich, a Slavonian at the Lightner mine in Angels Camp, crimped a blasting cap in the wrong place and set it off along with five other boxes of caps nearby. Carried out "badly mangled" but still alive, a reporter wrote dryly, "If blood poison does not occur the man has an excellent chance of recovery." Wade Johnston remembered three "close calls" in Calaveras mines, once when a blasting fuse misfired and twice due to human error while working in shafts under hand-cranked windlasses. Hayden "Buck" Stephens, a Sheep Ranch resident, worked underground until an unexploded charge went off near him. Escaping unharmed, he gave up mining in favor of storekeeping and raising a family. Others tempted fate, going back down the shaft year after year despite the poor odds. Without workers' compensation, when an accident did happen not only did the workingman suffer, but so did his family. In the most serious cases the usual response was to take up a collection for the widow and orphans.[58]

To men working underground, fire was the biggest fear, although fatal fires in metal mines were relatively scarce. In Calaveras the worst accident was a cave-in at the Utica that took 14 lives in 1889. Forty years later 5 men were killed in a cave-in at the Calaveras Copper Mine in Copperopolis. Both were far less than the 47 lives lost in 1922 in the Argonaut mine fire across the Mokelumne in Amador County, the deadliest single accident in California gold mining history.[59]

Rock falls were far more common, a greater contributor to mine fatalities as well as the leading cause of injuries in the three decades after 1890. Just one incident in Calaveras illustrates the universal problem. John Drury, an experienced miner and foreman of a shaft crew, stood near the hanging wall in the Crystal mine at Angels Camp one morning in 1903 "when a slab of slate of nearly three tons weight slid out of the wall. It caught Drury on the back of the legs, just above the knees, cutting one leg off and badly mangling the other." He bled to death before he could be rescued.[60]

The annual reports of the U.S. Bureau of Mines after it was organized in 1911 reflect these grim statistics. The death rate in California metals mines for that year was 38 out of 10,888 miners, or 3.49 per thousand. The rate steadily rose in the next four years, reaching 4.71 per thousand in 1915. In that same year nonfatal injuries in California mines totaled 2,928, or 281 per thousand, higher than for the United States as a whole. By 1920, with less than 5,000 miners still at work in California, the death rate had declined to 3.88, with a total of 1,138 injuries. County data cannot be extrapolated from these state figures, but it is reasonable to assume Calaveras rates were comparable.[61]

Miners also suffered from a variety of debilitating occupational diseases, some well known such as tuberculosis, others less prominent but still serious. A 1916 study by the U.S. Bureau of Mines, for example, found hookworm in 30.8 percent of Mother Lode miners. The most insidious killer, not well understood until well into the twentieth century, was silicosis, or "miner's consumption." Though medical research indicated by World War I that at least 25 percent of western miners were infected, the United States lagged far behind other industrialized nations in recognizing silicosis as an occupational disease.[62]

Despite the overwhelming body of evidence linking work-site hazards to employee illness, accident, and death, nineteenth-century corporate culture resisted any government encroachment on traditional property rights. Before the Populist-Progressive Era, legislation regulating the workplace seemed antithetical to the spirit of free enterprise. If corporations and the public were slow to accept a larger government role in protecting workers' health and safety, the courts were even slower. Mining corporations had powerful legal allies in the common-law doctrines of "assumed risk," "fellow servant," and "contributory negligence." These limited corporate liability by placing the burden of responsibility on the employee in case something went wrong on the job.[63]

In the face of this collective resistance, efforts to improve mine working conditions made slow progress before World War I. Though often divided by ethnic, cultural, and language differences, those who worked underground knew they had more to gain by uniting in common effort against implacable forces of opposition than by standing alone. In the words of one labor historian, "the forces drawing miners together proved stronger than those dividing them." Organized labor in the mining West had its first major success on the Comstock, where miners' unions took the lead in promoting health and safety measures along with better wages and shorter hours. Comstock miners also helped themselves by promoting hospitals, raising money for disaster relief, establishing burial funds, and other measures. The Comstock precedent spread outward by the late 1870s, but elsewhere labor gains were minimal and often accompanied by strikes and intermittent acts of violence. After more than a decade of internecine warfare, state governments reluctantly stepped in, prodded by labor lobbyists from the newly organized Western Federation of Miners (WFM). The result was a sequence of measures that overrode common-law liability doctrines, imposed a state regulatory regime, required workers' compensation, and gradually improved the health and safety of all mine workers.[64]

Labor issues that arose in Calaveras reflected trends repeated across the mining West in the three decades prior to World War I, yet union organization on the Mother Lode lagged behind other mining regions for a variety of reasons. First was the psychological legacy of the Gold Rush. The entrepreneurial spirit still affected miners and their families, especially those from American and British backgrounds whose roots in California mining went back to the days of '49. The press, the courts, the police, and the state government all considered unionization an alien concept, something un-American, antithetical to traditional ways of doing business. Even though most miners by the 1890s were wage earners, they still took pride in their craft. Many hoped to become mine owners, or at least rise to positions of power and leadership in the community.[65] To support unions was an admission of failure, an inability to make it on your own. The "California dream," to use Kevin Starr's evocative phrase, died hard in the Mother Lode.

A second reason was the slow development of lode mining in California, in contrast with the frenzy of bonanza districts such as the Comstock or Goldfield in Nevada. On the Comstock, as Richard Lingenfelter's study clearly demonstrates, unions gained and held power by taking advantage of critical economic junctures. In good times they imposed modest demands on vulnerable mine and mill owners, and in bad times managed to hold their own through effective organization and community support. Sally Zanjani recently suggested that Goldfield unions had a jump start over other mining communities because of the peculiar nature of Goldfield ore bodies, and the high incidence of leasing that developed after the initial excitement. Lessees lacked the management strength and conservative values so noticeable in older traditional districts. They were more concerned with rapid development and quick returns than more systematic development and cost control. Hoping to get in and get out quickly and quietly, they tended to settle labor issues expeditiously, thus encouraging union strength.[66]

These bonanza conditions did not exist on the Mother Lode. Though there were many spectacular discoveries of rich pockets, and a number of highly productive mines, for most of the formative lode years the hits were few and the misses many. Some mine owners did resort to contract mining, as we have seen on Carson Hill in the 1850s and 1860s, for example, but by the 1890s families or corporations controlled the best mining properties. Surface outcrops and rich pockets soon gave way to low-grade ore bodies requiring systematic mining and milling. Slow, steady development, and small family-owned or locally controlled lode mines did not provide a favorable climate for unionization.

Third, and perhaps most important, was the small-town atmosphere of California mining communities, a milieu not conducive to collective bargaining. California hardrock mining in the nineteenth century did not draw together great concentrations of capital and labor, as did the base-metal industries that emerged at Leadville or Bisbee or Butte. Most mines were small, employing few men and struggling to make ends meet. Before he became chief engineer in the U.S. Bureau of Mines, for example, Charles F. Jackson said that in one little operation he had "performed the duties of surveyor, assayer, sampler, draftsman, and geologist, and for a while, acted as night foreman." Even the biggest mines—the Empire in Grass Valley, the Sierra Buttes in Sierra City, the Kennedy in Jackson—had a workforce of only a few hundred miners, muckers, and surface staff. Some lived in company boardinghouses, though most found their own housing in towns such as Angels Camp,

Amador City, Sutter Creek, and Sonora. These were not transitory camps as in Gold Rush days, but permanent settlements that reflected the traditions and values of rural America. In this setting, management and labor were not dichotomous entities. Even if owners and general managers lived elsewhere, mine superintendents and other supervisory staff lived in the communities where they worked. They participated in community affairs and gave to local charities. They sponsored sports teams and community picnics, and joined in parades and patriotic rallies. Promoting goodwill locally reduced employee tension and helped keep creditors at bay. Unlike industrial centers with a long history of labor-management conflict, California's foothill communities treated local mining officials deferentially as leaders and benefactors. W. J. Loring, as general manager over several mining operations in the southern Mother Lode, recognized the value of company patronage when he applauded the community efforts of a local superintendent:

> I wish to congratulate you upon your taking an interest in the Baseball Team and Trap Shooting Club for the amusement of the employees, and I should like to suggest in this connection that the membership should not be confined entirely to the employees of our mine. Perhaps the Baseball Team may be confined to the Dutch Sweeney employees, but I should make it as general and as free as possible at the same time retaining the control in our hands.[67]

Union organizers had a hard time in such settings. Mother Lode managers were notoriously antiunion, and so were most of the old-time miners and their families in Calaveras and other mining communities. They understood the connection between mining activities and the region's economic health. Although town residents sympathized with hardworking miners and their problems, when the chips were down management ruled the heart and soul of the mining community. Even the rise of industrial mining after 1890 and the influx of miners from central and eastern Europe with more favorable union sentiments could not overthrow the conservative establishment that dominated mining life on the Mother Lode.

* * *

Hydraulicking, drift mining, and dredging were the technological successors to the shovel, pan, and rocker of the pioneer placer miner. For more than a half century after the Gold Rush the placer industry forged ahead, finding better tools and technologies to work massive but lower-grade deposits left behind by the earliest California argonauts. But even the application of more efficient methods could not overcome the fundamental geological and economic limitations facing major placer producers. By World War I the best placer ground was used up, and the costs of operating approached and in some cases exceeded the net income.

These same limitations, on a much smaller scale, confronted the marginal placer operators. What some have termed *remainder mining* or *sniping* carried on though the end of the nineteenth century and far into the twentieth. Whether a solitary miner, a family with inherited claims, or a small group of Chinese or Hispanics, these relics of the Gold Rush era carried on with traditional tools and modest returns, sometimes working elsewhere while they mined on weekends, eking out a living but not much more. David Drew's experience on the Stanislaus River in the late 1850s exemplifies the life of a post–Gold Rush placer miner. His diary for 1856 shows a gross income of $1,585, mostly from river mining and other placer operations. But he spent $1,576 during the same year on mining expenses, assessments, housekeeping, and other costs, leaving a net profit of $9. Later he gave up placer mining, married a New Hampshire girl, raised a large family, worked intermittently in the county's copper mines, drove teams hauling freight from Stockton and Milton to Copperopolis, and died in 1903 after a full but not very prosperous career.[68]

With placer on its way out by the 1880s, the mining scene in Calaveras shifted to the county's lode prospects. New technologies for mining and milling were already in place or just around the corner. The most promising leads had already been discovered, developed, and in some cases mined down to the deeper levels, but to optimistic mining men there was no shortage of good ore. Like many observers, both expert and amateur, they believed rich ore bodies lay deep below current operating levels. The way to wealth was downward, and to tap it meant better organization and management, bigger and better machines for mining and milling, better transportation and timber supplies, and increased productivity of the labor force. Though expectations often fell short, Calaveras lode mining between the late 1880s and World War I achieved its highest rates of production and profit. The county's golden years were about to begin.

# 6: Lode Mining in the Golden Years

The golden years of lode mining in Calaveras began with the Utica consolidation in the mid-1880s. It ended more than three decades later, when most of the big mines either had exhausted their payable ore or were forced to shut down because of rising costs and labor shortages during World War I. In the peak years between 1893 and 1920, Calaveras mines averaged more than 10 percent of the total annual statewide gold production. Lode mining contributed most of this, yielding nearly one hundred thousand ounces annually from 1901 to 1916.[1]

In this thirty-five-year period, hardrock mining in Calaveras completed the transition from exploratory pioneer to modern industry. In five districts — Angels Camp, Carson Hill, Lower Rich Gulch (Paloma), Sheep Ranch, and Hodson — major producers shifted from selective mining of shallow high-grade deposits to systematic mining of all developed ore, including deeper lower-grade deposits. The biggest mines attracted significant capital resources, consolidated the best properties, hired skilled managers, tapped into the largest ore bodies, adapted to newer mining and milling technologies, and reached levels of production and profitability unparalleled in county history.

Copper mining also came to life again in this period, revived as much by improvement in the price of copper as by technological advances. The late 1880s saw the reopening and modernization of the copper lode mines both at Copperopolis and at Campo Seco. By the turn of the century Calaveras was annually producing more than a million pounds of copper, a trend that continued, with peaks exceeding seven million pounds a year, until 1930, with two smelters in operation much of the time. Thus, copper vied with gold in importance to the local economy.

Consolidation was one of the keys to high production during the golden years of lode mining. Merging several adjoining claims into one consolidated property made possible the development and extraction of large low-grade ore bodies under one management. Before the 1880s most Mother Lode mines were small operations under individual ownership, with a low volume of ore processed each day. Owners could keep going only if the millhead values were high enough to offset the high unit costs of production, which before 1875 ran from six to ten dollars per ton.[2] By the late 1860s the trend toward consolidation in Calaveras had already begun, but it took another couple of decades

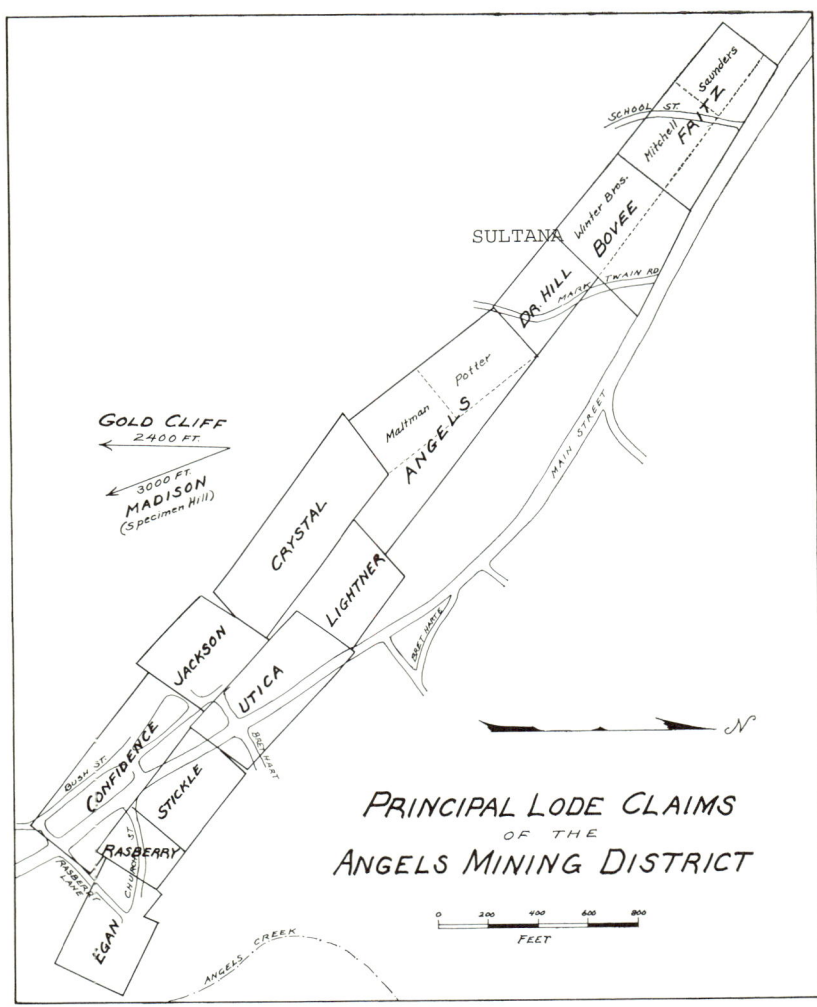

Angels Camp mine map. (Courtesy Calaveras County Historical Society)

to make significant headway so that large lower-grade ore bodies could be efficiently worked.

In the 1890s mining was an economic catalyst that helped Calaveras and other Mother Lode counties weather a severe national depression. Hard times in the Corn Belt did not translate into hard times in the East or West Belts of the Mother Lode. Depressed farm prices and foreclosed homesteaders in Iowa and Nebraska did not have much impact in Calaveras, though the Southern Pacific Railroad monopoly kept regional shipping prices high for mining companies and merchants alike. The national "Battle of the Standards," pitting advocates of a bullion-based, deflationary monetary policy against proponents of a credit-based, inflationary system, seemed remote and esoteric to most Mother Lode residents. Gold was clearly king in California, where productive mines kept foothill economies on an even keel throughout the turbulent nineties and into the first decade of the new century. The Golden State, unsympathetic to the idea of unlimited silver coinage, voted for hardmoney interests both in 1892 and in 1896, and applauded the adoption of the gold standard in 1900.[3]

## The Angels Camp District

In the thirty years since discovery, the mines along the Davis-Winter lode had gone through several periods of high activity and subsequent decline. By the 1880s most of these early claims had been consolidated into three principal properties, with a fourth emerging early in the next decade.

At its northernmost extreme, where the original lode was discovered in the early 1850s, the Sultana mine eventually emerged out of a collection of smaller properties first brought together by William H. Bovee in the 1860s. As noted earlier he worked the property until his mill burned in a spectacular fire in 1872, at which time he left the mining business and became a real estate promoter in San Francisco. For eight years the mine remained closed, but in 1883 it reopened as the Marshall mine under the ownership of Captain Volney Cushing, inventor of the Hooker steam pump, a popular alternative to the Cornish pump in that era. Upon Cushing's sudden death the mine came under control of the Mohawk Company that renamed it the Sultana, the last and most memorable of its various names. In the 1890s "General" G. W. Huddleston, whose nickname was allegedly earned in Colorado where he helped settle a miners' union dispute, purchased it. He was an internationally known mining financier who also held properties in South America. According to D. C. Demarest, Huddleston's cultivated tastes kept him in Europe much of the time following Italian opera and savoring good wine. Under his ownership the Sultana reached its deepest level, 700 feet, and kept a fortystamp mill in operation until 1905. Production is estimated to be around fifty thousand ounces.[4] It was the first of the consolidated mines of Angels Camp to open and the first to close.

Adjoining Sultana to the south was the Angels mine, consolidated in 1884 by James V. Coleman out of the Angels, Doctor Hill, Maltman, and Potter claims. With the Potter mine came a twenty-stamp steam-powered mill

The Sultana mill at Angels Camp, with its high trestle to deliver ore from the shaft. (Courtesy Calaveras County Historical Society)

known as the "big mill" because it was the largest in the district at that time. Demarest remembered it fondly. "The noise from the stamps of this mill, and the blowing of its whistle, morning, noon and night, put life into the air of an otherwise quiet camp," he said.[5]

A nephew of Comstock magnate William S. O'Brien, Coleman lacked the vision and the financial skills of his uncle. Demarest wrote that in the mine's early years Coleman gave too many incompetent friends management jobs, with predictable results. While miners still used old-fashioned hand tools belowground, on the surface an experimental ball mill had been installed to supplement the more traditional stamps, but the mill was plagued by technical and mechanical problems. In 1889 a chlorination works was added to process concentrates, but that failed also.[6] By 1900 the mine was only 500 feet deep but $650,000 in debt.

Coleman finally hired an experienced engineer to manage the property. Before moving to the Angels, Thomas Fullen was master mechanic at the Utica mine and Demarest's partner in the Altaville Iron Works. With a competent foreman, Charles Jacobs, Fullen set to work modernizing and deepening the Angels mine. They added an air compressor to eliminate hand drilling, built a sawmill to frame mine timbers on site, remodeled the blacksmith and

The Angels mine headframe and surface plant, ca. 1895. (Courtesy Holt-Atherton Library, University of the Pacific)

machine shops, junked the faulty ball mill, enlarged the old stamp mill, and added vanners to improve recovery. With the acquisition of the nearby Crystal mine shortly after 1900, Coleman and his crew were ready for sustained production. Operating almost continuously until the beginning of World War I, the Angels added nearly two hundred thousand ounces of gold to county production figures. By the time it closed, with its ore reserves exhausted, the main shaft had reached a depth of 850 feet, with a winze extended to the 1,050 level.[7]

After shutdown, the property deteriorated rapidly. Maintenance stopped, and the lower levels filled with water. Coleman tried to sell the mine, but prospective buyers were wary. In 1915 one offered to spend $100,000 on new equipment and development, but rejected Coleman's $300,000 price tag in the characteristic language of a hard-bargaining mine dealer: "I look at this property, as it stands today, as a failure. I think you will agree with me in this, whatever the fault may be, lack of capital or perhaps the management. . . . Consequently . . . I must have the most favorable terms to be able to induce my associates to go into it."[8] Coleman turned down the counteroffer, the deal fell through, and the mine never reopened.

Down the hill from the Angels mine, toward the central business district

The Lightner mill in the 1890s, showing the elevated tram from the headframe. (Courtesy Calaveras County Historical Society)

of Angels Camp, was the Lightner. Opened in the 1850s along with other claims on the same lode, it was unproductive in its early years and was finally abandoned in 1866. Nearly twenty years later Z. T. Carlow relocated the property but did very little work other than the minimal requirements to hold the claim. He sold it in 1888 to Joseph G. Eastland, who "for some reason" did not work it either, so the papers reported.[9] Had he done so he might have spared his heirs the inconvenience of having to prosecute a lawsuit against its neighbor, the Utica, for stealing ore below the Lightner's 60-foot level, the deepest it had penetrated before the mid-1890s. The suit, filed a year after Eastland's death late in 1895, claimed that for several years the Utica had been "running tunnels under the works of the Lightner Company" and had worked out "a great portion" of the Lightner's ore before Eastland discovered the fraud in June 1894 and demanded payment. The suit dragged on for years, but the Lightner heirs persisted and in 1904 won a judgment of $54,000. On appeal the state supreme court struck down the punitive damages, reducing the final award to $27,000.[10]

A year after his death, Joseph Eastland's daughter Alice sold the mine to a syndicate of Stockton and Oakland investors organized as the Lightner Mining Company. They had been attracted to the property by James Maltman, a local promoter and the son of William Maltman, an early Angels mine de-

The Maltman mine, with the Maltman family and friends. (Courtesy Calaveras County Historical Society)

veloper. Demarest remembered that the younger Maltman, who had a good "nose for ore," convinced the backers to deepen the mine in order to open a large low-grade ore body. But the new development was expensive, and as capital investment increased without dividends, so did the anxiety in Stockton. To protect their investment and try for quicker returns, the Stockton group sent one of their own members, Alexander Chalmers, to manage the mine. A shrewd businessman who had no previous mine training but considerable executive ability, Chalmers kept overhead to a minimum while he modernized the operation. He lived in a small cottage next to the mine with his wife, hired a bookkeeper, and kept an eagle eye on expenses. Most important, he employed a competent and trustworthy staff.[11]

In the 1890s the advent of cheap electricity gave the Lightner and other Mother Lode mines a welcome, new cost-cutting technology. Throughout the Lightner's operation, two-phase alternating-current Westinghouse motors took over from old steam- or water-powered equipment. A 100-horsepower motor drove the compressor that sent high-pressure air underground to the drills. A smaller 10-horsepower motor ran the crusher, and a big 150-horsepower machine drove the mill. The Lightner was the first mine in the county powered primarily by electricity.[12]

Converting to cheap electric power was perhaps the most important savings, but Chalmers and his managers found other ways to lower costs. For example, they saved an extra $0.50 per ton by expanding the number of mill concentrators by one-third over other mines in the district, thus reducing the amount of gold lost in tailings. By keeping operating costs to a bare-bones $2.25 per ton, Chalmers was able to work profitably a massive body of low-grade ore that averaged less than a half ounce of gold to the ton.[13]

The Lightner's impressive ore body, which in some sections widened to 120 feet, was the source of both its productivity and its demise. Huge underground stopes required heavy timbering. Each fall contract lumbermen unloaded long wagon trains of Sierra logs that "took up every foot of surface space available on the property."[14] A sawmill at the mine framed the square-sets, which timbermen installed in the great stopes underground. Despite these precautions cave-ins were a constant threat. In 1903 a tremendous roar shook the town like an earthquake. Miraculously, no one was working below at the time. When the dust cleared, townspeople and miners were stunned to see a massive surface depression where a few minutes before was level ground. All of the underground stopes down to the 600 level had collapsed, burying the ore and most of the equipment. Today part of the result of this settling can still be seen by comparing the surface level of the northern end of upper Utica Park in Angels Camp with the ground level of adjacent Highway 49.

Chalmers made a valiant attempt to recover the ore from the caved stopes, but two years later another disaster dealt the mine a crippling blow. A spectacular fire destroyed the mill and hoist, and nearly cost Chalmers his life. The mine superintendent was asleep when the fire alarm sounded at 1:05 A.M. He jumped up and jerked on his clothes. By the time he ran out the door the hoisting works were ablaze. The fire had started in the change room when a candle ignited clothing hanging near some blasting caps that had been prepared by the night-shift boss. He had descended to the 200-foot level to get ready for the next blast, leaving his caps behind. When they exploded, burning timbers were tossed in all directions. The fire quickly consumed the hoisting works and then the gallows frame. Within fifteen minutes the mill was in flames. Racing to the mill just minutes before it caught fire, Chalmers scrambled to the top in an effort to turn on several hydraulic nozzles. But the smoke was too thick, and before he could reach them he fell to the floor unconscious. The mill was engulfed when he finally recovered, and he stumbled out, almost falling into the ore bin on the way.[15]

Although the surface plant was nearly leveled by the fire, it did little damage underground. The night crew of sixteen men worked hard at the bottom of the shaft dousing the falling timbers before they could do much harm. The

Remains of Lightner compressor, now an artifact in Utica Park at Angels Camp. (Authors' collection)

men escaped through the Angels mine, which had opened a safety connection to the Lightner at the 600 level a few months before.[16]

Hardly had the ruins cooled when help began to arrive. Only 20 percent of the $100,000 loss was covered by insurance, but Chalmers was confident that he could rebuild without stock assessments. Most of the mine owners and merchants in Angels, mindful that the Lightner's demise would cripple the entire district, chipped in with supplies and equipment. Despite long-standing differences with his next-door rival, the Utica's superintendent, Fred Martin, was a picture of friendship and concern. When he heard the alarm he had rushed over to help Chalmers fight the fire, and now he offered to sell the Lightner an old headframe that still stood over the Utica's original shaft but was no longer needed. A deal was struck, and within a matter of days the reconstruction began. Lightner employees worked overtime and on night shifts to speed up the rebuilding. Chalmers, relying on short-term credit, erected a new mill about 100 yards west of the old plant, on better ground away from the shaft collar. He also installed twenty-four new concentrators and a new electric plant, confident that he could get back into production quickly enough to pay off the debt with returns from the mill.[17]

For the next five years Chalmers kept the Lightner afloat, but mine production never caught up with expenses. Eventually, he used up all his operating funds and his credit as well. When it became impossible to continue without extensive new development, the Lightner backers refused to put up the money and forced Chalmers into retirement. In 1910 a new company sank a 900-foot vertical shaft on the property in an unsuccessful attempt to get under the sector that had caved seven years before. The mine closed for a few years, but just before World War I "General" Huddleston tried to revive it. Hoping to repeat his earlier Sultana triumph, he returned to Angels Camp and brought Chalmers out of retirement long enough to secure an option on the Lightner. The "general" secured $50,000 from New York investors and turned it over to the aging Chalmers, but it was not enough to dewater the old works and produce paying ore. In 1915 the mine closed again, this time for good. Total production was about 176,470 ounces of gold.[18]

By far the most important mine in the district was the Utica, located south of the Lightner in what is now the main business district of Angels Camp. This property attracted at one time or another some of the more important names in western mining history. James G. Fair's connection with the mine in the 1850s and 1860s has already been mentioned. Robert Leeper located the abandoned property in 1869, renamed it the Invincible, and did a small amount of development. In 1884 the workings were acquired by Charles D. Lane, a Missouri native who had come west to Knights Ferry with his parents in 1845. He and his partners restored the original name.

Lane was one of the last pioneer engineer-managers in the mining industry. Lacking a formal education, he relied on experience, hard work, personal attention to detail, and luck to reach the top of his profession. Characteristic of Charley Lane was his oft-repeated formula for locating pay dirt: "I am digging all the time and praying like hell." A practical miner rather than a trained engineer, he was also mystical and superstitious. When spiritualism swept the country after the Civil War, it attracted Lane and at least one other California mining notable, Alvinza Hayward. A Vermont attorney and Michigan miner before joining the Gold Rush in 1850, Hayward began his western mining career at Independence in Calaveras County. Success there bankrolled his foray into the fabulous Comstock, where he teamed up with John P. Jones at the Crown Point mine, fought off the Bank Ring, and emerged with a reputation as one of the wealthiest and shrewdest of western mining men. Later he bought controlling interest in the Eureka mine at Amador, which ultimately earned him a reported $2 million. Most of his mining profits went into Bay Area real estate development, exemplifying the capital transfer from mines to cities.[19]

Charles D. Lane. (Courtesy David
Clarence Demarest Collection,
Holt-Atherton Library, University
of the Pacific)

Both unconventional, Hayward and Lane ventured into spiritualism just
as they might speculate in mining. How deeply they delved is hard to say.
At any rate séances became almost a regular routine at Lane's Angels Camp
mansion. Demarest recalled Lane once investing in a hardrock venture that
cost him $40,000 without producing enough "to pay for the grease on the
cam shaft." When Demarest inquired why he risked so much, Lane replied
that he "did the usual thing, in contacting the spirits of the miners who had
worked in the Mine. In this case, as very seldom happens, they lied to me."[20]

Lane's spiritualist friends served him well in acquiring the Utica. Before
purchasing the property, he took an ore specimen to a San Francisco medium
who promised that "millions were to be had where that rock was found," and
predicted that "Charley" would also discover a "splended vein of oer [*sic*] of
which you now know nothing—you will soon reach it however, and it is indeed rich." With that assurance, reinforced by other such "tests" of Utica
samples, Lane and four other investors bought the mine for $10,000.[21]

Under Lane's guidance most of the early Utica production went into consolidation and capital development. Acquisition of the important Gold Cliff
property in 1884 on a flatly dipping vein to the west gave the Utica developers
a substantial low-grade ore body just a thousand feet away. Three years later
they added the Madison mine, which Lane's sons operated at first. In 1888, to

The two Utica mine shafts at Angels Camp in the 1890s, with the original mill in the center. After purchasing the Stickle mine and sinking the Cross shaft, this mill was dismantled and the equipment added to the Stickle mill. Today this site is part of Utica Park. (Courtesy David Clarence Demarest Collection, Holt-Atherton Library, University of the Pacific)

secure water for its expanded mill and later its chlorination works, the Utica contracted for all the Union Water Company's output, and two years later purchased the company.[22]

Topping off their acquisitions, in 1890 the Utica investors bought out the adjacent Stickle mine. Since 1885 it had been in the hands of Captain William A. Nevills, who had operated it as the Union. An eccentric investor who went on to become the "Big Mogul of Tuolumne County," Nevills was notoriously antilabor and at one time mounted a Gatling gun on his property to protect nonunion workers—"scabs" in labor parlance—during a strike. While he had the Stickle-Union, Nevills installed a forty-stamp mill and apparently did well. Production figures are lacking because he kept all his records a secret.[23]

Building the financial structure and the surface plant at the Utica took both engineering skill and money. Even though Utica ore production since 1885 paid most of the operating expenses, by the end of 1887 Lane and his fellow investors found themselves almost $1 million in debt. But Lane's close friend Alvinza Hayward and another mining financier, Walter Hobart, came to the rescue. Both were powerful capitalists in the mining world. Hayward had parlayed his spectacular success on the Comstock and in Amador County into a financial empire that centered in the San Francisco Bay Area. Hobart was less well known, but together their Calaveras venture stirred financial circles and helped put new life into the lagging Southern Mines. They bought out the Utica's minority investors, paid off the creditors, and pumped in sufficient capital for major expansion and development. Although Lane now shared the Utica ownership, he still controlled mining operations and did not welcome the advice of his partners on mining matters. On one occasion, soon after the South shaft on the Utica had been sunk, mill cleanup fell so low that Hayward and Hobart sent Lane a panicky wire telling him to shut down the mine. Lane exploded. His mill operator later recalled his defiant words: "To hell with them! I own a third of this mine, and I will take no such orders from anyone." Two weeks later the Utica made one of its richest strikes.[24]

In the nineties the Utica set national production records, turning out in one thirty-month period alone more than 244,000 ounces in gold for a net profit of $2.5 million.[25] Much of this money left the community in the pockets of its absentee owners, but the $40,000 monthly payroll of the combined mines in Angels Camp was spent locally and kept the town prosperous during a period of national depression.

The Utica was the "envy of the mining world" in this period, according to Demarest, but the five hundred men employed did not always share this opinion. Before the advent of effective health and safety laws, the rewards of industrial mining came at a high cost to the miners.[26] As the Utica shafts deepened, so did the anxiety of underground crews who had to work under very hazardous conditions. Like the Lightner, the Utica's large stopes required extensive timbering. Lane and Hayward adopted a modified square-set system, using peeled pine logs, round instead of square, up to 30 inches in diameter and usually 8 feet long, although some were 16 feet in the early 1890s. The longer of these awkward and heavy beams were lowered into the mine by chains hanging from the bottoms of ore skips. Timber crews drove dogs deep into the timbers to firmly anchor the chains. The thought of a half-ton timber careening down a 2,000-foot shaft was a considerable deterrent to carelessness. Even so, handling accidents did occur occasionally.[27]

Caving was a much more serious threat to Utica mine safety than careless timber crews. Hardrock mining under normal conditions is dangerous; in the shifting ground beneath Angels Camp underground crews virtually took their lives in their hands. The demands of mine investors added to the danger by pressuring managers to give higher priority to production than to safety. William H. Storms, a state mining engineer, condemned the "antiquated" and "objectionable" methods employed by such men as Alvinza Hayward, who opened huge stopes in order to return quick profits despite the caving perils. Even worse, Storms found timbering methods along the Angels lode inadequate. At one unidentified mine after a cave-in, despite the superin-tendent's insistence that the stope had been "carefully timbered" and filled, Storms spotted gaps of up to 6 feet between the tops of the square-sets and the stope "back" or roof. The filling had also been improperly placed, extend-ing only partway up timber sets and as much as 15 feet below the back of the stope. In Storms's words: "The great mass of ore, being undercut over a large area, began to settle, the process being aided by the talcose walls, and a cave resulted as a natural consequence."[28]

Unstable ground combined with expedience, parsimony, and perhaps some incompetence as well all contributed to the caving problems at Angels Camp. At the Utica the first serious incident occurred late in December 1889, when miners at the North shaft refused to go below because of the spongy condition of the ground above the 330-foot level. Lane had told his partners of the condition earlier, and he ordered a timbering crew to reinforce the shaft. Despite the foreman's protests, seventeen men descended to begin the repairs. Suddenly the shaft collapsed, killing all but three. Lane oddly escaped criticism since he was the manager, but miners openly blamed Hayward, who they said believed that "men are cheaper than timbers."[29]

In spite of this accident and others, miners in Angels Camp were slow to organize. Elsewhere, mine unions had made considerable headway. Com-stock miners had unionized as early as 1863. Homestake miners had founded a union in 1877, one year after the lode was discovered. Mining districts in Montana, Colorado, Idaho, and Arizona had all experienced union move-ments before Calaveras. But outside the Comstock district, which set the pat-tern for the entire industry and made unionization a major force in the West, mine unions spread sporadically. A recent scholar has found that unions were generally more successful in long-established camps but failed in the boom camps where there was less community solidarity, less personal contact be-tween miners and nonmining residents. However, California mining towns he found to be an exception. In districts such as Amador and Grass Valley, for example, miners were an integral part of the community, but the issues were more complex, the levels of violence greater, and the resistance to union efforts more difficult to overcome, in part because California mine owners were more conservative, as has been mentioned earlier.[30]

Dangerous working conditions provoked the first talk of unionization at the Utica, but the results were feeble. A miners' Protective Association had been established early in 1885 to improve underground working conditions, but this was more of a men's club than a union. Neither the 1889 accident at the North shaft nor the broken cable in 1891 that killed nine men stirred up major union agitation. Ernie Vogliotti, a Utica miner from Turin, Italy, barely escaped both disasters, but his reminiscences are silent on labor issues. He remained at his job for another decade, working a ten-hour shift at three dollars per day. In the 1891 incident Lane's personal attention to the problem, and his bravado, helped avoid a labor crisis. The victims were riding a cage when the hoist cable suddenly gave way, dropping them 350 feet to the bottom of the shaft. To address the problem Lane called on his former master mechanic, Thomas Fullen, who equipped a cage with his improved "Fullen Safety," apparently a modification of the "safety dog" developed in 1866 by William N. Shaw, a blacksmith at the Union mine in Copperopolis. The mechanical attachment closed automatically when the hoist cable went slack, catching on the shaft guides and stopping the descent. Lane tested it the first time himself, bravely climbing aboard a modified cage that was suspended over the main shaft by a hemp rope. With dozens of men watching, at his signal one of them cut the rope, and the device stopped the cage after dropping only a foot. No one ever accused Charley Lane of cowardice. The Fullen Safety was soon installed as standard equipment at the Utica.[31]

Not until 1894, during the national labor unrest following the panic of 1893 and the Pullman strike, did a major labor crisis occur in Angels Camp. Organized miners led by Slavs and Italians struck the Utica and blew up a boardinghouse in protest to dangerous underground conditions and lack of medical facilities. The strike ended after mine owners agreed to build a hospital, but not before a vigilance committee had run the most militant union organizers out of town.[32] Union sentiment was still so low by 1898 that two Tuolumne union officials left in disgust after a futile one-day organizing effort at the Utica.[33]

Working conditions improved at the Utica after 1900, although the Lightner and other mines continued to be troubled with shifting ground. However, after the turn of the century union support increased more because of demands for shorter working hours than because of dangerous conditions underground. Nationally, the eight-hour movement had gained ground steadily since Comstock miners first introduced the concept in the 1860s. By 1910 it had replaced the ten-hour day in every major western mining district. Hence, agitation in Angels came in the midst of a rising national labor outcry. In 1907 a strike shut down the Utica, Angels, and Lightner. Four months went by before owners compromised on a nine-hour day.[34]

Despite a sullied labor reputation, the Utica continued its deep development in the search for more ore. Work halted temporarily in July 1895 when a dangerous fire in the support timbers of an old stope on the 800 level burned several days. Lane got his crews out without loss of life, but the mine had to be flooded, and it took more than a month to pump out and rehabilitate.

Within a year Utica miners began sinking the Cross shaft, which, combined with underground drifts and winzes, eventually reached 3,050 feet, the deepest of any mine in the district. By 1910 more than 200 miles of Utica workings extended under the town of Angels, whose residents obligingly had relinquished all mineral rights below 500 feet in return for title to their homes. By this arrangement Lane protected the Utica from trespass lawsuits by surface owners. To save haulage and hoisting expenses as well as to shore up and stabilize the heavy ground, almost all mine waste was gobbed underground, leaving the surface unusually free from waste dumps.[35] Operators even used surface gravel on occasion to fill the huge underground stopes.

In the golden years of its mining history, the Angels district produced some 1.5 million ounces, nearly three-fourths of it coming from the Utica group. Total Utica-Stickle production from 1887 to 1918 was over more than 800,000 ounces of gold. Adding in the Gold Cliff's 167,000 and the Madison's 60,000, the Utica proprietors earned a $4 million operating profit.[36] It is important to note that the Utica group remained in the hands of a three-man partnership throughout its latter history. Without stockholders to share profits, Hayward, Hobart, and Lane rank first among Calaveras County mining investors in total earnings, if not total production.

By the time it shut down in 1918, the Utica was no longer under the supervision of Charles D. Lane. The Yukon rush had caught his eye at the turn of the century, and he left for Nome, Alaska, to pioneer gold development there, leaving the Utica in the hands of Superintendent Fred Martin. Unlike the Utica venture, however, Lane spent much more money than he earned. D. C. Demarest, who knew him well, said failing eyesight prevented him from distinguishing good claims from bad. Other mining interests in Alaska, Arizona, Nevada, and Mexico also drained his capital resources. By the time of Lane's death in 1911, most of his fortune had evidently been lost.[37]

## Angels Camp Miners and Families

The flow of local gold during the early 1890s insulated Calaveras from the national monetary crisis and depression that followed. In 1894, the year of the Pullman strike and Coxey's Army, when business slumped and the jobless rate zoomed elsewhere in the country, Angels Camp heard "no cry of hard times," in the words of W. H. Hibbit, a Stockton tailor who visited the town in March of that year. "Everybody is spending money . . . with the merchants and business men." Monthly payrolls from the major mines, he reported to a Stockton paper, keep "plenty of money in circulation, and therefore times are good."[38]

A month later Hibbit wrote again, this time as an Angels resident making four times the money he had in Stockton. He liked the business but was less sanguine about the town's prospects. Too many idle people were on the streets, and the mining camp atmosphere reminded him of the perilous frontier. "Shooting scrapes are frequent, and no notice is taken of them unless some one is killed or unless the man shot insists on his assailant being arrested." At a local dance hall, he said, the piano player was nearly killed when a bullet "passed unpleasantly near" his head. He kept on playing, though, while the first violinist "descended from the platform and disarmed the shooter."[39]

Incidents such as this grew more frequent in the spring of 1894 as men from valley towns rushed in to find jobs. Hibbit's darker view may also have reflected the rising labor troubles at the Utica, which brought out the vigilantes. As the *Mountain Echo* reported early in May, "Angel's Camp has more idle men looking for work than any other town in the State, and a public mass meeting was held by the citizens of that burg last Sunday night to take steps to drive bad characters out of the town." Other regional papers, however, inspired either by the booster spirit or by the subsidies of the prominent mine owners, disputed these alarming reports and blamed the incidents on a few "ten-cent gamblers" and members of the "undesirable class."[40]

The disruptions of 1894, though troubling to mine owners, did not permanently alter the boomtown atmosphere that pervaded Angels Camp during the nineties. While the mines were active business boomed in the building trades, clothing, real estate, banking, transportation, lodging, and other services. Boardinghouses filled up, and so did the hotels. Grandma Rolleri's Calaveras Hotel, the largest in town, expanded rapidly, eventually covering an entire block, with fifty rooms and a couple of apartments. Single miners were the biggest customers. They could take meals prepared by Chinese cooks in one of two dining rooms, and could also have their dinner pails filled with sandwiches, pie or cake, coffee, milk, or cooked fruit. Cornelia Barden Stevenot, granddaughter of Olivia Rolleri, remembered lines of miners coming off shift from the Utica, swinging their pails as they walked down the hill from the mine collar to the hotel, still in their "diggers," or working clothes. Only later did mine managers add change rooms in an attempt to cut down on "high-grading," a practice discussed later. Even after the golden years of mining ended the Rolleris continued to run the hotel until fire destroyed the building in 1938.[41]

For children Angels was an exciting town in the boom years, filled with strange people, exotic smells, awesome machinery, loud noises, and lurking dangers. The streets were crowded with timber and freight wagons, and the sky dramatic with smoke and steam. Kids played games in the mine yards and hoisting works. The bolder ones occasionally climbed the headframes when no one was looking, or jumped up to catch and briefly ride the bucket carriage on the aerial tram that passed overhead carrying concentrates to the Utica mill across Angels Creek. They had to drop off quickly or find themselves hanging 150 feet in the air.[42]

Ruth Harper Lemue, six years old at the turn of the century, lived on Democrat Hill overlooking the Stickle and Utica mines. Her father, James Harper, was mill superintendent at the Sultana and later the Lightner, just up the road from the Utica. From her house she could watch Utica miners stepping off the cage at the shaft collar as they came off shift. For townsfolk the day began at 5:30 A.M., when the mine whistle first called the men to work. It sounded again a half hour later, when the shift commenced. Hotel guests a few blocks away might complain about disrupted sleep, but to residents the blasts were taken in stride, the sound of progress and jobs.[43]

The lode mining boom altered the ethnic mix in Angels Camp and other Calaveras towns. The Chinese had represented a sizable community in the 1860s and 1870s, working as cooks, laundry workers, vegetable peddlers, day laborers, and in other occupations that served both the white community and themselves. But the Chinese population had steadily dropped after 1880 because of the exclusion laws, the decline of river placers, and the continuing opposition to their urban enclaves. By 1910 only forty-nine were reported in the Calaveras census. With few exceptions, the evidence of their presence also rapidly disappeared. Of the two Chinese houses of worship in Mokelumne Hill, the first burned in 1897, and the second was stripped to the bare walls and incorporated into a hog pen. The flood of 1909 destroyed most of the Chinese quarter along Angels Creek. Only a few structural remnants have survived to present times. Almost all trace of Chinese cemeteries and burial plots in the Mother Lode was erased after families of those who died reclaimed the bones of their ancestors and returned them to family plots in China. The Joss House in Angels Camp stood until the 1950s, but ultimately gave way to the community high school. Recently, one brick building in Angels, an 1860s-era store once owned by Sam Choy, has been restored and used for law offices.[44]

Though the Chinese did not work in the lode mines, their decline opened opportunities for immigrants and women to expand into service and supply trades once occupied by Chinese. White or Hispanic women, often miners' wives or widows, took over much of the laundry trade. For example, in West Point, Mexican-born Antonia Balensuela, a widow at age sixty, was listed in the 1880 census as a "washer woman." She kept a boardinghouse with five other occupants. In the same district Hester Hepburn, age forty, married with husband "absent," also ran a boardinghouse. She was from Ireland, and

The ethnic diversity of mining crews in the southern Mother Lode is well illustrated in this shift at an unidentified mine before World War I. Note the young "nipper," or miner's helper, at center. (Courtesy Holt-Atherton Library, University of the Pacific)

lived with her three daughters, two of them teenagers, all born in California. Doubtless, she also did laundry for extra money. Often, taking in laundry was the only available source of additional income for older women.[45]

Italians by the 1890s, though only 4 percent of the county population, provided most of the fresh vegetables and fruit in Calaveras towns and villages. Frank Canepa, on his Vallecito farm, each morning loaded a wagon and brought produce to Angels Camp, Murphys, Sheep Ranch, and Copperopolis, leaving his wife, Sara, at home to care for their ten children. He made trips as far as Calaveras big trees to serve sawmills along the way. Others used pack animals for the same purpose. The presence of these hardworking, nonmining villagers from the Old Country earned grudging respect from more affluent Americans. With a hint of disdain one mining superintendent described the Italian "peasants" who occupied the hills and canyons of the southern Mother Lode, converting old mining claims to gardens and vineyards, trading fruit and vegetables in town for basic necessities, building "mere huts," living on "next to nothing," and poor in everything "except children."[46]

In the mines of Amador and Calaveras, a Serbian influx was noticeable by the 1890s. A rising clash of ethnic and religious minorities in the Austro-Hungarian Empire after 1880 accelerated this Mother Lode immigration.

St. Sava's in Jackson, the oldest Serbian Orthodox church in the United States. (Authors' collection)

Census records unfortunately do not distinguish Serbians from other "Austrians" in these years—despite their ethnic pride and devotion to *Srpstvo* (Serbianism)—so it is not clear how many Serbs composed the 11 percent Austrian foreign-born in Calaveras in 1910, or the 21 percent in Amador. Serbian miners in Angels Camp numbered six to eight hundred in the period 1910–1915, according to one estimate.[47]

Though most immigrant Serbs were single males living in boardinghouses, Serbian families added to the ethnic mosaic of the southern Mother Lode. The spirit of *Srpstvo* brought Serbs together for social, religious, and fraternal purposes. Family patriarchs often played the *Gusle* and recited epic poetry, preserving in folklore the deeds of classical Serbian heroes such as Marko Kraljević and Miloš I Obrenović. At Angels Camp many Serbian young men belonged to *Srpsko Druztvo,* a benevolent and patriotic society active during the Balkan crisis prior to World War I. To serve the Serbian population, St. Basil's Orthodox Church was established at Angels Camp in 1909, following

Corporate executives welcomed hardworking southeastern Europeans during the golden years of mining. These "Slavonians" helped build the road to the Stanislaus Electric Power Company's plant at Camp Nine. (Courtesy Calaveras County Historical Society)

the founding of St. Sava's in Jackson in 1894. The Angels Camp congregation declined soon after the local lode mines closed, yet the Serbian spirit still lingers today among the active parishioners of St. Sava's.[48]

Mining managers welcomed eastern European minorities, both for their work ethic and for the ethnic diversity that they hoped would offset the threat of unionization. Keeping the workforce divided by language, religion, and ethnicity was a favorite management tactic, dating back in the United States at least to the Henry Clay Frick era in the steel industry. California had generally escaped serious labor trouble before the 1890s. As one mining editor explained, "Our ores are for the most part, free and easily treated, while labor is almost everywhere cheap and manageable." But as earlier discussed, union talk in Angels Camp rose in the early 1890s and reached the confrontational stage after 1894. Adding Serbs and Italians to a workforce dominated by Irish and Cornish was no mere coincidence, but union sympathizers countered by seeking common bonds among workers. Health and safety issues were the most important, although not very effective in Calaveras County, as we have seen. Efforts to pass mining health and safety laws made slow headway, but after 1900, Calaveras unionists found some common grounds of unity in mutual-aid societies, which assisted workers and families brought down by illness or accident. In 1909, for example, W. S. Reid of Local 55 in Angels Camp described the funeral after a horrible mine accident:

> The weather was fine and there were about seven or eight hundred men in the procession headed by the local band. Two of the deceased were Austrians [Serbians?], and one was an Italian, but it was really gratifying to see the way in which all thoughts of nationality were swept into the background—a happy indication of the fast disappearing artificial barriers which the exploiting class would be glad to maintain between the workers of different nationalities.[49]

The Angels mines avoided serious labor trouble during the golden years, but nagging health and safety concerns continued to drive a wedge between labor and management right down to the close of the major mines during World War I. After that, mining jobs were too scarce in the Mother Lode to provide much of a platform for labor agitation.

## The Gwin Mine

Mining activity virtually ceased in the Paloma district after the Gwin shut down in 1882. Rather than underwrite a modernization program, William M. Gwin Jr., son of the senator, sold out in 1893 to the Gwin Mine Development Company headed by Frederick F. Thomas, a professional mining engineer with a successful record in Amador County and New South Wales. Hiring as assistant manager, David McClure, a trained engineer and friend of Herbert Hoover, Thomas in 1894 began sinking the North shaft farther down on the gulch from the earlier workings. They opened up a large low-grade ore body that they followed eventually to the twenty-eight hundred–foot level. Even though averaging only about 0.2 ounces per ton, the large volume of ore available made it possible to work the mine profitably under skilled management.

Thomas and McClure were equal to the task. To take better advantage of the waterpower available from the nearby Mokelumne Hill and Campo Seco Canal, they reorganized the surface works and fed the discharge water from the main hoist and a new forty-stamp mill, later expanded to one hundred stamps, to a nineteen-foot diameter Pelton wheel at the bottom of the gulch. It turned a huge compressor that furnished enough high-pressure air to power the hoist on the old South shaft, two pumps in the main shaft that discharged more than one hundred thousand gallons per day, three drills, and a smaller hoist at the fourteen hundred–foot level. They also dismantled the expensive chlorinator and shipped sulfides to the Selby smelter instead. By 1903 these cost-effective measures had reduced operating expenses to $2.42 per ton.[50]

Frederick F. Thomas, manager
of the Gwin mine. (Courtesy
Calaveras County Historical
Society)

The Gwin's successful management cannot be attributed solely to technological expertise. Like other good mining engineer-managers, Thomas and
McClure also knew how to get the most out of their men. By paying attention
to their needs, listening to complaints, even eating in the same mess hall, they
kept morale and productivity high, while at the same time defusing potential
labor troubles. Such rapport stood in sharp contrast to that of men such as
Captain Nevills of the Rawhide mine near Jamestown or Kemp van Ee of the
Royal, whose callous insensitivity to worker complaints recoiled against them
and increased labor agitation.

The superior tactics of the Gwin management were amply demonstrated
in 1903, during widespread labor unrest in Amador County that spilled over
into Calaveras. A massive strike in the heart of the Mother Lode grew out
of a union organizational movement that had swept mining districts from
Colorado to California after the turn of the century. Led by the Western
Federation of Miners, union organizers clamored for union recognition, no
member discrimination, and shorter hours—the "bread and butter" issues of
American labor. Miners at the Kennedy and Argonaut across the Mokelumne
had turned out in droves to hear "Big Bill" Haywood, later cofounder of the
radical Industrial Workers of the World, demand the eight-hour day underground and union recognition. Fired up by Haywood's visit, Jackson miners

The Gwin mine, ca. 1905. Note the new steel headframe and the steam traction engine of Captain Hiram Ashley Messenger, a Civil War veteran. He used it to haul mine timbers and other products. (Courtesy Calaveras County Historical Society)

formed a local branch of the WFM and posted notice that all nonunion miners in the district would be treated as scabs.[51]

To counter rising labor unrest, Mother Lode mine owners and managers resorted to a familiar tactic. They organized the California Mine Operators' Association, vowed not to recognize the hated WFM, and fired fifty men who were known union members.[52] In the Coeur d'Alene, at Cripple Creek, and elsewhere across the mining West this kind of economic pressure had been used repeatedly to undermine union efforts. But the violent confrontations of the 1890s, as Richard Peterson suggests, "made many entrepreneurs realize that uninterrupted production was preferable to reduced labor costs." After 1900 enlightened owners and managers strove for more accommodation and less confrontation. By 1903 Utah, Colorado, Montana, and Idaho had joined the ranks of labor reform, but California was the least organized and the most resistant of all the mining states.[53]

The vindictive action by the mine owners' association in Amador, however, caused a backlash. Instead of stopping the union movement dead in its tracks, it aroused the rank and file and garnered public sympathy. By early April more than one thousand miners had walked out, shutting down all the mines in Amador and some in Calaveras. A scornful prolabor paper in nearby Tuolumne condemned the mine owners for firing "fifty good miners, family men, old residents and tax-payers," just for the "crime of belonging to the Miners Union." Even traditional management allies sounded surprisingly sympathetic, especially as the strikers kept their cool and avoided violent confrontations. The Amador sheriff, noting that miners were peaceful and "had made no threats," decided that it was not necessary to "put the county to the big expense of appointing a large force of deputies." His action was seconded by the district attorney, who said "the strike is just."[54]

The Amador strike quickly spread across the Mokelumne to the Gwin mine, where 75 percent of the miners joined the union and the strike after a WFM rally. With only fifty men left at work, Gwin managers planked over the shaft and stopped work except for essential maintenance. A few days later the impasse ended almost as abruptly as it had begun. With public sympathy building for a shorter workday, the Gwin owners offered a major concession. David McClure, acting as representative and chief negotiator for the owners, accepted an eight-and-a-half-hour day underground and agreed to rehire the strikers without recrimination. Though they did not win union recognition, strike leaders had cracked the management hard core. By May 1 most Gwin miners were back to work.[55]

The Gwin settlement became the basis for a general agreement that reopened the affected mines in Calaveras and Amador. The California Mine Operators' Association, still refusing to recognize the Western Federation of Miners, accepted almost the same terms McClure had offered his men at the Gwin. Miners went back to work, hailing the agreement as "one of the greatest victories ever won by union labor." In reality it was something much less. Rather than signal the "ultimate unionizing of the entire county," as predicted by a prounion editor, labor organization stalled in all but the major mines, and union strength was more chimerical than real. Later that year the collapse of a WFM-led strike at Cripple Creek in Colorado illustrated the differences of race and class that weakened mine unions in the face of determined forces of opposition.[56]

After the failed organizing effort in 1903, union membership went into a long period of decline, reflecting in part the inability of the WFM to withstand a protracted factional dispute, and in part the progressive effort to address fundamental labor issues through regulatory legislation. Moreover, as Alan Derickson has shown, after 1910 mine safety became an important issue among industry leaders who recognized the correlation between workers' health and industrial productivity. Corporate paternalism was slow in coming to the Mother Lode, but by providing worker benefits elsewhere it helped undermine union strength everywhere. As a result union leaders had trouble keeping even the largest mines organized. As late as 1922, for example, while frantic fire crews tried to rescue forty-seven men trapped underground in the Argonaut mine in Jackson, American Federation of Labor organizers used the tragedy as a rallying cry to sponsor a new membership drive.[57]

The Gwin closed in 1908 after producing more than two hundred thousand ounces of gold under Thomas's management. Water shortages to power the hoists and mills, lack of development capital due to a too generous dividend policy, but primarily rising costs and lower grades and tonnage of ore contributed to the shutdown. Demarest claimed a directors' disagreement about future development played a major role, but Thomas's son denied the charge. A mining engineer later wrote that Thomas had been preparing to sink the main shaft below twenty-five hundred feet to develop promising ore bodies explored earlier when "the telephone rang and they were instructed by the directors of the company meeting in San Francisco to shut down. All those at the mine were completely taken by surprise and thought that the shut-down would be for a short time only."[58]

During World War I a San Francisco scrap dealer bought the idle surface plant and removed the machinery. An enterprising second party purchased the empty mill building, tore it down, and ground-sluiced the site, earning a $15,000 profit on the amalgam that had dropped through the cracks in the mill floor. Perhaps the former stockholders did not begrudge the loss. They had collected $496,500 in dividends during the mine's working life.[59]

## The Royal Mine

Study of Gwin-mine operations makes clear that low-grade ore could return substantial profits if efficiently worked. F. F. Thomas and David McClure, both trained engineers, protected the profit margin by maintaining a close watch on ore assays, and by keeping tonnage high and operating costs low. In contrast, John Charles Kemp van Ee, an untrained mining promoter, managed the Royal unprofitably because he overbuilt the surface plant for the amount of ore in sight and was unable to maintain profitable ore grades. He also failed to contain labor troubles that contributed in the long run to the mine's closure.

We have seen how in the late 1870s Henry Botcher started up a small but successful mining and milling operation on the Pine Log lode, one of the earliest claims in the Madam Felix district. In 1881, Isaac Wilbur and the Castle brothers of Stockton bought the Pine Log Gold Mining Company and continued operating the mine until 1884. Then, believing all the ore had been mined out, the new owners decided to investigate an interesting development at the nearby Royal claim. This was a promising outcrop that had been prospected off and on for fifteen or twenty years but never really developed. Wilbur and the Castles moved in slowly but decisively, acquiring the Royal and five adjoining claims by 1885. They initiated small-scale mining the following year, with encouraging results. By 1890 a new stockholder from Stockton, J. D. Peters, one of the region's prominent grain dealers, injected new enthusiasm and capital into the operation, and development of what proved to be a major ore body was rapidly advanced.[60] Within a year the Royal Consolidated Mining Company, into which had been merged the Pine Log mine and mill, was paying dividends.

In 1891 the company hired an experienced mining engineer, Daniel Jutton, who soon took over as superintendent. He sank a new shaft, built a more complete surface plant, and increased the size of the old Pine Log mill to twenty stamps. Dividends increased substantially as a result, and by 1896 the Royal, although still a small operation, had attracted enough interest to stimulate vigorous prospecting throughout the district.

Early in 1897 the Royal prospects increased considerably when John C. Kemp van Ee, a mining promoter and operator, hired William P. Miller, a

J. C. Kemp van Ee. (Courtesy
Calaveras County Historical
Society)

San Francisco mining engineer, to examine the Royal and submit a detailed
report. This led to a buyout of the Royal company late in the year by English
interests.

Who was this Kemp van Ee? A New Jersey native of Dutch ancestry, he
came west in 1869 while still in high school, lured by exciting mining news
from Nevada and California. Changing his name to Jack Kemp, he took up
photography as a profession, but added mining promotion after a few years
in the mining camps. At Bodie he acquired a hotel and ran a studio with a
partner on the second floor, and doubtless promoted mines in the lobby or
in the nearby saloons, which frequented that colorful town.[61]

This fledgling promoter's first real experience with a major mine project
came in 1880, when he was hired by the Great Sierra Mining Company to
open up the Sheepherder silver lode in the high Sierra just north of Tioga
Pass. Although the mine never came into production, Kemp van Ee, as he
now styled himself, had been able to equip the mining site and drive the great
Sheepherder tunnel. He also had constructed one of the first telephone lines in
the area, and had laid out the original Tioga Road down to Crocker's Station.
By the time he left the project in 1883, he had gained considerable experience
and a local reputation as a talented promoter and developer.[62]

Kemp van Ee spent much of his time in London in the late eighties and nineties working on other promotions and inventions. His mining pursuits eventually caught the eye of an English investor, John T. Hodson, who hired him as an agent and sent him back to the United States to look for promising mining properties in the American West. Late in 1897, after reading Miller's detailed and enthusiastic report on the Royal, Kemp van Ee optioned the mine on the spot for $400,000. Then he raced back to London with a proposal that required $60,000 of the option price to be paid in cash, with the balance to be paid by an 80 percent royalty on all production revenue over $4.32 per ton, the Royal's current operating costs.[63] At that time all the underground headings appeared to be in ore averaging some $8.00 per ton in gold, with at least 105,000 tons in sight. This was a truly rosy picture for both buyer and seller.

Now as general manager, Kemp van Ee, drawing on his high Sierra experience, set to work with characteristic enthusiasm to exploit the property and upgrade its surface plant. Over the next several years he established the company town of Hodson, hired nearly one hundred men, installed steam-powered pumps to bring water from Salt Spring reservoir, increased the old Pine Log mill's capacity to 40 stamps, and deepened the inclined shaft to nine hundred feet. For a time the investors were full of hope and excitement under Kemp van Ee's management. But by 1902 most of the high-grade ore had been mined out, and after that the run-of-the-mine ore grades declined steadily. Undaunted, Kemp van Ee moved ahead with plans to expand. Funded by an additional $375,000 from his English backers, he rebuilt the hoisting and crushing plant at the shaft collar, constructed a magnificent new 120-stamp mill several hundred yards away, and connected it to the mine by an electric tram over a trestle. The mill opened on July 21, 1903, before a crowd of one thousand, mostly Calaveras residents. The featured speaker, Percy L. Shuman, told onlookers that for years investors like Rockefeller, Delamar, and Poniatowski had scorned the Royal. But the big mill, expected to process 30,000 tons a month, seemed to vindicate both Kemp van Ee and his English investors.[64]

The Royal mill had the largest number of stamps under one roof ever assembled on the Mother Lode. With twelve separate batteries, each designed to run separately or stop for repairs or maintenance while others were running, the mill could operate continuously seven days a week if ore was available. At a distance the noise probably resembled one old miner's description of the din from the Treadwell mill in Alaska: "It was just a very low rumble . . . low enough . . . that around town it became unnoticeable, you know. Except . . . when it stopped. Then everybody got out of bed to see what the

The Royal surface plant, with the hoist house on the left, and the high trestle connecting the shaft collar to the 120-stamp mill in the distance. (Courtesy Calaveras County Historical Society)

trouble was."[65] But inside, imagine the roar of 120 stamps, each weighing a half ton, dropping one hundred times a minute! And this in a day without mandatory ear protection for mill men.

Despite the expansion and the impressive surface works, things soon began to go wrong. Ore grade continued to drop, and mill returns fell far behind operating costs. In order to keep both mills running, Kemp van Ee activated a series of small glory holes in the hanging-wall mineralization, and at the same time increased stoping on the main vein. In a procedure similar to shrinkage stoping, the low-grade hanging-wall ore was dropped down glory holes and pulled out through chutes, then trammed to the shaft and hoisted to the crushing plant. Selective mining was abandoned in the push for more and more tonnage to keep the mills busy, but this further lowered ore grades by diluting good ore with waste from the glory holes. Even the better ore, developed from time to time in the main vein, was insufficient to keep the average millhead values profitable.[66] Installation of a chlorination plant in mid-1904 came too late to affect the final outcome.

Kemp van Ee had still more problems, even before the completion of the big mill. First was a lawsuit instigated by a former partner that dragged on for two years before it was dismissed. In the interim it tied up assets and strained the resources of the London investors. They eventually sent over an investigator who threatened to force out Kemp van Ee. Then came a strike late in 1903

Underground crew at the Royal, ca. 1902. (Courtesy Calaveras County Historical Society)

by members of the Independence Miners Union, organized in the district and encouraged by unionizing activities elsewhere along the Mother Lode. But Kemp van Ee, unlike F. F. Thomas and David McClure at the Gwin mine, had little patience for employee relations. The strike hit with brutal force that he returned in kind. Refusing to negotiate, he locked out the strikers, hired strikebreakers from San Francisco to keep the mine open, and won a court restraining order against the local union. In retaliation, the strikers attacked the scabs, threatened Kemp van Ee's life, and blew up the new underground pumping plant. In December federal marshals arrived to enforce the court injunction, and the mine reopened. A month later, facing criminal charges as a result of the violence, the organizers called off the strike and left the district.[67] But it was a Pyrrhic victory for Kemp van Ee. What little capital funds remained went into repairs for putting the mine back into full production. For a few months new ore discoveries kept both mills running, but the ore grade continued to drop.

The end came in November 1905, when Wilbur and Peters, the previous owners of the Royal, having received royalties only in 1898 and 1899, demanded the balance due and took the matter to court. That finished Kemp van Ee and his London backers. The creditors foreclosed, and the court appointed as receiver Daniel Jutton, the former mine superintendent. The mine was in such bad shape that he ordered it to close. Although it had produced more than fifty thousand ounces of gold during Kemp van Ee's regime, his English investors took a large loss.

The court eventually exonerated Kemp van Ee and awarded title to the London investors, but continuing legal problems kept the Royal mine closed for nearly a decade. In the interim its nearly new surface plant rusted away, and its lower workings filled with water. Herbert Hoover, back in the States in 1905 after spending eight years in Australia and China as a mining engineer and consultant, traveled to Hodson on behalf of a London friend and spent a day talking with Kemp van Ee. Later he sent his brother, Theodore, to look it over, suggesting they might "lease it or buy it or something." Evidently, "Tad" was not impressed, for nothing came of the venture.[68] Just before World War I another group of investors tried to reopen the mine and mill but shut down in two years, unable to make a profit. Not until the thirties did the mine return to profitable production, when Frank Tower established a small but highly successful operation, largely with block leasers, selectively mining only the remaining higher-grade parts of the main vein.

## The Sheep Ranch Mine

In Calaveras County, the Royal was the largest gold producer on the West Belt of the Mother Lode. Its counterpart on the East Belt was the Sheep Ranch, which yielded 350,000 ounces in seventy years of intermittent production. Much of this output came in the Hearst era. George Hearst, one of the West's most successful mining promoters and developers, had earned his first big money on the Comstock after learning the business as a lead miner in Missouri before coming to California in 1850. A financial setback in the 1860s— about the time his son William Randolph was born—drained him of investment capital, but he formed a partnership with two prominent developers to offset the loss. Lloyd Tevis and James Ben Ali Haggin, as Gertrude Atherton once wrote, "were clever, shrewd, ruthless men . . . and highly respected in San Francisco." The same might have been said of the elder Hearst, who relied on his partners to provide the capital while he contributed a remarkable ability to ferret out mines of high potential. In 1872 they purchased the Ontario mine in Utah, which by 1875 had developed into a major silver producer, yielding enormous profits to its owners. That same year the three associates acquired the Sheep Ranch—Hearst's first gold venture but certainly not his last, for two years later the trio obtained control of the fabulous Homestake mine

The Sheep Ranch mine in the 1890s. Note the steam venting out of the powerhouse at center, fueled by the piles of cordwood in the left foreground. (Courtesy Holt-Atherton Library, University of the Pacific)

in South Dakota. From that one property alone came 35 million ounces between 1878 and 1962, equivalent to nearly one-third of California's total gold production.[69]

Compared to the Homestake, the Sheep Ranch was a very modest mine, but while it was in Hearst's hands it performed well. Soon after purchasing the property in 1875 from A. P. "Cap" Ferguson and William A. Wallace, the Hearst interests bought up the adjoining Chavanne mine to control the entire vein. Their manager was W. H. Clary, a Calaveras old-timer, born in Kentucky, who had once owned the Quail Hill mine near Telegraph City. Deepening the mine to fourteen hundred feet, Clary uncovered three relatively narrow but high-grade ore shoots that kept a twenty-stamp mill busy for eighteen years. Emile Guidici, born at Sheep Ranch in 1905, remembered the barren hills for miles around that had been denuded by woodcutters to fuel the three steam boilers that powered the mill. When the local supply ran out, teamsters like Manuel Swmegar [sic] hauled cordwood by ox team from Indian Creek and other foothill regions to feed the boilers. Hearst's death in 1891 hastened the mine's closure, for the other partners by that time were busy expanding their more promising property, the Homestake.[70] The Sheep Ranch remained open for another two years, but declining ore prospects and

The Sheep Ranch mine during the Hearst years. George Hearst is the bearded man at left center, standing next to his manager, William Clary. (Courtesy Calaveras County Historical Society)

heavy rains that overloaded the pumps gave the owners an excuse to close rather than to put more money into the property.

Five years later Clary organized a group of San Francisco investors who financed the mine's rehabilitation. They modernized the surface plant and worked the property until 1907.[71] It was idle several years before another syndicate tried its luck during World War I, without much success. The mine evidently had yielded most of its good ore long before, but was to have at least one last burst of production in the late thirties.

## The Carson Hill Mines

The pocket hunting and small-scale haphazard mining that had characterized early Carson Hill history gave way in the 1880s to the first major corporate development. Two principal vein systems, the Calaveras vein and the Bull vein, formed a concave lens set on edge that extended from the south side of Carson Hill northwesterly to the town at Carson Flat. Outcrops along each vein had attracted early miners, although smaller veins and pockets in the ground between them contained the richest ore, including the 195-pound mass of native gold found in 1854.

An early sluice and amalgam operation on Carson Hill, apparently reworking tailings from a mill above. (Courtesy Segerstrom Family Collection, Holt-Atherton Library, University of the Pacific)

Dozens of claims dotted the hill, but the key to development was consolidation, especially along the main veins. Gabriel Stevenot had realized this at an early date as he strove to acquire additional claims. In 1875 he and other claim owners incorporated the Melones Consolidated Mining Company to put under common ownership many of the claims on the Bull vein zone, including the Stanislaus mine down on the river.

Gabriel Stevenot died in 1885, leaving his son, Emile, to carry on at Carson Hill. A year later Emile sold most of the remaining Stevenot claims on the Calaveras vein on the west side of the hill to a group of British and American investors that included John W. Mackay of Comstock fame—the third of the bonanza kings to be associated with Calaveras mining. Organized as the Calaveras Consolidated, they acquired other properties along the vein, drove an adit one thousand feet below the summit of Carson Hill, and collared just north of Robinson's Ferry. The tunnel headed northwest toward the Relief claim at the other end of the vein. In 1889, to improve ventilation, British manager Theodore Allen commenced sinking an "air shaft" some distance ahead of the tunnel face on the Santa Cruz claim. Not far from the surface his crew struck a large shoot of low-grade ore. The practical Allen at once changed plans. He discontinued work in the tunnel, started blocking out ore in the shaft workings, and decided to build a mill at the shaft collar rather than on the river. The development of this low-grade strike marked the beginnings of a new era at Carson Hill.[72]

Allen's careful plans soon went awry. In December 1889 the directors told shareholders that to process this low-grade ore, a mill would be built and ready for operation by the end of 1890. However, a year later stockholders were informed that the management had failed to raise the necessary additional capital for mine development, and that no progress could be reported. At the same time the directors proposed to raise funds through a bond issue. Thus, the downward financial spiral began long before the Calaveras Consolidated stockholders could reap the rewards of their investment. Although Allen managed to complete a twenty-stamp mill, it was in operation only a short time before the mine closed.[73] He left there a frustrated man, but stayed in Calaveras County as mine superintendent of the Utica at Angels Camp.

In the meantime another part of Carson Hill was undergoing significant organizational changes. Eight years after they had incorporated the Melones Consolidated, Gabriel Stevenot and his fellow stockholders sold out the controlling interest in the Reserve mine on the Bull vein zone. The purchasers were George W. Grayson and Archibald Borland, San Francisco financiers who had also become major stockholders in the Union Water Company. They hired Percy S. Buckminster to manage the Reserve operations. The old Reserve mill was rebuilt and expanded to forty stamps. However, the ore was of such low grade, with substantial amounts of the gold values in the refractory sulfide or "gray ore" type, that in a little over two years the uneconomical operation closed down. With the sulfide mineralization also a serious problem in the Angels district, Grayson and Borland commissioned Buckminster to construct a large chlorinator at Angels Camp. This successful plant was soon taken over by the Utica Company, as noted earlier, and continued in operation until the Sierra Railway completed its tracks to town in 1902.

Although their Melones-Reserve operation had failed, Grayson and Borland still held the consolidated properties that included every important claim on the Bull vein zone southeast of the Morgan, Melones, and Reserve mines except the South Carolina, whose complex ore had frustrated earlier milling attempts. Their disappointing venture turned into a new opportunity under Grayson's son-in-law, William C. Ralston Jr. Son of the flamboyant San Francisco banker and Comstock magnate, young Ralston was one of the first trained engineers at a major Calaveras lode mine. A few years after graduat-

ing from the University of California in 1887, he took control of his father-
in-law's Carson Hill properties. Following in the elder Ralston's footsteps,
he raised capital by using his name and connections to float a nine hundred
thousand–dollar bond issue in Boston. In 1897 his backers incorporated the
Melones Mining Company in West Virginia, elected William Ellery Chan-
ning Eustis of Boston as president, and turned the mine over to Ralston as
general manager. But political aspirations and a job as appraiser for the Port of
San Francisco occupied much of Ralston's time in the 1890s. To guide Carson
Hill development he hired a brilliant New York consulting engineer, Walter B.
Devereux. A graduate of the Columbia School of Mines, Devereux had spent
twenty vigorous years in the mining business as an engineer, manager, and
owner before poor health forced a partial retirement in the 1890s. He had
moved to New York, set up a consulting business, and built a reputation for
engineering successful methods to mine and treat low-grade ore—the kind
that Ralston faced at Carson Hill.[74]

In the late 1890s, Melones miners had sunk a shaft on the Reserve claim,
which Ralston had held under bond from Grayson's Melones Consolidated
Mining Company. This work uncovered a large low-grade deposit in the footwall of the Bull vein. To mine it the Melones Company took an option on the
South Carolina mine and reactivated its old tunnel some 450 feet below the
bottom of the Reserve shaft. By 1898 it had been extended into the Reserve-
Melones claims, but it was abandoned after the South Carolina interests refused to renew the lease and option. The investment was not entirely wasted,
however, for the advance had confirmed the Reserve's footwall ore body.[75]

Ralston next turned to Devereux for advice. The New Yorker engineered
a masterful plan, which the Melones syndicate adopted, to connect all the
company claims by driving a long adit at the 1,100-foot level that would start
near the river on the Stanislaus claim and extend north between the Calaveras and Bull veins. Collared in 1899 and completed in 1902, this 5,000-foot
tunnel became the main artery of traffic between the footwall ore body of the
Bull vein and the new sixty-stamp mill that was built just east of the portal
near the town of Melones. The mill opened in 1902, and three years later it
was increased to one hundred stamps. A dam on the Stanislaus River provided water for hydroelectric power to operate the mill, the surface plant, and
the underground facilities, including an electric trolley in the main haulage
tunnel.[76]

Devereux left before he saw his handiwork completed, but Ralston knew
how to pick good men. In 1901, the same year he was elected to the state assembly, he briefly enticed William Loring to Carson Hill as superintendent,
but the next year Loring was lured away to Australia, where a British firm

hired him to manage a major group of gold mines. Frank Langford later ran the Melones, and took over general management after Ralston himself lost favor with his Boston backers in 1904.[77] Still later, Devereux's son, W. G. Devereux, ran the mine until it shut down in 1919.

To keep operating costs as low as possible, Ralston's miners glory-holed the footwall ore by means of vertical ore passes connecting with the 1,100-level haulage tunnel. Waste from the glory hole reduced the values of the already low–grade ore, just as had happened at the Royal, but the large and efficient concentrator at the new town of Melones recovered a very high percentage of the gold values. A cyanide plant was erected sometime after 1902 to process the concentrates. The well-run mining operations and low-cost tunnel haulage, plus the milling efficiency, kept operating costs at a remarkable $1.08 per ton before 1910, the lowest in the Southern Mines, although not as low as the incredible 50 cents per ton achieved in the early 1890s at the Dalmation mine in El Dorado County, the first California mine to use electricity as a major power source.[78]

Under Ralston's regime the Melones Consolidated Mining Company also discovered the lower end of an important new ore body in the hanging wall of the Bull vein. Near the end of the haulage tunnel, the company collared and sank a winze. At the 1,350 level miners drove a crosscut through the Bull vein to the hanging wall, locating an ore shoot that they followed down rake to 3,000 feet. Mined by shrinkage stopes the undiluted hanging-wall ore was higher in grade than the footwall ore and kept the mill busy in the winter when the glory hole was too wet to mine.[79]

In drifting toward the hanging wall the Melones engineers crossed into the idle Morgan property, evidently unclear as to who controlled the apex of this ore body. Had the Morgan owners protested a legal challenge might have stalled any further operations. However, as we have seen earlier, the Morgan had been in dispute for years as a result of the suit between James G. Fair and his former manager, William Irvine, who also claimed ownership. Eventually, the bonanza king won clear title but died before he could make use of it. Irvine had also claimed the entire 143 acres of the Carson Hill town site, using a homestead entry. Local residents fought eviction for years. In 1909 the Morgan Mining Company bought out Irvine's claim for $500 and ended the dispute. By that time Fair's heirs had hired a British manager to reopen the Morgan, but he could not turn a profit. The mine soon closed again and was not reworked until the Loring interests took over during World War I. In the meantime the Melones Company had mined out a large segment of Morgan ore.[80]

The Melones mine was Carson Hill's main producer for nearly a quarter century. By the time it closed in 1919, the mine had yielded 265,000 ounces of gold. Ralston, as noted above, was not in control at the close, for he fell victim to financial pressure just as had Kemp van Ee at the Royal and for nearly the same reasons. Stockholders waited in vain for dividends that never came despite the high production figures that in the early years, at least, had more than paid the costs of operation. But as the glory hole on the footwall widened, ore values diluted to the point that mill revenue was insufficient to cover production costs, low as they were. Bondholders then forced Ralston's resignation but had little better luck under Frank Langford or W. G. Devereux, his successors.

The Melones Consolidated Mining Company had been the first at Carson Hill to master three essentials of modern mining: technology, finance, and management. Despite this achievement, measured by traditional investment standards the Melones Company was a financial failure because it never earned sustained profits for its stockholders.

## The Copper Mines

Lode mining in the West Belt copper districts underwent a modest revival in the 1880s, after the disastrous fall in copper prices twenty years before. The Copperopolis mines had been shut down since 1867, except for a very brief period of production in the early seventies.[81] During this period of inactivity Frederick Ames from Massachusetts had acquired control of the Union and Keystone mines, operating them as the Union Copper Company with corporate offices in Boston. In the late 1880s the company constructed an improved smelter and reopened the mines. However, due to difficult market conditions, as well as problems in treating the low-grade sulfide ores encountered in the deeper levels, the company shut down again in 1892.

Ten years later, under the management of G. McMillan Ross, or "MacRoss" as he was often called, the mines were dewatered and copper production resumed. Ross put a new concentrator on line in 1905, and for the next two years the mine produced well. Within a year after the smelter had been rebuilt and reactivated in 1908, the mine was sold to Calaveras Copper, a new operating company, who turned over its management to Samuel M. Levy. He added a flotation circuit to the concentrator, discontinued the on-site smelter, and shipped the concentrates to a smelter in Tacoma, Washington. During World War I, with the rise of copper prices, the local smelter was reactivated, and the Union and Keystone mines entered the most productive period in their history.[82]

Technology advanced mining and milling in Calaveras during the prewar

The Union-Keystone concentrator one mile south of Copperopolis. An electric tram delivered ore from the mine over the high trestle; the concentrates were then fed to the adjacent smelter. (Courtesy Holt-Atherton Library, University of the Pacific)

years, but life changed little for the average miner. Edward Lee Riggs provided a revealing vignette of his years as foreman at Calaveras Copper from 1915 to 1921. He lived with his wife and three children in a concrete company house with inside plumbing and electricity supplied from the company generator. Power was so irregular, however, that only three of the twenty-five-watt bulbs that served each room could be used at any given time. Sometimes even those gave off no more than a dim red glow. Water was piped in from a nearby creek, but no one dared drink it. The main supply came from the sump at the bottom of the mine. The Riggs family used it to wash dishes and clothes on days when it was reasonably clear, and to flush the toilet. Wastewater from the house drained back into the creek. To supplement the food supply, Riggs raised chickens and ducks, and hunted squirrels, rabbits, and other small game animals. Store-bought goods, as well as mining supplies, arrived by wagon from the Milton railhead twelve miles away. To navigate the muddy roads in winter required a string of eight or ten horses per wagon. Passengers could ride to Milton on a horse-drawn stage until 1918, when it was replaced by an old Ford touring car remodeled into a "jitney" by cutting away the body and adding a platform with bench seats. Thereafter, horses were used mostly for tow service during the rainy season.[83]

Mining at Campo Seco paralleled developments at Copperopolis in this

The surface plant at the Union shaft at Copperopolis as it appeared ca. 1903. (Courtesy Holt-Atherton Library, University of the Pacific)

period. Although copper had been opened up here within a year after the Copperopolis strike, production during the Civil War boom from three small mines (Campo Seco, Lancha Plana, and Copper Hill) was only a fraction of that at Copperopolis. A small smelter was in operation for a short time in the late sixties. Apparently, there was little activity in the Campo Seco copper district from 1868 to 1883, when H. D. Ranlett reopened the Lancha Plana, renaming it the Satellite mine, selling it three years later to the San Francisco Copper Company. Ranlett then moved on to Copperopolis to reopen the Union-Keystone-Empire group for Frederick Ames. About 1887 the Penn mine reopened under the proprietorship of A. C. Harmon, who acquired it and its parent, the Penn Chemical Works, from Christian Borger. Shortly afterward Harmon purchased the Satellite and adjoining claims and operated the entire property as the Penn mine. After a few years of very expensive and discouraging explorations, mining was temporarily discontinued. In 1898 the mine again reopened, and the Penn Chemical Works built a smelter that was "blown in" late in 1899. Thence ensued a twenty-year period of profitable copper production until the mine's closure in the post–World War I slump. About 1910, Harmon sold out to Loughride and McIlvane, who reincorporated as the Penn Mining Company. They made a number of improvements

Distant view of the smelter at Campo Seco during its heyday, ca. 1906. Note the un-scrubbed sulphur dioxide fumes venting in the open air and the proximity of the Moke-lumne River. The mine was successful, but later generations had to confront the toxic legacy. (Courtesy Calaveras County Historical Society)

that improved production and reduced costs. As a result, the Penn mine out-ran the Union-Keystone to become the most important copper mine by 1919, not only in Calaveras County, but also all along the West Belt. As will be seen in the next chapter, between the wars there were a few short-lived attempts to reestablish the operation, with a final boom at the Penn mine coming in World War II.

* * *

Improved technology and corporate financing, modernization of mining and milling equipment, consolidation of significant properties, better manage-ment, and systematic production—these were the key elements that acceler-ated the development of Calaveras lode mining between the 1880s and World War I. Mining in the golden years at the turn of the century was not only the county's principal industry, but also its raison d'être. It energized the eco-nomic life of the county's towns and villages, stimulated trade and travel, provided employment and support for most of the county's population, and reinforced a cultural milieu that emphasized hard work and material growth.

But even though mining culture in the golden years still championed the

Tapping floor of the Penn smelter, showing pots for drawing off molten matte and slag from the eighty-ton blast furnace, 1900. (Courtesy Calaveras County Historical Society)

values of rugged individualism and personal achievement, mining was no longer a pioneer business open to anyone. The formative years of individual or small-company operations had given way to modern industrial mining, where individual miners were not entrepreneurs but hired hands working in shifts for a weekly paycheck.

In this period Calaveras led all other counties of the southern Mother Lode in modernizing the mining business. As W. H. Clary said in 1894, mining in Calaveras County was a corporate enterprise with little room for the old-time prospector who would be better off "in a pocket country such as Tuolumne or Mariposa counties." The remark was overtly provincial and biased, since Calaveras had had more than its share of little mines and pocket hunters, some of whom did very well indeed. Even as late as 1908, Vic Lagomarsino and Dave Queirolo of Angels Camp came to town with a load of quartz averaging one hundred dollars a pound from the Gobbi Ranch near Fosteria, which "has

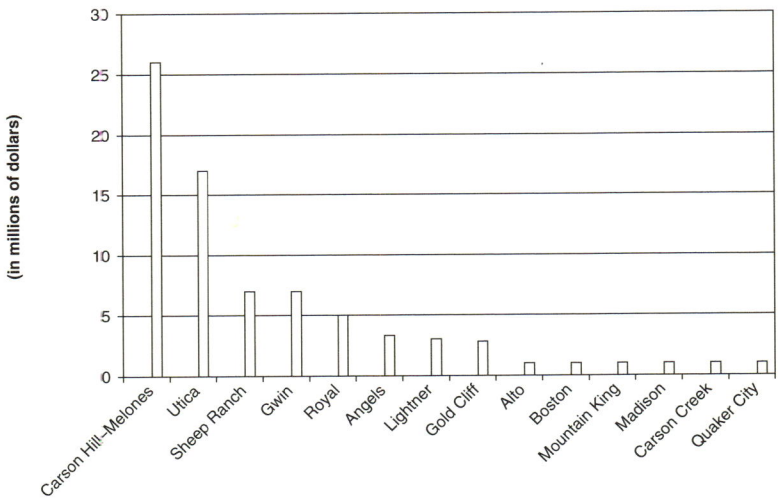

Calaveras Lode Gold Mine Production, 1848–1968. (Authors' database)

always been regarded as a pocket section of the country," said the newsman who reported the story.[84] Yet Clary's point was clear: Old-fashioned methods could no longer sustain a modern industry.

Even the smaller mines recognized the need to modernize after the 1880s, but raising capital was difficult for small and marginal operators. Some simply mined only the high-grade ore for quick cash, as the Tanner mine operators did at Murphys between 1908 and World War I. Others were more sensible, trying to find technological alternatives that could prolong production and still save money. In 1896 at the G.A.R. mine in Angels Camp, for example, W. G. Drown, to remedy a power shortage, installed a gasoline engine to run a new rock crusher and other surface equipment. A few years later at the Alto mine near Copperopolis, Tommy Lane, brother of the Utica owner, expanded operations by glory-hole mining to feed his new forty-stamp mill powered by electricity from the new Tulloch dam and power plant at Knights Ferry on the Stanislaus. Despite a low ore grade, Lane kept operating costs below fifty cents a ton.[85]

Through these and other efforts by progressive mine owners, large and small, Calaveras mining came of age by World War I. The Utica and Carson Hill–Melones mines emerged as the two largest gold producers, with the Gwin and Sheep Ranch not far behind. As the comparative graph in Table 6.1 indicates, between 1893 and 1920, the high point in their history, Calaveras mines produced nearly 2.5 million ounces of gold, more than 10 percent of

TABLE 6.1    California vs. Calaveras Gold Production, 1848–1995 (in millions of ounces)

| Period | Calaveras | California | Percent County to State Total |
|--------|-----------|------------|-------------------------------|
| Before 1880 | 4.24 | 49.67 | 8.5 |
| 1880–1892 | .047 | 9.52 | 4.9 |
| 1893–1920 | 2.43 | 23.98 | 10.1 |
| 1921–1933 | .044 | 7.62 | 5.8 |
| 1934–1958 | 0.7 | 16.26 | 4.3 |
| 1958–1995 | 0.45 | 8.99 | 5.0 |

California's total output. The Penn and the Union-Keystone copper mines rivaled the Utica and Carson Hill operations in employment, if not in value of metals produced. In a deflationary era, with operating costs and wages low, the largest Calaveras mines earned significant profits for their owners.

The combined impact of these modern mining operations gave an enormous lift to the regional economy, but at high cost to employees and the environment. Modernization in this era did little to improve the health and safety of underground crews. Better working conditions remained a high priority to workers' families and union leaders, but they made little headway against entrenched forces of resistance in the conservative districts of the Mother Lode and elsewhere in the mining West. Throughout the nineteenth century and well into the twentieth, the mine fatality rate in the United States remained higher than in Europe, where mine regulatory legislation had been imposed much earlier. In the West state-sponsored mining safety laws were adopted as early as 1896 in Utah. Other states passed similar legislation and established state mine inspectors to implement the new rules, but most of these early efforts were piecemeal and lacked rigorous implementation. Not until the U. S. Bureau of Mines was established in 1910 did the federal government begin a systematic campaign to improve mine safety, and not until after World War I did "safety first" become a ubiquitous and mandatory corporate policy underground.[86] Environmental concerns came much later, primarily after the 1960s, when state and federal regulators began to call attention to heavy metal residues and acid mine drainage problems on the Mokelumne and other regional waterways.

Despite its emphasis on tools and technology rather than on personnel relations and safety, modernization was one important consequence of industrial mining in Calaveras and other Mother Lode districts in the years before World War I. Yet modernization came too late for many mining companies. The impetus for continued growth did not last for two main reasons: the steady depletion of Mother Lode ore reserves, and the effects of war and its economic aftermath. Both eroded the incentives that for more than three decades had energized the regional mining industry.

# 7: The Unstable Twenties and Thirties

In the half century after World War I, Calaveras County witnessed economic changes that radically altered its prewar status. Before the war the mining industry dominated the county economy and provided the major impetus to urban growth and development. In 1912, for example, Angels Camp's five thousand inhabitants, most of them directly or indirectly tied to the mining industry, welcomed the incorporation of their city as a sign of progress and prosperity. That optimism faded as gold mining slumped during World War I and continued to fade through the twenties, despite a few noteworthy exceptions. Rising costs with a fixed gold price, depletion of known ore bodies and fewer discoveries of new ones, dried-up sources of capital, and labor problems all took their toll.

Some believed the traditional conservatism of Mother Lode operators also contributed to the economic malaise. Writing in 1921, Rau Roesler, a mining engineer, blamed the problem on obsolete, old-fashioned mining and milling methods that limited productivity and drove away investors. He was convinced that outside capital and systematic mass-mining and -milling of low-grade ore bodies could keep costs well below the profit margin and restore the Mother Lode to "the head of the gold-producing areas of the world."[1]

Others pointed to the credit crunch and postwar inflation that weighed down mining companies already saddled with heavy indebtedness left over from the war years. W. J. Loring, in lobbying for a government bailout in the early twenties, claimed that spiraling operating costs were ruining the industry, forcing gold mines into selective mining using the shrinkage stope method. This, he argued, ruined the mine for future underground operations and ended any chance of economically recovering low-grade ore that might otherwise be worked if gold prices rose.[2] The ultimate solution, of course—an idea close to the heart of gold miners and classical economists alike—was to end government regulation of gold as a monetary unit and let the yellow metal float on the open market.

Despite disagreement over causes and solutions, the gold mining industry suffered in the immediate postwar years. By the early 1920s all the major gold mines in the county had closed except for Carson Hill. The only bright spot in extractive industries during the decade was the opening of the Calaveras Cement plant in 1926. Even in its early stages, the cement plant was a godsend to a county economy already in decline before the national collapse in the early thirties. Because of its major impact during and after World War II, the cement industry in Calaveras will be discussed in the last chapter of this narrative.

During the Depression, however, precious-metals mining bounced back. Calaveras gold production peaked at 123,500 ounces in 1939, though state-wide production continued to grow until 1941. America's entry into World War II brought an end to this new bonanza period for gold miners, but the base-metals industry enjoyed a vigorous, if brief, revival as part of the nation's strategic resource development. Overall, the twenties and thirties were transitional decades in the Mother Lode, with unpredictable external forces adding to the normal uncertainties associated with mining.

## Precious-Metals Mining in the Twenties

Although the twenties were years of prosperity to many businesses nationally, most of the communities along the Mother Lode suffered sharp declines in both business and population. Calaveras County's population dipped to its lowest since the Gold Rush—a direct consequence of the mining slump attributed largely to the depletion of profitable ore reserves confronting a fixed gold price.[3] In Angels Camp, for example, the Utica Cross shaft and winze by World War I had been sunk to the 3,050-foot level, but relatively little ore had been found and mined below the big ore bodies in the upper levels. All the major mines in the Angels district had shut down by the end of World War I except the Gold Cliff. It survived until 1920.

One indication of the extent of the decline was the changing status of the old Altaville Iron Works, which had been the county's chief builder and installer of mining machinery since the 1850s. In 1928, reforming as the California Electric Steel Company, it ceased production of mining equipment and converted entirely to the manufacture of steel castings for industrial use.[4]

As miners left to find jobs wherever they were available, mostly outside the county, some mine owners sold their used equipment and surface plants to scrap dealers. Others made a valiant but mostly futile effort to reopen and mine successfully. For instance, Frank Tower attempted to restart the Royal mine several times in the twenties but without significant development or production. In Murphys the Bullion Mining Company had a brief run. It leased the Washington mine, erected a five-stamp mill in 1921, soon increased it to ten stamps, but within two years shut down, in part due to a costly lawsuit.[5] All across the county similar cases attested to the economic slump that affected Calaveras and other mining regions after World War I.

## The Boom and Bust at Carson Hill

The only major gold producer in the twenties was the Carson Hill Gold Mines, Inc., under the management of William J. Loring and his superintendent, Archie Stevenot. Loring was one of the most important figures in Calaveras mining history. A self-taught mining engineer and developer, he attributed his success partly to hard work and experience, and partly to "his willingness to take a chance."[6]

Untrained American engineers in Loring's day were the exception rather than the rule, especially in the major mines. The term *mining engineer* is here used in a broad sense to indicate anyone engaged in mining engineering regardless of education. Before the 1890s trained engineers were rare, but by 1915 they dominated the industry. Most were of British ancestry like their financial brethren. Like them also, they were from middle-class backgrounds, beginning their careers at the operations level and gradually working upward. By 1900, as mining became more specialized and diversified, trained engineers moved rapidly up the corporate structure to assume executive positions, even to become company presidents. Many covered the globe as consultants, directors, or managers. Their most important contribution was to modernize the mining industry through new technology. "With the application of technology," explained one specialist on the subject, "mining became a modern industry, with increased production per unit of capital and manpower." The point is embodied in a mining engineer's slogan: "increased tonnage at decreased costs."[7]

Except for his lack of formal education, Loring typified the men of his profession. He was a native Californian of middle-class Anglo-American parentage. His father had emigrated from Illinois during the Gold Rush, eventually settling in Amador County. There Loring grew up, attended the local schools, and began his mining career as a handyman at the East Keystone for fifty cents a day. Hired by Alvinza Hayward at the Plymouth mine when Loring was fourteen, he learned every aspect of the miner's trade as he worked his way up from mucker to oiler to mill man to assistant amalgamator. In 1888, after the Plymouth mill burned, Hayward transferred him to the Utica mill at Angels Camp. There he remained for the next thirteen years, working under Charles D. Lane and his brother Thomas, rising in the organization, and making a name for himself as a creative technician with a sound business mind. As superintendent of the Utica mills in the bonanza 1890s, he improved stamp-milling techniques that increased the capacity of the Utica's 160 stamps by nearly one-third. Loring's published account of his Utica operations is an excellent technical description of gold milling procedures in that day.[8]

Carson Hill's postwar boomlet attracted important figures in the regional economy. *Left to right:* David Clarence Demarest, William J. Loring, Lawrence Monte Verda, and Percy Wood of Robinson's Ferry. Behind them is the mine's main adit. (Courtesy David Clarence Demarest Collection, Holt-Atherton Library, University of the Pacific)

In 1901, W. C. Ralston hired Loring as superintendent of the Melones, where he started construction of the 60-stamp mill and water delivery system that powered it. Early in 1902 he caught the eye of Theodore J. Hoover, later a dean at Stanford, whose brother Herbert Hoover was a partner in the British-based Bewick Moreing Company (BMC), a consulting, investing, and management firm. This was the organization that launched Herbert on his successful mining career, and it would do the same for Loring. After some negotiation Loring accepted a $9,000 per year offer to become superintendent of the Sons of Gwalia mine in western Australia, a BMC enterprise. For the next seven years he revamped BMC mines and earned the respect of both miners and managers in that arid country. By 1908 he had charge of sixteen different mines and had saved enough money to buy out Hoover's partnership in the parent organization.[9]

As a full partner and general manager of all BMC mining operations around the world, Loring traveled widely between 1908 and 1913. He covered every mining continent and every major district, investigating silver mines in Burma, exploring gold mines in northern Ontario, and developing new tin mines in Cornwall. His work took him to New Zealand, Nicaragua, West

Africa, the Far East—to every corner of this far-flung British mining empire. An international figure and a distinguished engineer by the early twenties, Loring was honored by the American Mining Congress when they elected him president for two years in a row.[10]

While he was still with Bewick Moreing, Loring returned to California in 1911. Pumping British capital into the Mother Lode, he reopened the Plymouth Consolidated, Alvinza Hayward's old mine in Amador County that had shut down when the main ore body pinched out at sixteen hundred feet. Upon Loring's recommendation, BMC picked up the option on the Plymouth despite the prevailing London opinion that it was "not good policy" to invest in a mine that had only 110,000 tons of low-grade ore in sight after extensive investigation. Loring, however, had studied other Mother Lode mines and had seen a pattern: In "mine after mine," he later wrote, mine developers had found "better ground by deeper sinking." By 1920 the Plymouth was at thirty-one hundred feet and had generated more than $1 million in profits despite wartime disruptions and rising costs. Eventually, gold production totaled 470,000 ounces below Hayward's old workings.[11]

Early in 1914, Loring left London on a business trip and was in the Far East when World War I erupted. Rather than risk a European crossing he returned to the United States and remained in California to investigate other Mother Lode properties for his British firm. Sometime during the course of the war he took leave of BMC to strike out on his own—perhaps because the war cut off British capital flowing to the United States.[12] At any rate because of his experience and reputation his services were in demand regionally, and soon he was deeply involved in mining enterprises both in Nevada and in California.

His first major step as an independent entrepreneur was to take an option on the Dutch-Sweeney complex in Tuolumne County, a property under the control of the Segerstrom family of Sonora. Late in 1914, W. G. Devereux, son of the consultant who had engineered the master plan for the Melones development more than a decade before, had optioned the same mine and had worked it a year, but failed to raise the $250,000 purchase price. Loring in the fall of 1915 saw his chance and drove a hard bargain. Putting up $100,000 of his own money, he secured a ten-month option for $50,000 less than the price to Devereux the year before. His negotiations brought him into contact with Charles H. Segerstrom, the family spokesman. A well-known regional banker and investor, Segerstrom was an important player in Calaveras mining history, as well as in many different western mining districts from California to Utah. With Segerstrom's approval Loring took control of the Dutch-Sweeney. He assumed the role of general manager and promised big returns,

Archie Stevenot, "Mr. Mother Lode," an active member of E. Clampus Vitus, was honored at the dedication of the ore car on Carson Hill in 1961. The bridge over the Stanislaus River at Melones that bears his name was dedicated posthumously in 1976. (Courtesy Richard Coke Wood Collection, Holt-Atherton Library, University of the Pacific)

but development stalled, in part due to his inability to raise additional capital for modernization and exploration, and in part because underground work did not find enough good ore. In 1916 he was granted an extension on his option, but America's entry into World War I foiled his Tuolumne County plans, and soon afterward bigger developments just north of the Stanislaus River occupied much of his time and attention.[13]

Loring was no stranger to Carson Hill. He had worked at the hill as the Melones superintendent in 1901–1902, and he was well acquainted with the managers and mill men. He also recognized the opportunity presented right after the war. Most of the mines appeared to have good potential, but were shut down or stalled for a variety of reasons. What Carson Hill needed most was adequate capital and better organization, and Loring thought he could provide both. At the very time he was negotiating to operate the Dutch-Sweeney in neighboring Tuolumne County, Loring made a bold move at Carson Hill. The Calaveras Consolidated seemed the most likely immediate prospect, and to secure it Loring, in 1916, turned to a Boston friend, Edward A. Clark, president of the American Zinc, Lead, and Smelting Company. With a substantial loan from American Zinc to help capitalize his newly organized Carson Hill Gold Mines, Loring secured an option on the Calaveras Consolidated properties. He had his eye on the large low-grade ore body that Theodore Allen had located and blocked out more than twenty years before, in the Santa Cruz claim. Instead of using Allen's shaft and mill site at Carson Hill, Loring reasoned that the way to mine this ore successfully was to build a new, efficient mill at the portal of the Calaveras tunnel, located near the Stanislaus River in the town of Melones. With his hands full overseeing the Plymouth mine in Amador and the Dutch-Sweeney in Tuolumne, Loring hired Archie Stevenot to manage the Calaveras Consolidated. This Calaveras native son, scion of Gabriel Stevenot, Carson Hill's early mine developer, wasted little time. Soon he had built a ten-stamp mill, rehabilitated the Calaveras tunnel, and begun advancing into the Santa Cruz claim to get under the Allen ore body.

Up and down the Mother Lode excitement mounted as the mining fraternity anticipated big developments under Loring's direction. They were not disappointed. Loring brought a refreshing concept to Carson Hill: Unite the entire district under one management. That proved to be the key to successful exploitation of the Carson Hill deposits after sixty years or more of fragmentary and desultory efforts. Ralston, with Devereux's help, had gone part of the way twenty years before, but after Ralston's departure the Melones Consolidated Mining Company had lacked the vision and the resources to work Carson Hill successfully. Loring had both, and proceeded swiftly to carry out his plans to consolidate the district.

Luck also played a role in Loring's success. In 1918, obtaining control of the old Morgan mine with a lease and bond, he suddenly changed priorities. Development work on the Calaveras Consolidated stopped temporarily, and Stevenot's crew shifted to the Morgan. Exploring the three hundred–foot level, Stevenot ran into high-grade "gray ore" in the hanging wall of the Bull vein. This proved to be the unworked upper part of the hanging-wall ore body that the Melones Company had mined far below. Subsequently locating the apex by trenching the surface and thus confirming their ownership, Loring and Stevenot arranged with the Sierra Railway to ship Morgan ore around the hill and down to the Calaveras mill. Soon the proceeds from this good ore generated the cash flow needed to expand the mill to thirty stamps.

In 1919, after floating a new bond issue in Boston to pay off American Zinc's loan of three years before, Loring and his backers reorganized with a new name, the Carson Hill Gold Mining Company, with Boston as the corporate headquarters for the expanding operations. Another plum fell into his basket early in 1920, when the Melones Consolidated Mining Company, which had stopped underground operations in the fall of 1918, agreed to an option that gave Loring's company complete control of Melones property, including use of the eleven hundred–level haulage tunnel. Stevenot soon connected the upper Morgan workings with ore passes to the eleven hundred level, making it easy to mine out the rest of the hanging-wall ore body from the Melones mine, and to tram the ore to the Calaveras mill just west of the main portal. Production increased rapidly, reaching nearly 13,000 tons per month by 1921. Following the hanging-wall shoot eventually down to four thousand feet, the Loring interests extracted 412,000 ounces of gold from 1 million tons of ore.[14]

As president and general manager of the consolidated mining operation, Loring by 1921 was the undisputed king of Carson Hill, but his kingdom was built on a financial bog. Relying on an installment smorgasbord of options, leases, and other purchase agreements, he had consolidated all the major mines. In spite of substantial cash flow from mining the high-grade hanging-wall ore body and other ore shoots that had been opened up, there was still not enough cash flow available for Loring's ambitious plans. He kept afloat by using up all his mill returns, supplemented by loans from the Segerstroms. For all of Loring's achievements, the mine paid no dividends under his management. Lease payments, option installments, and the rising cost of daily operations due to postwar inflation ate up most of the revenue. It was the same old story: Undercapitalized for the amount of work needed to mine at deeper levels, Loring robbed Peter to pay Paul. He soon exhausted his credit and used up all his liquid assets just to pay the most pressing bills, with no money left for contingencies and not enough ore in sight to keep going much longer without new exploration. His objective was to "open up the lower levels," which he thought would uncover significant new ore bodies like those discovered below five thousand feet in Amador County mines. Without deep exploration, Loring well knew what would happen next. As he exclaimed to Clark, in reply to criticism from an unhappy shareholder who wanted to cut expenses by stopping any more development,

> [Doesn't the investor] . . . quite realize that this mine is producing 15,000 tons monthly, and that it is our desire to increase this tonnage if we can, and that 15,000 tons is a devilish big pile of ore, and it requires development work to produce ore at this rate? Does he not know that development work is the soul and life of a business of this kind? . . . The idea of curtailing development work is suicide, absolute suicide, in a few months.[15]

The first crisis came in the summer of 1921, at the height of his fame. While publicly accepting the plaudits of his peers as president of the American Mining Congress, privately he was pleading with his Boston backers for more time and more money. With local creditors clamoring for payment on long-outstanding accounts, the last straw would be to default on the payroll: "Should labor trouble overtake us," he wrote, "we would be simply put out of action because of no cash and no real credit." The only local creditor that seemed to understand the situation was Charles H. Segerstrom, a "simply wonderful friend . . . [who] is carrying certain matters that no one else on earth would think of doing." Why Segerstrom was so considerate is not clear. Perhaps he appreciated Loring's managerial skills and his engineering abilities. Perhaps he was impressed by Loring's achievements. Perhaps he wanted Loring's help in managing other mining properties. There were plenty of opportunities to reciprocate, and Segerstrom's papers show that Loring provided substantial assistance to Segerstrom properties in Nevada and elsewhere. Whatever the reason Segerstrom remained a good friend and an understanding investor even after Loring's fall from grace at Carson Hill.[16]

Loring's last major achievement before the fall was to absorb the few significant properties that remained on Carson Hill. In 1922 he acquired the South Carolina in a deal Frank Solinsky worked out with the Bank of Sonora, which held a judgment against the mine for defaulting on debt payments. That same year he picked up full title to the Melones. Since 1919 it had been under lease to Loring's company, but in 1922 he exercised the purchase option and floated a $600,000 five-year bond issue to pay for it. Justified to the stockholders as a means of developing new ore below the haulage level and at the same time reducing annual operating expenses by eliminating expensive lease and royalty payments, Loring's decision brought Carson Hill under one ownership but at considerable increase in the bonded indebtedness.[17]

After the death of Edward Clark in 1922, Loring had few friends left in Boston. The directors called him to account in 1924, setting strict conditions for continued operations under his management. Faced with the redemption of $450,000 in short-term gold bonds issued in 1919, on April 8 the remaining Boston directors sent him a "protocol" demanding, among other things, that processed ore value must be raised to a minimum of $6.00 per ton. Stevenot was forced out as superintendent, and his successor, R. C. Eisenhauer, was told in no uncertain terms that his job depended on raising mill revenue and cutting costs. One Boston board member said that when the Carson mine boss had "succeeded in raising the grade of the mill heads and in securing greater efficiency in development and operations generally, your next aim should be to work for greater economy by cutting down expenses, ruthlessly, in every direction, 'without fear or favor,' whenever and wherever possible." Eisenhauer, following orders, stopped all further development, bandaged together old equipment, and managed to raise tonnage value for a time by processing the best ore in sight. Loring chafed under the stipulations and wrote long letters to Boston trying to explain why his program for systematic development was sound, and why the high-grading and cost-cutting were shortsighted. But it was a waste of words: The investors insisted on greater efficiency and higher revenues. Loring stayed on for a few more months, and then resigned in disgust. Segerstrom, still friendly, tried to get him to remain as president of the Pacific Tungsten Company, a Nevada firm Loring once had heavily invested in and managed. But Loring had turned all his tungsten mine shares over to Segerstrom as collateral on several loans, and he wanted no more to do with either that "phantom dream" or Carson Hill. Titles made him a "target," he told the Sonora banker, and for a change he decided to "let somebody else have stones thrown at them, rather than myself."[18]

Loring's later career never matched his earlier triumphs. For a time in the late twenties he was in Arizona, trying to revive lost fortunes by taking charge of a copper prospect, but his luck had run out. As he explained to Segerstrom in seeking another loan extension, "It has been a very hard struggle indeed, and I been at this camp alone with my crew for weeks at a time without going to the outside, with absolutely no relief at all, and it gets very trying at times to continue the life. . . . I need money so badly that I nearly suffer at times."[19]

Times got worse in the thirties for Loring, just when gold mining began to revive. One venture after another failed. His biggest fiasco was at Virginia City, Nevada, where an Arizona syndicate under his management leased several historic properties on the Comstock, hoping by open pitting to process low-grade deposits missed earlier. The ore values were too low for profitability, however, and in the meantime his syndicate ran into trouble with the newly organized Securities and Exchange Commission over some dubious stock transactions. Discredited and heavily in debt, he stored his personal things in a Reno warehouse and went on the road as an itinerant mining consultant. In the late thirties he eked out a minimal existence, with his second wife helping out by working menial jobs in San Francisco. During World War II, however, restrictions on "non-essential" mining hurt many freelance operators like Loring, who fell even further into debt. But he struggled on for a few more years. Declining health in his last few years added to his troubles, as he described to Archie Stevenot in 1952: "I have been very unwell for the past ten months, having overworked myself in climbing mountains here in Nevada, the result being that my heart gave out and . . . I have been under the care of my doctor ever since." Considering he was then eighty-three years old, his illness was not surprising. He died that October in Tonopah of a heart attack. His wife in San Francisco had to borrow the forty dollars for bus fare to see him interred.[20]

Loring's departure from Carson Hill in 1924 provided only temporary reprieve for the Boston investors who wanted more efficiency, better ore grade, and lower costs. For a time the new management resorted to "selective mining" by working the highest-grade ore in sight and leaving the rest, a practice that had caused the demise of many good mines in the past. This time the results were no different. By early 1926 a consultant predicted the mine would fail within three months. He missed by a few weeks: The mine actually closed in December after the board of directors failed to raise new development capital through assessments. Lawrence Monte Verda, describing the faulty mining practices of the interim management, wrote later that in some cases "the material used for back-filling carried higher values than the selected ore milled." The Bank of Boston, a major creditor, foreclosed in 1929, and bondholders sold the remaining company assets. Stockholders were left high and dry, with two hundred thousand worthless shares still outstanding.[21] Though more than four hundred thousand ounces of gold had been produced during the 1919–1926 period, foreclosure at Carson Hill brought lode mining to a new low in Calaveras.

## The Mining Revival of the Thirties

While the Great Depression hurt almost every large industry, gold mining increased significantly after 1929 and especially after 1933. Most of the lode activity was on a modest scale. A number of small lode mines started up again, many the result of a new wave of promotional fever during the period. They were mostly exploration and development operations on a limited basis, and the total production was small.

Some of the bigger mines came back into production after the Roosevelt administration, in the first year of the New Deal, abandoned the gold standard, devalued the dollar, and raised the official price of gold from $20.67 to $35.00 per ounce. For years, gold producers had been clamoring for a change in policy. While he was president of the American Mining Congress in the early 1920s, Loring had lobbied for a special government subsidy to offset rising costs. His hopes rose briefly with the election of his old mining boss, Herbert Hoover, but the Hoover administration clung to the gold standard

Frank Tower. (Courtesy Calaveras County Historical Society)

and made enemies of many former friends. During the frenetic Hundred Days after Roosevelt's inauguration, one Mother Lode mining man told another that FDR "is doing now what that numbskull Hoover should have done a year or more ago."[22]

Anticipating a boost in gold prices, Frank S. Tower reopened the Royal mine in 1932. He had been associated, one way or another, with the Royal and the Hodson district since the golden years before World War I, although before the 1920s his main business was meat, not metal. His grandfather Jacob S. Tower had arrived in Calaveras in 1850, but after mining unsuccessfully he took up a homestead in the Salt Spring Valley. After the elder Tower died in 1880, one of his sons, Jacob F. Tower, started a butcher business on the ranch, delivering meat to miners and merchants throughout the Salt Spring Valley and beyond. A successful businessman, Jake had several retail outlets, including a butcher shop at Hodson, near the high-grade enrichment his younger brother Frank later mined. Frank worked for Jake several years, and then took over the butcher business after World War I. He was also a successful businessman, but his main interest was mining. A shrewd practical miner, he explored every gulch and stream in the Salt Spring Valley, and became thoroughly familiar with the land and its resources.[23]

After several unsuccessful attempts trying to help others reopen the Royal during the twenties, Frank Tower obtained full control of the property and started up on a very modest scale, working only one ten-stamp battery of the big mill and principally processing ore from block leasers in the mine. He also mined on his own account, and did some custom milling for other small mines nearby.

Tony Dutil, a native of Angels Camp who grew up on his family's ranch adjacent to the Royal property, worked for a time at the old Royal mill in the 1930s. As a teenager just entering high school he was hired as a handyman to George Blazer, an elderly practical engineer who ran the mill for Tower. Dutil remembered watching with fascination as Blazer loaded the copper plates with quicksilver. To test whether the amalgam was "soft" enough to absorb more gold, he scraped a small portion into a pile and delicately rubbed it between his fingers. If more mercury was needed, Blazer measured the dosage by pouring it drop by drop through a hole bored through a tree branch cut especially for the purpose.[24]

Keeping everything on a small scale, and closely watching milling costs, Tower treated some 16,000 tons from 1932 to 1934 with an average recovery of $8.04 per ton. After the price rise, he milled 61,788 tons, recovering $5.81 a ton, and averaging a net return of $3.15 per ton. Before it closed in 1942 the Royal had provided a ten-year livelihood for about a dozen employees and leasers, and a comfortable living for Tower.

Another old mine in this district, the Mountain King, also reopened in the thirties. Byron Rowe, a mining engineer, organized the Jumbo Mining Company, dewatered the mine, rebuilt and modernized the mill, and began an active underground operation. Despite this effort the project was a financial failure, primarily because the ore was too low grade to be profitable. After the company folded, Stewart and Nuss, a Fresno firm, bought out the assets and began mining. They hired Tony Dutil, by then an experienced miner and machinist, to run the mill, using ore from the Hobo pit on the property. Not long afterward, however, they were forced to shut down along with most other gold operators as a result of government restrictions after the United States entered World War II.

The Gwin mine never reopened after its 1908 close, despite several attempts over the next two decades. After Thomas's death David McClure continued to hold the properties for the Gwin Mine Development Company, occasionally taking bids and option proposals but rejecting most. In 1928 he agreed to lease the mine to a new syndicate organized by two Mother Lode residents, Jeffrey Schweitzer and Jack McSorley. A contract was drawn up and signed, but neither party realized at the time that the attorneys who drafted the documents had received mixed instructions. The error was not disclosed for five years. In the interim the stock market crashed, and the new syndicate folded before it even started to work. One morning in 1933, while working in San Francisco near McClure's office, Schweitzer received a frantic call from the former Gwin boss who demanded an immediate meeting without saying why. Schweitzer hurried over, and McClure told him that the lease and bond agreement they had signed in 1928 was actually a deed giving Schweitzer sole ownership of the Gwin mine! Dumbfounded, the Amador engineer thought a moment, and then told McClure to have papers drawn conveying the mine back to its former owner. As Schweitzer told the story later, "Mr. McClure looked sternly at me and putting his face near mine, said in a deep determined voice, 'Do you mean that?' I replied a simple, 'yes.'"[25] McClure got his mine back, but it remained moribund.

Aside from the Carson Hill, the biggest Calaveras lode mine to revive in the thirties was the Sheep Ranch, which reopened in 1933 but shut down soon afterward. Four years later, the St. Joseph Lead Company of New York, a major firm with sufficient capital to properly rehabilitate the mine and renew production, leased the property. Dewatering alone took nearly four months of steady pumping. By 1938 the mine had been deepened to thirty-one hundred feet and a new flotation circuit had been installed to improve the recovery of the fine-grained auriferous sulfides. Occasionally, high-grade ore shoots were uncovered that excited even the scoffers. In January of that year, for example, management reported that a recent "round of shots broke into a jewelry shop" deposit of high-grade gold that assayed up to $100,000 per ton. A year later, at the thirty-one hundred level, muckers turned out a "5-car sample of ore [that] averaged 2 ounces of gold to the ton." In this period the Sheep Ranch was second only to Carson Hill in total lode gold production, but first in profits. Right up to U.S. entry into World War II it was paying quarterly dividends of 25 cents and more a share, compared to Carson Hill's meager 1.5–2 cents.[26] The mine closed early in 1942 as war priorities turned attention to base metals and cut off gold mining supplies. Some of the mine's two hundred workers were drafted; others headed for better-paying defense jobs in the coastal cities. Most of the equipment and staff and a few of the remaining employees were transferred to the company's lead-zinc-silver mine at Hughesville, Montana.[27]

## The Rejuvenation of Carson Hill

The most productive, if not the most profitable, lode mine in the thirties was Carson Hill, which reopened in 1933 after a seven-year hiatus. The man primarily responsible for bringing it back into production was Charles H. Segerstrom, the Sonora banker who knew the mine intimately and who never lost faith that it could be made to pay well under proper management.[28]

His chief ally in the new venture was Lawrence Monte Verda, a Calaveras County native son and twenty-five-year veteran of Mother Lode mining. He started in the business as an office boy for the Calaveras Consolidated Gold Mining Company at Carson Hill, and later became a bookkeeper with the Altaville Iron Works. Acting on Monte Verda's advice, D. C. Demarest took an option on Calaveras Consolidated just before World War I, then transferred it to Loring during the latter's heyday on the hill.[29]

In 1927, while Carson Hill was moribund but still in the hands of the Boston syndicate, Segerstrom and Monte Verda made their first move. With some local financial backing, including commitments from Utica mine interests under M. H. Manuel, son of the Calaveras lumberman and a principal investor in Utica properties, the two promoters made a bid to acquire and mine Carson Hill. Segerstrom committed more than $100,000 in personal assets to absorb the bonded indebtedness of the previous owners, and Monte Verda went east with a proposal to reorganize, consolidate and recapitalize, placate old shareholders with new stock issues, and raise $800,000 in new development capital. The deal was still under consideration when the First National Bank of Boston foreclosed in 1929, wiping out the old syndicate and absorbing all the remaining assets. The California interests then regrouped, and after some months they worked up a new proposal. In 1930, Monte Verda returned to Boston, negotiating for a purchase option on the mine for $200,000, to be paid in installments over ten years. A fourth of the money would come from western investors, the rest to be raised in the East. However, it was not a good time for speculators. Monte Verda managed to raise $35,000 from eastern capitalists, but they backed out after a disagreement over how the corporate structure would operate and who would be in control.[30] By that time the rippling waves from the crash of 1929 were eroding away the financial strength of the nation, drastically shrinking the pool of investment capital.

Without eastern backing the western group had enough money to take up the option but not enough to develop the mine. They went ahead anyway, organizing the Carson Hill Gold Mining Corporation in April 1930, with Charles H. Segerstrom as chairman of the board and managing director, M. H. Manuel as president, and Lawrence Monte Verda as vice president and general manager. These three and four other California investors, primarily from Fresno, constituted the initial board of directors. Incorporated under Nevada law, the company capitalized at $2.5 million, issuing that amount of common stock with a $1.00 par value. The directors controlled 600,000 shares; another 1.5 million went to secure the Carson Hill properties, leaving the remainder to be offered on the open market.[31]

For the next three years the new company struggled to restore the mine to minimal working condition, and to raise development capital. But no matter what financial difficulties faced the Carson Hill promoters, Segerstrom remained confident. Indeed, he had reason for optimism, since what little exploratory work the new group was able to accomplish showed very promising potential. Monte Verda sampled one untapped ore body that assayed nearly $60 a ton. With prices for goods and services declining both nationally and locally, which would translate into lower operating costs and higher net yield, the Sonora banker thought it would just be a matter of time before investors woke up to this golden opportunity. "Personally," he exclaimed to a Chicago stock salesman, "I feel if this is not a good gold mining property, there is none to be had in America."[32]

Segerstrom's confidence grew bolder by the month, as Monte Verda and a skeleton crew worked to rehabilitate the properties, secure adjacent holdings still not acquired, and explore new ground. Early in 1932, John A. Burgess, a consulting engineer with Newmont Mining, one of the industry's most respected firms, was brought in to assess Carson Hill's mill capacity and estimate the reserves. His report rekindled local enthusiasm by concluding that the mine should yield big dividends from massive low-grade ore bodies. Under current economic conditions, he wrote, operating costs could be kept to $1.50 per ton. Later he revised the cost estimate downward to less than $1.00 per ton. Even the Morgan's exposed $3 ore, estimated at 500,000 tons that could be easily mined through the glory hole, would look good at that rate. Echoing the "deeper the better" philosophy of Loring and other Mother Lode pioneers, Burgess also recommended probing below the three thousand–foot working levels. "What I am trying to say," he told a Visalia investor, "is that no geological bottom has yet been found to small shoots of high-grade ore in the Mother Lode belt." His conclusion was a shot in the arm to both the Carson group and his clients: "It is my opinion that the Carson Hill Mine can be operated at a very satisfactory profit on ores now available, and that future developments will enable the mine to continue profitable operations for many years."[33]

A few months later the Carson investors received a second opinion, equally positive, from Oscar Hershey, a San Francisco consulting engineer. Emphasizing the strong geological evidence of massive low-grade ore bodies waiting for development, he described Loring's achievements and concluded with a challenge to investors: "What is now needed is the nerve to continue the process."[34]

Important financial figures gather around an unidentified young visitor at Carson Hill. Lawrence Monte Verda is second from the left. Charles Segerstrom stands at far right, behind Sol Grossbard, representing the Anglo-American syndicate that took control of the mine in 1937. The lace-booted engineer at far left is probably John W. Burgess. (Courtesy Segerstrom Family Collection, Holt-Atherton Library, University of the Pacific)

Bolstered by positive field reports and by upticks of interest from a variety of investing firms, the Carson Hill group shrewdly changed their financial strategy. Instead of trying to raise funds on the open market while the mine was inactive, they withdrew the unsold shares and sought financing from a selected group of interested Nevada investors. This would provide enough operating capital to go into full production, while at the same time stimulate broader investment interest. As Segerstrom told David Maltman, a New York broker, evidently related to the Maltmans of Calaveras, keeping the public guessing while producing good ore would "make a market for the stock." When the property is in operation, he said, "the stock will probably be very active and I feel sure it will give every one connected with it a chance to make some money."[35]

By the spring of 1932, despite depressing economic news all across the country, at Carson Hill the mood was decidedly upbeat. Financial circles were buzzing with rumors, and mail poured in from agents and potential investors wanting copies of the prospectus and any other information available. One broker in Exeter, California, who caused "quite a stir" by distributing the prospectus, wanted twenty more "pamphlets" to distribute to those "who want to invest a little that have a small savings as well as the thousand dollar man and five hundred dollar man." Another salesman from the San Francisco investment firm of Bennett and Company telegraphed a hurried message from Santa Monica: "Rush me fifty pamphlets whether Bennett's name is on or not."[36]

The reports and rumors from Carson Hill flushed out some of the biggest names in mining. In October—a month before the nation's only mining engineer–president would be voted out of office—Segerstrom learned that John Hayes Hammond, a friend of Hoover and one of the most respected professional engineers in the country, would be in California all winter and wanted to inspect the hill properties. His informant said that Hammond "is still looking for a large gold property that has reasonable assurance and promise of being a great mine when put in production." Hammond may or may not have followed through—the record is not clear—but it probably would have made no difference, for Carson Hill was no longer available. With little cash but holding nearly nine million dollars in assets, including lands, claims, options, plant, and equipment, and an estimated six million dollars in ore reserves, the company was in an excellent bargaining position.[37] All it needed was enough cash to start mining and meet the weekly payroll.

That winter the Carson Hill company made a deal with the Anglo American Mining Corporation that led directly to full-scale development. After tying up the loose financial ends, Segerstrom announced the "new deal" in March 1933—complimenting, unintentionally no doubt, the "New Deal" in Washington. The Anglo American firm, newly organized by a syndicate of London and New York investors, purchased a block of 250,000 shares of Carson Hill stock, providing an operating fund of one hundred thousand dollars to rehabilitate the mine and bring it into full production. Anglo American also had an option to purchase another block of 300,000 shares at fifty cents apiece, half the par value. At the same time, the directors voted to wipe out the corporate indebtedness by letting Segerstrom have nearly 260,000 shares. He also held 1.5 million shares in trust for the incorporators. The remaining 450,000 shares available were to be traded publicly on the New York and London exchanges.[38] Privately, Sol Grossbard, the New York agent for Anglo American, advised one of the Carson Hill directors to adopt Segerstrom's strategy of waiting until the mill had operated for six months and had "actually shown a minimum profit" before distributing stock to public traders. "Due to the [new Roosevelt] Administration's inflation policy," he wrote, "and due to the speculative wave which is existing at present in New York," he thought the remaining shares could bring two dollars apiece as starters.[39]

Segerstrom traveled east during the early summer of 1933 to discuss Carson Hill finances and to encourage the new administration's "inflation policy" by lobbying for an increase in the price of gold. He wrote Grossbard that with the help of "my friends in Washington," and "without any question of a doubt, there will be a free market for gold at an early date." His most important political ally was Harry L. Englebright, California's congressman from the Second District, who in August happily announced to his "Friend" in Sonora that Roosevelt had issued a proclamation giving the "gold mining industry the right to export its newly mined gold and obtain the world price thereon."[40] The current market price was thirty dollars, but shortly the administration fixed the official price at thirty-five dollars. It was not quite what the gold interests wanted, but even a fifteen-dollar increase automatically upped the value of Carson Hill's estimated reserves by four million dollars. In 1933 dollars that was not a bad bargain.

Meanwhile at Carson Hill renovation and preparation for full operation proceeded rapidly under John Burgess, earlier that year hired as general superintendent with orders to start up by August 1 if at all possible. Burgess did his best, but the work was slow. Real mining could not begin until the working levels could be retimbered and cleared, the machinery cleaned or replaced as needed, the compressor and air lines renovated, and a thousand other details attended to. Beginning with a handful of men, by May he had thirty-eight employees at work, and by September another forty-two had hired on. On September 4, at the mine headquarters at the base of Carson Hill, Burgess and his men took time out to join the corporate officers and directors, representatives from the Anglo American syndicate, a few stockholders, and about fifteen hundred happy citizens from Calaveras County and environs in grand-opening ceremonies. The men went back to work the next day, not mining but still preparing. Regular mining and milling did not begin until mid-September, and it took several more weeks before the mine became fully operational.

By October enough preliminary work was completed to begin full-scale mining. The immediate objective was to work the developed low-grade ore above eleven hundred feet in the Morgan and Calaveras mines, but at the same time to explore virgin ground in the most promising places, both in those mines and elsewhere on the eleven hundred–acre consolidated property. Development at deeper levels would have to await dewatering below the two thousand level. As Burgess told his bosses, the new operating plan was a 180-degree shift from the objective in the interim between 1924 and 1926: "The proposed operation is based principally on mining the large low-grade ore bodies above the haulage levels at low operating costs. This is contrasted with the earlier operations, and higher costs, of deep mining on higher grade ore."[41]

Through the mid-thirties Burgess stuck to the plan, keeping ore production apace with mill capacity, slowly building the surface plant to handle greater volume, adding men as needed, cleaning out old levels, and preparing for deeper exploration. Occasionally, there were moments of excitement, as in March 1934 when a high-grade pocket in virgin ground near the surface on the Morgan claim was uncovered with a power shovel. Local headlines screamed the news: "$80,000-TON GOLD ORE STRUCK IN MORGAN MINE AT MELONES," and Lawrence Monte Verda told reporters that "the strike serves to blast the belief that Calaveras ore is of low grade and all the good gold has been worked out."[42] But this was mostly publicity, an effort to stimulate investor interest and "create a market" for the company stock when it was released. A selected crew picked the pocket clean in a matter of hours, and it hardly made a blip on the gross production records. The real business at Carson Hill was low-cost mass production.

Keeping production costs down placed a heavy burden on the employees. Cost-cutting at Carson Hill in the thirties was a job ordered by the board and supervised by the local mine management, but the miners and mill men and all the auxiliary workers felt the daily impact. Unions had never gained a foothold at Carson Hill, and workers had little recourse against indifferent bosses or hazardous working conditions. Jobs were scarce in the Mother Lode, and those who complained were few and far between. As Burgess told Segerstrom in 1932, the competition "results in much greater efficiency in labor than for many years."[43]

Carson employees in earlier decades had no better conditions, however. Milo Bird, a mill man, remembered a host of hazards confronting Carson Hill surface workers in the Loring era and beyond: silica dust that fogged vision, clogged noses, and left a quarter-inch deposit on the mill floor every twenty-four hours; deafening noise that made talk impossible; water sprays and pools that soaked mill crews within minutes of going on shift; narrow, unprotected catwalks over and around moving machinery and agitation tanks; and timber and scrap metal and slash piles everywhere. Workers who were injured had to pay their own doctor bills, and their paychecks were docked for lost time. The workday often ran into overtime without extra compensation.[44]

These conditions did not improve much in the thirties, although by the latter years of the decade federal and state safety boards had been upgraded

Two views of the Carson Hill mill in the 1930s. At top is the new ball mill installed to fine-grind discharge from the stamps. At bottom is the revamped mill, with the Stanislaus River bridge in the foreground. The mill later burned, and the bridge is now deep below the surface of New Melones reservoir. (Courtesy Segerstrom Family Collection, Holt-Atherton Library, University of the Pacific)

to monitor mining operations. But management balked at every turn. One letter from an insurance salesman after the death of J. R. Risley, a twenty-four-year-old miner who fell from a skip, was indicative: "[A]s this fatality happened so soon after we were discussing Group Insurance for the Mine, it naturally attracted my attention. It is too bad that Group Insurance was not in force, for if it had been, we would have been in a position to hand a $1,000 check to Mr. Risley's beneficiary."[45]

As late as 1941 the Mother Lode Mining Association, with Burgess as president, passed resolutions opposing the first two safety orders issued by the newly created California Industrial Accident Commission, which tried "to force mine operators to supply helmets and goggles to mine workers." The association also fought enforcement of new federal minimum-wage and maximum-hour regulations on grounds that rules based on interstate-commerce power "cannot be legally binding on California gold mines operating wholly within the state."[46] The wage scale was the one real difference in labor conditions between the two decades, since wages had declined 40 percent during the Depression. Minimum-wage guidelines did not go into effect during Carson Hill's operating life, so only a corresponding decline in the cost of living made life manageable for working-class miners.

By 1935, Carson Hill had boosted its workforce to more than two hundred men and its daily output to nearly five hundred tons, but that was just the opening phase of a management plan to increase production to thirty thousand tons per month. Accomplishing this objective required extensive new development and new equipment. On the surface, power shovels pushed deeper into the open cuts on the Morgan and Santa Cruz claims. Underground, haulage levels were extended, and new winzes and raises reached into newly developed ore. Additional pumps were installed to dewater the deeper levels. At the mill a new crusher and several new cyanide tanks were added, and plans were made to add a flotation circuit. Operating costs rose appreciably because of these measures, and to pay for them the directors pumped all the net earnings back into the mine. Segerstrom, as president, announced the plan in a spring progress report. Withholding dividends to expedite mine development in the short run, he assured stockholders, would result in lower unit costs and greater operating profits in the long run.[47]

As a development strategy for low-grade ores the plan had its merits, but as an inducement for investment it had serious drawbacks. Investors of all types—big and little, actual or potential—were impatient as well as short-sighted. Given a choice most preferred dividends over development regardless of long-range consequences. Indeed, as the thirties wore on, Carson Hill directors found themselves confronting a paradox: Where mine revenues were limited, long-range development was incompatible with investor demand for quick returns. Without dividends brokers could not sell Carson stock unless it was heavily discounted, no matter how impressive the production figures or the development potential.

It took nearly three years for this lesson to sink in, however. Segerstrom, as president and CEO between 1933 and 1937, carried on the original plan with unflagging determination despite dismal market results for Carson stock distributions. In the meantime management plunged ahead with systematic development both on the surface and underground. By 1936 the mine's full-time employees totaled more than 250. Daily production averaged twelve hundred tons of ore at a cost of $2.71. The directors authorized $110,000 in mine development and another $90,000 for new equipment, including two ball mills to supplement the stamps, new classifiers, and still more improvements to the cyanide plant. Lawrence Monte Verda, Carson Hill's assistant general manager who had separately acquired Demarest's old Altaville Iron Works, got the contract to install the new equipment. The improvements were expected to lower operating costs to $1.50 per ton and to increase daily capacity to twenty-two hundred tons, making Carson Hill the largest producer in California.[48]

All of this new development impressed the directors but not the investing public. Segerstrom complained bitterly to Sol Grossbard that "we have done everything in our power to keep the stock in a pretty fair position and about all we ever got out of it was a lot of letters asking why we didn't pay the dividends and why we didn't do certain things that had been promised." The mine's president blamed brokers for talking down the stock, but that was only a symptom of the underlying problem Segerstrom himself described in the same letter: "[W]hen the stock was [first] sold, promises were made . . . that dividends would be paid within a reasonable time, and now that probably an extra year has gone by, [investors] . . . are fearful no dividends will be paid and are in a mood for selling."[49] The net result was that operating costs continued to absorb all the available revenue despite the increased production, and without stock sales the company could not continue major development.

By the end of 1936 Segerstrom and the California directors faced a financial crisis resolved only when Anglo American, the eastern syndicate that had bought enough heavily discounted shares in 1933 to restart the mine, decided to exercise its option and take control. This was a mixed blessing for the original investors. They had held off Anglo American's takeover, hoping the market value of Carson stock would rise to par value or above. As Anglo American president Walter Lyman Brown explained to a New York investment firm,

Segerstrom and his California associates "are not particularly happy at having to sell their stock to us at the option price [of $0.40–0.50 per share]." But by the beginning of 1937 the western investors had no other choice. Anglo American was the only interested party. With the concurrence of the other western directors, Segerstrom sold the syndicate the bulk of the remaining shares outstanding, including his entire personal Carson Hill portfolio, at a heavy discount. The easterners took control at a special meeting on May 26, 1937, when most of the old directors resigned and new ones were elected. Burgess remained general manager, but the upper echelon was swept away. The old office of president gave way to a new title, chairman of the board, which W. L. Brown assumed commensurate with his control of Anglo American. Segerstrom was given a unanimous vote of thanks, small consolation for his long efforts and his financial losses. No longer tied financially to Carson Hill, an outsider without board status or managerial responsibility, he turned to other mining interests. Family members attributed this change in part to foresight: With World War II approaching, Segerstrom became a spokesman for the strategic-metals industry and one of the nation's leading tungsten producers. But the Sonora banker remained emotionally tied to Carson Hill, and until it closed continued to answer stockholder inquiries and to act as spokesman for Mother Lode miners.[50]

Now in the driver's seat, Anglo American pressed ahead with Burgess still in charge of development and with but one significant financial change. In the fall quarter of 1937, for the first time in the mine's long history, Carson Hill declared a dividend. That was a calculated move to stir investor interest. The dividends were small, amounting to only 2 cents per share for the first three quarters. But Grossbard thought the regularity of dividend issues was more important than the amount. Late in 1938 he told Segerstrom: "If we can maintain a dividend of 2c per quarter we should be able to market a block of this stock around seven to eight times its earnings or probably ten times its annual cash dividend."[51] That turned out to be wishful thinking; the market never improved much for Carson Hill stock despite continued development under Anglo American.

By the end of the 1939 fiscal year, ore production was up to nearly 33,000 tons per month. Despite expanded open-pit operations along the Calaveras vein, more than half of the ore was still coming from low-grade shoots and stopes in the Morgan mine that could be easily dropped by gravity to the haulage level. For the year the operating costs, including all development, diamond drilling, maintenance, and improvements, were $1.86 per ton, slightly higher than the year before but still well below the figures prior to 1935. With

Open-cut mining with a power shovel at Carson Hill in the 1930s. Note the 1–2-yard dump trucks. Huge trucks with 100–150-yard capacity are standard equipment in modern pit mines. (Courtesy Calaveras County Historical Society)

average ore grade running $2.37 per ton, the mine was profitable, if just barely at $0.51 per ton. At least its approximately 750 stockholders, predominately Californians holding one hundred shares each or less, were receiving regular dividends.[52] Total dividend distribution for the year was $132,000.

In its last three years of operation, Carson Hill continued the pattern established nearly a decade earlier. Ore production averaged more than 32,000 tons per month, operating costs remained below $2.00 per ton, and ore value held steady at about $2.15. Stockholders continued receiving small quarterly dividends, and Anglo American persisted in believing stock values would rise so long as the checks arrived regularly. But disaster struck early in June 1942 when a fire "of undetermined origin" burned the mill to the ground, stopping all production and throwing more than 200 men out of work. By that time any chance of rebuilding died because of war "priorities." The announcement of the War Production Board (WPB) later that year effectively shutting down all "nonessential" mining made official what Carson Hill management knew months before. The impact of that order will be discussed in a later chapter. In 1943 what remained of the surface plant at Carson Hill, and all the salvageable underground equipment, was removed and sold at auction.[53]

Anglo American, the parent company, at first contemplated reopening the mine after the war with all new equipment, but by war's end the company plans had changed, and the mine remained closed until the 1980s.

Carson Hill literally went up in flames after rising to the top among California's major gold producers between the golden age before World War I and the new era after 1970. Total production over the nine-year period between 1933 and 1942 was nearly 220,000 ounces of gold.[54] Over its lifetime it was one of the biggest gold lode producers in Calaveras County. It still had one more chapter to write after nearly a forty-year hiatus following the 1942 fire.

## Placer Mining in the Thirties

The hardrock revival was perhaps more glamorous, but a resurgence of placer mining in the Calaveras foothills and elsewhere during the depression was no less significant. In the early thirties a new type of gold dredge made its appearance and began to recover gold from stream placers long abandoned, many of which had been reworked by Chinese. These little dredges—some constructed in backyards out of used equipment and scrap metal, others built by large equipment companies—consisted of a track-mounted dragline that excavated stream gravel and dumped it into a portable washing plant usually mounted on a shallow-draft pontoon boat moored close to the dragline. In some cases the plant was mounted on wheels and towed behind the dragline. Capable of dredging large tonnages in narrow and shallow creek bottoms, and working upstream in gravel too shallow for bucket-line operations, they were popular devices in Calaveras County. Like their larger bucket-line cousins, these little machines stacked the washed gravel tailings behind them as they worked, leaving trails that resembled the residue of a common insect. Soon they were everywhere known as "doodlebugs."

Costing far less to operate, highly efficient with good recovery of fine-gold values, and using only a handful of men, doodlebugs could profitably work ground that assayed as low as 20 cents per yard at the new gold price. For example, from September 1935 to August 1937, the Milton Gold Dredging Company dredged 1,234,908 yards at an average cost of 12.01 cents per yard. Farther south, on the Tuolumne River below Moccasin, Glenn Bump ran a low-maintenance, cost-efficient doodlebug on electric power that dredged 125 to 150 yards per hour around the clock, seven days a week. By 1938 doodlebugs contributed 25 percent of California's placer gold production, operating in areas primarily considered "worthless" by larger dredge companies.[55]

In Calaveras the most extensive draglining was carried out at South Gulch between Milton and Jenny Lind, and near Camanche. However, doodlebugs

Doodlebug dredgers came in many different shapes and sizes. The dredge at top operated at Coyote Creek in Vallecito. At bottom is the Moundwell dredge at Lancha Plana in 1952. (*Top:* Courtesy Calaveras County Historical Society. *Bottom:* Courtesy Holt-Atherton Library, University of the Pacific)

also worked most of the North and South Forks of the Calaveras River and
its major tributaries. On the Mokelumne River and at Jenny Lind the bucket
dredge was still king, larger and improved from older models. At least five
were working on the Mokelumne and Calaveras Rivers by the end of 1937.[56]
By government order as a war measure, all types of dredge operations halted
in 1942. Dredging in the thirties, though perhaps more efficient, was no less
destructive to the riparian environment than in earlier decades. Following
the example of the bucket-line dredges downstream, doodlebugs completed
the destruction of the narrow streambeds and bottomlands of the upstream
tributaries.

Drift mining also revived in the thirties, although on a much smaller scale
than dredging or lode mining. At least two dozen small or exploratory operations were concentrated in the Vallecito-Altaville and Mokelumne Hill districts. However, most drift mines were marginal operations, and more often
than not production costs exceeded revenue.[57]

The one major exception was the Calaveras Central, working near the site
where the Calaveras skull was "discovered." In the thirties it was California's
largest Tertiary gravel mine. Harry Sears, who had acquired the property in
1926 from previous owners, spent $300,000 over the next five years modern-
izing the surface plant and improving the underground workings. His engineers electrified the hoist, added new pumps, enlarged the drifts to run two
electric-powered trams on parallel tracks, expanded the compressing plant to
operate mucking machines and drag scrapers, and increased the capacity of
the mill to handle 250 tons per day of gravel that averaged nearly 0.16 ounces
of gold per ton. The result was a highly efficient, fully mechanized operation
that reopened in 1931 accompanied by a wave of favorable publicity despite
the depressing economic climate. Through the thirties the Calaveras Central
remained a model mine, continuing to make improvements, keeping production costs below $2.00 per ton, and demonstrating the utility of modern
mass-mining techniques in drift gravel operations.[58]

## Miners and Families in the Depression

For Calaveras families dependent on mining for a livelihood, the Depression
was a mixed blessing. Some benefited from the rise in gold prices after 1933
that stimulated a mining revival, but not the majority. The base-metals industry continued to slump, and much of the service and support industries
formerly tied to mining were too far gone by the mid-thirties to feel the effects
of a gold resurgence.

In Copperopolis sliding copper prices and a major underground disaster

that took five lives late in 1929 closed the last mine and took with it most of
the business that had kept the town alive. Many Mexican families that had
come to Calaveras in the 1920s to work in the mines or in supporting services
moved away, leaving the town poorer and less able to sustain the few shops
and stores that remained. The Stone family garage, for example, survived for
a few more years, but the end was near. It had served the community since
1916, when James Stone and his son Charles added a garage adjoining the
blacksmith shop James had acquired in 1890. When James died in 1924 the
blacksmith business closed, but Charles had kept the garage open through
the twenties despite the mining slump. It could not sustain the desolate years
of the Depression, and closed in 1937.[59]

Many foothill families eased the hard times by reworking old diggings.
They controlled most of the mining ground, often as unpatented family
claims that dated back to the nineteenth century. Holding a claim normally
required a modicum of labor on weekends or during vacations to complete
the one hundred–dollar annual assessment work required under the 1872 mining law. As the Depression deepened, however, old claims saw new life. South
of Mokelumne Hill, for instance, a number of drift mining operations revived in old Tertiary gravels that had been intermittently mined since the
1850s. One of several owned by the McSorley family was the Green Mountain
mine, better known for the high quality of its quartz crystals, used in lenses,
jewelry, crystal glass, and other specialty products. The Tiffany Company
alone had ordered twelve tons of Calaveras crystal; in the 1920s the United
States Navy purchased fifty tons for crystal radio sets. The McSorley brothers
continued a family mining enterprise that had begun under their father in
the 1860s. As second-generation Californians born and raised in Calaveras,
Hugh, Tom, and John McSorley had "spent most of their youth mining" in
Alaska, Nevada, Mexico, and other parts of the world as well as at home. By
the 1930s, as crystal demand waned, the McSorleys shifted to other mining
operations.[60]

Some Calaveras families eased hard times by panning the surface streams.
Nancy Ellen Doster Shelby and her husband, Thad, for instance, collected
gold dust in Bull Durham sacks and traded it for food and staples at Attilio
Domenghini's general store at Mountain Ranch. By 1932 they and other local
residents found themselves competing with an influx of dirt-poor refugees
from the cities known as "snipers," a pejorative term frequently heard in the
Depression, when sniping was widespread throughout the West. In the 1930s
hundreds of men and a few women down on their luck literally headed for
the hills to hunt for gold, usually along the streambeds and in the gulches,
but sometimes in the pits and adits of old hardrock prospects and work-

Three McSorley brothers—John (with pan), Tom (on horse), and Hugh (far right)—
with John Burton (holding dipper) and another employee processing Green Mountain
mine quartz crystals, 1897. (Courtesy Calaveras County Historical Society)

ings. Jesse Coffey remembered one friend finding "several hundred dollars"
of rich, crumbled, iron-stained quartz pieces while looking over an "old prospect hole."[61] Some of these new gold seekers took up abandoned claims and
had them recorded according to the rules of the district; others did not bother
with standard procedure but worked any unoccupied site until they ran out
of ore or were driven off by the claim owner or the authorities.

For a number of unemployed and homeless men and a few families, sniping was better than living in a Hooverville or joining the army of job seekers
in San Francisco or Los Angeles. Most had little or no mining experience,
but they learned by doing, as had their placer predecessors. They worked old
crevices and streambeds and built sluice boxes and rockers, sometimes trying
novel ways to recover gold. One sniper fashioned a homemade diving bell and
worked the crevices in deep pools along Mother Lode streams. Their earnings were meager, but they remained independent and self-supporting, living
on a diet of "bacon and beans," supplemented with small game, fish, and a
few garden vegetables if they could stay in one place long enough to cultivate
them. For most snipers, compared to the urban alternative, camping in the
foothills was not a bad existence.[62]

Guy Castle. (Courtesy
Calaveras County
Historical Society)

   At Carson Hill, Milo Bird, an experienced miner and assayer, discovered
the workings of one of these unconventional miners by accident. His reminiscences provide a fascinating glimpse into the latter days of Calaveras lode
mining. He recalled an old and scruffy prospector named Charlie Allen whose
weekend antics were familiar to Angels Camp residents. Every Saturday, Allen
came to town with a small vial of "freshly crushed quartz" gold to buy a week's
supply of groceries. He wrapped the purchases in a gunnysack, tied it to a tree
in back of the store, and then headed for the local bar to drink up the change.
By Sunday afternoon he was sober and on his way back to his mysterious
mine. No one knew the source of his gold until Bird stumbled on it one day
while inspecting the south end of the Melones Consolidated Mining Company property near the Stanislaus River. On an old quartz outcrop beside a
bush he found a small coyote hole along with a single-jack hammer inscribed
"C. A." Charlie later admitted he had been sniping on private property, but
said he "knowed the company had gave up on that outcropping years ago."[63]
   For Chris Porovich and Guy Castle, boyhood friends who had grown up
in Angels Camp with mining and miners, the Depression had little impact on
their jobs at the Calaveras Central mine, which had reopened in 1930. Harry
Sears, president and general manager, was more politician than miner. One

of the state's most visible, and voluble, mining men during the 1930s and 1940s, he lobbied for favorable mining legislation, wrote dozens of mining articles, led the Mining Association of California, and had a major role in the gold mining revival. But politics and promotion did not make for good mine management, and Sears left daily operations at the Calaveras Central to his superintendent, W. H. Warick. Porovich and Castle gained valuable experience at the Calaveras Central but made different decisions when new opportunities arose. Castle left for Melones when the Carson Hill reopened in 1933, but Porovich stayed at the Calaveras Central for three more years, first as a driller, then as a shift boss, the youngest in Calaveras.[64]

Working in a gold mine proved too much of a temptation to some employees faced with earning a living during the Depression. Even though living costs had declined since the twenties, wages and salaries had dropped as well, and there was still no social safety net for workers and their families. To supplement meager weekly earnings, miners often resorted to high-grading, or stealing ore. It was an ancient practice, found wherever gold and greed came together. As Otis Young has suggested, gold miners who could clandestinely reward themselves a pay raise were less likely to join radical unions, unlike fellow workers in base-metal districts where radical organizers had greater success. In hard times the pilferers rationalized greed with need. They cached high-grade ore for later retrieval; walked off with it in tobacco tins and lunch pails; or carried gold flakes and dust in their cuffs, under their fingernails, and even in body cavities. Porovich recalled a man with a "mouthful of gold—I got him laughing and he spit gold all over the floor."[65]

Converting stolen gold to cash was always a problem, but nearly every mining camp had an underground economy. Black market gold was highly discounted, but lucrative enough to keep high graders going back for more. Before the 1933–1934 rise in gold prices Guy Castle remembered a San Francisco fence making regular buying trips to Calaveras, picking up stolen gold at ten or twelve dollars an ounce and redeeming it at the mint for eighteen dollars. Castle also recalled a familiar method of disposing of stolen gold from the Calaveras Central. To mask a fencing operation, said Castle, two other men "started a mining operation down on the South Fork of the Calaveras River. Pretty soon they began showing people the beautiful pieces of gold they said they had mined. We all knew where it came from. It was high graded gold from the Calaveras Central they had bought at $10 an ounce." At the Calaveras Central, as Porovich discovered, theft was so widespread that management installed double change rooms for miners coming off shift. The men dropped their "diggers" or work clothes and showered in the first, then walked naked from the shower to the second to dress in street clothes. Following traditional high-grading countermeasures, the company also employed detectives and undercover agents, but the results were dubious. "I don't think they ever caught very many," said Porovich.[66]

At a different level were the tramp miners or "10-day miners," experienced hardrock men with honed skills that were in greater demand after gold prices rose in 1933. They traveled from mine to mine, bidding jobs on contract, or hiring on as the pay scale and conditions warranted. As Guy Castle found when he encountered these men at Carson Hill, they had little respect for local miners or their managers, but they were "real professionals" who "knew every mining method and had encountered every kind of mining problem. You could put them anywhere and they knew exactly what to do." Castle "learned more from them than anyone, but the problem was they wouldn't stay. . . . They knew they could get a job at mines anywhere and after a couple of paychecks they'd be gone."[67]

Tramp miners and snipers were transitional figures in Calaveras mining. Both were visible during the Depression years, but by the forties they had largely disappeared, leaving little in their wake but a few colorful stories and fading memories.

\* \* \*

Between the two great wars of the twentieth century, mining struggled to reassert its leadership as the predominant industry in Calaveras County. The prosperity that many Californians felt in the twenties did not extend to the rural counties or those where traditional mining had been the economic mainstay. Farmers and miners in the twenties were both hurt by the lingering dislocations of World War I. By the early 1930s, however, while farmers dropped even deeper into an economic quagmire, mining benefited from the national conditions that reduced costs, inflated the gold price, and put thousands of miners back to work. Ironically, while the nation struggled to shake off the effects of the most severe depression in its history, miners in Calaveras and other gold districts reached production levels not attained since Gold Rush days.

Despite bigger production and higher employment, the gold industry in the 1930s remained an uncertain enterprise, far from the powerful economic engine it had been during the golden years at the turn of the century. Rau Roesler's prediction that a new golden era would return to the Mother Lode with modern methods and outside capital did not bear fruit. During the Depression decade the major mines had brought in outside capital, invested heavily in new machinery, introduced mass mining on a scale never before seen in the Mother Lode, and broke tonnage records. Yet the regional impact of all this effort was limited by the lack of sustainable ore reserves. It was a familiar story. Even though gold prices had risen 61 percent, as war began in Europe major companies could not maintain profits in the wake of rising costs and declining ore grades at deeper levels. Neither could the smaller mines. Though spending money locally and contributing a fair share to the local economy, small marginal operators abandoned their claims when new war-related jobs opened up in the cities. Had the government not closed the remaining Mother Lode gold mines when the United States entered the war, the industry might soon have collapsed on its own.

# 8: A Half Century of Change

Gold mining is a fickle business that often defies mainstream economic trends. In peacetime, hard times on Main Street usually mean good times in the gold districts. That was the case in both the 1890s and during the Great Depression, as we have seen. But wartime is a different matter. Most basic industries—agriculture, transportation, lumbering, manufacturing, and the like—feed on wars like kids on candy. But the wartime impact on the mining industry depends on the mineral in question. Anything deemed "strategic," or potentially helpful to the war effort of a belligerent nation or its allies, is in great demand. During World War II, for instance, mineral extraction was a booming business in the nation's oil, coal, and gas fields. Base metals such as iron, lead, copper, aluminum, and zinc reached unprecedented production plateaus, and all-out efforts were put into the exploitation of deposits of strategic minerals such as tungsten, molybdenum, cobalt, and boron. But before the space age, gold had little immediate utility to nations at war. During the great wartime emergency following Pearl Harbor, those whose livelihood depended on mining the precious yellow metal had to be sacrificed on the altar of national security. The gold mines shut down, and the gold miners went to war or found jobs in copper and other strategic minerals.

Wartime brought mixed responses in Calaveras. The gold mines closed, but copper blossomed, only to wither again soon after the fighting ended. Gold partly revived after World War II, but the respite was only temporary. Yet the war and its aftermath changed the Mother Lode in more profound ways. Economic diversification after the war, combined with a steady population increase, softened the impact of upturns or downturns that in earlier days might have meant widespread dislocation and hardship. By the 1950s lumbering and cement manufacturing had replaced mining as Calaveras County's principal enterprises.

In more recent years mining has given way to other forms of economic development. Since the 1970s tourism, agriculture, real estate development, trade, and service industries have all expanded significantly. Modern technology has also substantially influenced the scope and direction of the county's economic development, since it has reduced the size of the labor force in traditionally labor-intensive industries such as mining and logging. Moreover, modern road systems have provided greater mobility for county residents and have enabled large numbers to live in the county but work elsewhere. No longer dependent on a single resource industry, but not yet fully diversified, Calaveras is still in economic transition.

## Government Wartime Mining Policy and the Base-Metals Industry

Old-timers in the mining industry can still be heard to complain about L-208, the War Production Board order that in 1942 closed all mining activity not needed for the war effort after October 15 of that year. Aimed primarily at gold mining, which was considered a nonessential industry, the order did not affect base-metals mining even if gold and silver were produced as byproducts. Western mine officials and their political allies vehemently protested the order, citing the adverse affect on regional economies, the unfairness to gold producers (since foreign and base-metal producers were not affected), and the faulty data used by the War Production Board to justify the closure. As the mining representatives noted in hearings prior to the closure, WPB figures indicating that 15,000 miners "could be made available for copper mining through a shutdown in gold mining" were based on old information from an outdated labor department report. The real number, said the mining representatives, was closer to 750 men, not enough to significantly boost copper production, but economically devastating to the targeted mines and communities. These and other arguments fell on deaf government ears. The order stood, and the mines closed until 1945, when L-208 was finally lifted.[1]

Though they gained few new miners in the aftermath of L-208, Calaveras copper mines revived for a brief period during and after World War II. Government subsidies helped offset production costs. In 1943 the Eagle-Shawmut Company leased the Penn mine. It reopened the underground workings and mined substantial tonnages of copper-zinc ore, which was trucked to the Eagle-Shawmut mill in Tuolumne County for processing. The company produced large amounts of copper and zinc as well as some precious metals, but the subsidies were phased out over a period of several years after 1945. Postwar inflation added to the Penn's woes, although it lingered for a few more years. American Goldfields installed a mill after that company took over in 1948, but declining revenues and increased costs painted a dim picture. The Penn closed in 1953 but reopened briefly in 1955, then closed again and the underground workings were abandoned. Cement copper was produced from the mine waters for a short time afterward, then the last operators pulled out, leaving a scarred landscape and deteriorating tailings ponds and leaching mine dumps that became the focus of statewide controversy nearly thirty years later, as will be seen. This was a sad ending for one of Calaveras County's

Dismantling the Sheep Ranch mill in the fall of 1942. (Courtesy Calaveras County Historical Society)

best mines, credited with a total production of nearly eighty million pounds of copper, twelve million of zinc, and six hundred thousand of lead, as well as sixty-two thousand ounces of gold and about two million of silver.[2]

At Copperopolis, where the Union-Keystone had been shut down since 1930, wartime demand for base metals resulted in the opening of the old North Keystone in 1942. The next year Otto Schiffner, of the Lava Cap gold mine at Grass Valley, which had been closed by L-208, formed the Keystone Copper Corporation and took over the Copperopolis operation. The North Keystone produced a substantial tonnage of high-grade copper sulfide ore that was trucked for processing to Stewart and Nuss's Mountain King mill in Hodson. At the same time the Pacific Mining Company reprocessed a large tonnage of Union-Keystone tailings.

Tony Dutil was foreman at the Mountain King mill during the copper interlude, and recalls the difficulty finding the necessary manpower to operate. The ore coming from Copperopolis had to be crushed, then hand-sorted before it was fine-ground and fed to newly installed flotation cells to recover the concentrates for shipment to smelters outside California. Hand-sorting improved the mill-run ore grade from 45 percent to 75 percent, but it required a crew of a dozen or more pickers day after day standing alongside a long conveyor belt tossing fist-size low-grade rock onto the waste pile. It was a dirty, monotonous, back-straining, and low-paying job, the kind usually handled by immigrants or minorities, not one that would attract gold miners even if they had been thrown out of work by L-208. They found other jobs in Calaveras or elsewhere, or joined the armed forces.

With no other immediate options the copper company first tried using women. They were entering the war effort by the thousands in the shipyards and airframe factories along the coast, but at the Mountain King mill, recalled Tony Dutil laconically, they "didn't work out." But he had another idea. Since copper was vital to the war effort, Tony suggested asking for soldiers. His boss called the western regional defense command, and soon a detail of some two dozen men in uniform came to Calaveras. With no place to stay in Copperopolis—the boardinghouses had long since closed—they found lodging in Angels Camp and were hauled daily to the Hodson mill on a flatbed truck that also served as a bus to take Copperopolis-area children to school in Angels. This pragmatic arrangement worked well, at least well enough, and kept both the mine and the mill in business until the end of the war.[3]

The end of federal metal subsidies terminated this phase of Calaveras copper mining, and the only additional activity at Copperopolis was minor production of cement copper from mine waters of the old Union mine. Leach boxes containing shredded tin cans lined the surface near the portal. Mine water loaded with sulfuric acid and dissolved copper sulfates was then pumped in, causing a chemical reaction in which the copper replaced iron. This "cement" dropped to the bottom and was sluiced out, then dried, bagged, and shipped to market. Leaching was a common practice in old copper sulfide deposits and had been known since antiquity. In the late 1920s, Charlie Stone and other kids in Copperopolis got a dollar a sack for cans collected for the leach boxes, which were about three feet wide and two or three feet deep. The boxes had to be raked occasionally to keep the solution active, as another Calaveras leaching crew discovered right after World War II. They dumped scrap iron down the main shaft of the Quail Hill mine, but without agitating the salvage operation recovered only 15 percent of the potential copper values.[4]

Total production of the combined Union-Keystone–North Keystone operations is estimated at more than one million tons of ore, resulting in more than seventy million pounds of copper as well as some silver and minor quantities of gold.[5] The large output and the substantial employment over many years put this mine in the top ranks of Calaveras producers. Yet, ironically, it had only brief intervals of profitable production.

## Postwar Gold Mining: The Law of Diminishing Returns

Even without the government regulations that ended in 1945, the rising costs of labor and supplies, coupled with the fixed price of gold at thirty-five dollars per ounce until 1971, severely retarded mining activity after World War II. A few mines reopened for limited runs or to prospect new areas, but production remained small. The Mountain King mine in the Hodson district, for example, reopened in 1945 and was worked by open-pit methods until 1948 when it closed once again. Prospecting began on the Gwin mine in 1946, but nothing came of the venture. In 1947 the Calaveras Central drift mine at Altaville resumed operations. Harry Sears, energetic president and leading critic of federal gold policies, spent his development money erecting a new steel headframe and fixing up the old mill. He had kept the mine dewatered through the war years, fuming at the shutdown and hoping to recover once mining resumed. But postwar economic realities soon dispelled these hopes. He had to shut down again after only a short period of production.[6]

The Royal mine at Hodson also reopened for a limited run after the war, but in 1948, Frank Tower sold it to Stewart and Nuss, a sand and gravel company at Fresno. In the interim three major mining companies prospected the site with diamond drills but did not locate ore of sufficient grade to justify further large-scale development. For the next forty years it languished, but as will be seen later, it came back to life for probably the last time in the late 1980s.

Disposing of the Royal did not end Frank Tower's remarkable mining career in Calaveras. Before giving up the older mine he negotiated with the buyer to mine out an enriched zone located by exploratory drillers just thirty-seven feet below the surface in the old town of Hodson. He called it the Butcher Shop mine in memory of the little retail store his brother operated back at the turn of the century. In just two years this little mine produced some twenty-five hundred ounces of gold. For a while it was the talk of the county, but Tower did not let the sight of crystalline gold shining like jewels disturb his equanimity. He lived comfortably but not extravagantly in a stately ranch house he had built on land that had been in his family since 1852. Contract miners and relatives worked his mining properties, while he managed finances and handled ore shipments. Once he gave Dufo Gualdoni, a bus driver on the Stockton–Angels Camp route, a heavy package wrapped in brown paper, instructing him to deliver it at the railway-express office in Stockton. The package contained a gold ingot worth thousands.[7]

Calaveras doodlebug operators and the bucket-line dredgers shared the same fate as the underground gold miners. They could not cope with continuing inflation and a fixed price of gold. A few draglines and one bucket line resumed dredging at war's end. Some held out for nearly a decade despite wage hikes, declining profits, and increasing complaints from landowners and environmentalists who wanted the topsoil restored. Several California counties with more extensive dredging fields actually passed resoiling ordinances, but adverse court decisions prevented enforcement of these laws until passage of California's Mine Reclamation Act in 1975. In the long run rising costs and falling profits did more than legal battles to end California dredging.[8]

## The Cement Industry in Calaveras

The wartime shift from precious to strategic minerals spurred the growth of a Calaveras industry that had started in the twenties and grew slowly until the 1940s. The seeds were sown as far back as the Gold Rush, when enterprising merchants and artisans, alert to the fire danger in mining camps, began quarrying stone for building materials, manufacturing lime for mortar and stucco, and sawing and shaping dimension stone from marble and volcanic tuff. The market for most of these building materials was limited until California's population growth and urban development began to mushroom after the turn of the century. By the twenties California's concrete market was expanding faster than the state's fledgling cement industry could supply.

In Calaveras County, William Macnider was one of the first mining promoters to recognize this dynamic imbalance. An industrial salesman living in northern California, he was well acquainted with Calaveras geology and its abundant raw materials for the manufacture of cement. Early in the decade he proceeded to option several limestone deposits as the first step in the development of a local cement industry. His activities caught the attention of William Wallace Mein and George B. Poore, two prominent San Francisco mining engineers newly returned from the Witwatersrand in Africa, where they had been active in developing and operating gold mines. They formed a syndicate with several other mining engineers to take over Macnider's options and ideas. In 1924, after exploring regional market conditions and identifying the prime limestone deposits, they chose a plant site two miles south of San Andreas. Vigorous, knowledgeable, and experienced in mine development, they soon had a quarry open and a two-kiln plant under construction. To meet essential bulk transportation needs, company engineers and work crews worked jointly with the Southern Pacific Railroad to extend a spur from Valley Springs to Kentucky House, an old stage stop on Calaveritas Creek, two miles from San Andreas. By May 1926 production of high-quality cement

Dedication day at the Calaveras Cement plant, 1926. (Courtesy Calaveras County Historical Society)

was well under way. With its first big order from Pardee Dam contractors a few months later, the little plant was to become one of the county's most productive and profitable mining enterprises over the next fifty-six years. The company managed not only to stay alive but even to double its production capacity during the long, lean Depression years. By the end of the thirties Calaveras Cement was poised to provide a strategic material to help meet the insatiable demands of World War II.[9]

Calaveras Cement thrived during the war and expanded rapidly thereafter. Wartime price ceilings were finally lifted in the late forties, and the exploding demand for industrial concrete launched the company on an accelerated growth spiral. A series of expansion programs in 1946, 1951, and 1956 made it one of the county's most important and busiest modern industries, with five kilns capable of producing 650,000 tons of cement a year. The raw and finish mills as well as the kilns operated around the clock, seven days a week. A group of subsidiary aggregate and ready-mix operations and several rail-supplied cement transfer stations facilitated the distribution of cement and concrete to customers throughout the company's market area.

Accompanying these changes in the physical plant was a change of management. Closely controlled by William Wallace Mein and his associates for nearly thirty-five years, the owners in 1959 decided to accept the terms of a merger with the Flintkote Company, a nationwide building-materials manufacturer. Flintkote operated the plant for twenty years, expanding the distribution system, adding a new cement plant at Redding in 1961, and calling upon managerial and technical staff of the Calaveras Division to assist other Flintkote divisions throughout the country. Then in 1979 the Genstar Corporation, a Canadian company, absorbed the entire Flintkote operation.

The growth of the Kentucky House plant required expansion of the existing quarries at San Andreas and at Calaveritas to provide adequate supplies of raw material. By 1971 these quarries, operating on a two-shift, five-day-a-week schedule, had become so deep that waste-stripping ratios were too large for profitable operation. A new quarry was then opened at Cataract Gulch on the north side of Stanislaus Canyon, on the Camp Nine Road near Murphys. The company connected this quarry and its crushing and grinding facilities with the five-kiln burning plant at Kentucky House by a seventeen-mile slurry line.

For nearly a quarter century the Calaveras plant provided a very satisfactory cash flow to both the Flintkote and the Genstar companies. But by the early 1980s it became infeasible to continue due to constant increases in labor, fuel, and electricity costs. Conversion from oil to coal in 1977 brought temporary relief, but the plant needed much upgrading and modernization. California cement producers began importing Asian cement clinker and finished cement in the late 1970s because of increased market requirements, and the cheaper costs of the foreign product. This became a major factor in the final closure of the Calaveras plant. Newly imposed environmental constraints likewise affected the decision not to modernize. In December 1982 a union dispute over the use of outside contractors brought matters to a head. The parent company, to prevent further operating losses, closed the Kentucky House plant. Thus ended fifty-six continuous years of successful and profitable cement manufacture that had provided more employment, more tax revenue, and, indeed, a bigger impact on the regional economy than any other mining enterprise in the county's history.

## Transformations in Logging and Railroading

Calaveras County's lumber industry came of age in World War II. Before that time most county lumbering was regional in nature and in corporate development. A number of the local companies had found themselves financially over committed in the twenties, and few survived the Depression intact. Even the Pickering Lumber Company, a large firm that had acquired an extensive sugar pine stand within Calaveras boundaries in the twenties, failed in the next decade and had to be reorganized before it reopened in 1937.[10]

Timber-holding companies as well as logging enterprises changed hands because of the Depression. Charles F. Ruggles, a Michigan timberman known as "the last of the lumber barons," had first entered the Calaveras timber business in 1909 when he purchased the Solinsky holdings. By the twenties he had joined Frank Solinsky Jr. in organizing the Calaveras Timber Company. Like James L. Sperry a half century earlier, Ruggles wanted a railroad to develop his timberlands. When Calaveras Cement Company president W. W. Mein announced plans for a rail extension from Valley Springs to Kentucky House, Ruggles proposed to lay track from the cement plant upriver to his timber properties. However, hard times hit Ruggles in 1927, two years before the crash. His Michigan enterprises collapsed, forcing him to raise funds by mortgaging his Calaveras Timber Company. Neither he nor his timber enterprises could survive such financial burdens after 1929. He died in 1930 at the age of eighty-four, and soon after the Calaveras Timber Company fell into the hands of bondholders who organized the Calaveras Land and Timber Corporation. By the end of World War II this syndicate had pooled its timberlands with those of the Winton Lumber Company, which had acquired the Brown Brothers Lumber Company's tracts in Calaveras and the Amador Lumber Company's holdings across the Mokelumne. The merger made Calaveras Land and Timber one of the largest privately owned timber tracts in the southern Sierra, with sixty thousand acres of prime forestlands in Calaveras and another sixty thousand or so in Amador.[11]

Rising war clouds in Europe coincided with the revival of the Calaveras timber industry. In 1938, with privately owned prime timberlands in Sierra County running out, Frank N. Blagen, president of the Davies-Johnson Lumber Company, operators of the Calpine mill, secured a contract with the Calaveras Land and Timber Corporation to harvest part of its holdings. Using the trucks of his contract logger, S. D. "Doc" Linebaugh, who had been with Calpine since the early 1920s, Blagen dismantled his Calpine mill and reassembled it at White Pines, one mile from Arnold. A planing mill on rail at Toyon just above Valley Springs was completed at the same time, and the renamed Blagen Lumber Company began operations in 1939. Just before the United States entered the war, Blagen ran into financial difficulty and sold out to the Stockton Box Company, a subsidiary of the American Forest Products Corporation (AFPC).[12]

World War II thus marked the beginning of a new Calaveras lumber era. Wartime demand for all types of wood products led to rapid expansion and consolidation. Harvesting logs from both private and government lands, Doc

Linebaugh and his truckers delivered 35 million board feet annually to the White Pines mill that turned out 180,000 board feet per day in two shifts. By California standards this was a medium-size plant, but the White Pines mill surpassed all previous production records for Calaveras County.[13] As an example of the frenetic wartime pace, when the planing mill at Toyon burned to the ground in 1942, it was completely rebuilt in twenty days. The following year AFPC built a new mill a few miles south of West Point in prime sugar pine country. It also erected a company town there, named Wilseyville after Frank Wilsey, manager of the White Pines mill from 1939 to 1954. By the end of 1945, American Forest Products employed more than eight hundred people in Calaveras County alone, making it the county's largest single employer. For three years following 1945, Calaveras ranked among the top ten counties in statewide lumber production. By 1947 its mills were producing more than 150 million board feet annually, nearly equaling neighboring Tuolumne, whose total forest acreage is twice that of Calaveras.[14]

Despite growing annual production figures, Calaveras's percentage of total state production steadily diminished in the 1950s. By 1950, Calaveras had increased its output to 186 million board feet and had expanded to twenty-two mills, most of them small family-size operations. However, the county's share of state production had dipped to little more than 4 percent. That proportion had shrunk to 2.1 percent by 1955. By 1960 it was only 1.4 percent of the state figure.[15] AFPC, one of the nation's largest lumber producers in the mid-fifties and the major producer in Calaveras, estimated in 1958 that 40 percent of Calaveras residents depended, directly or indirectly, on the lumber industry. That was twice the number estimated for the county in 1939.[16] In spite of this optimistic outlook the fate of the Calaveras timber industry depended on circumstances beyond local control, including the corporate health of its biggest producers, the changing climate of forest rules and regulations governing timber sales on public lands, and the environmental conditions that affected the use of the resource. All of these external factors turned negative after the 1950s, and the result was an overall decline in timber production and a diminished role for the wood products industry in Calaveras over the next three decades.

The county's lumber industry continued to slide in the 1960s, averaging only 72.2 million board feet annually, down 57 percent from the previous decade. AFPC purchased the Calaveras Land and Timber Corporation holdings in 1961, but the next year closed the White Pines mill after twenty-two years of continuous operation. The Wilseyville mill shut down a year or so later. Still later AFPC acquired the Winton Lumber Company and its Amador County holdings, thereby nearly doubling its forest reserves and obtaining the large, diversified plant at Martell. The Toyon operation was then relegated to a very small assembly plant. Martell, in contrast, newly modernized and expanded, became the center of lumber activities in both Calaveras and Amador Counties. After the Bendix Corporation acquired AFPC in 1970, it built up Martell into one of the largest integrated forest-products plants in the state. Meanwhile a little plant at Toyon, owned by the Bohemia Company of Grass Valley, closed soon after. The Thornburg plant at Wallace, rebuilt and expanded by its new owner, the Snider Lumber Products Company of Turlock, survived until 1993.

In recent years lumber production in the central Sierra Nevada has been whipsawed by conflicting economic and environmental interests. On one hand are rising concerns over habitat protection and restoration in the wake of the spotted-owl controversy in the Pacific Northwest, as well as an accelerating and highly vocal opposition to traditional methods of logging, especially clear-cutting. On the other hand are pressures for increased timber production on private timberlands to offset the continued decline of logging on public lands. Complicating the picture are changing economic conditions that have accelerated the pace of corporate restructuring in the lumber industry, as well as demographic shifts that have begun to affect public perceptions about the importance of traditional industries in the Mother Lode.

The corporate restructuring that had begun in the 1960s increased dramatically in the uncertain economic climate of the 1980s and 1990s. In 1981, Bendix shed its AFPC subsidiary after a leveraged buyout by a private group affiliated with buyout specialists Kohlberg, Kravin, Roberts, and Company. Seven years later Georgia Pacific acquired the limited partnership that controlled AFPC's assets, including 125,000 acres of Sierra timberland and the sawmill, particle-board plant, and molding and millwork plants at Martell. The molding plant soon closed, but the others remained open until the mid-nineties. In the meantime Sierra Pacific Industries, a private family-operated northern California timber products company based in Redding, suddenly assumed giant status in the central Sierra. In a bold move to increase his private holdings to offset the decline of timber harvesting on public lands, in 1988 Archie Aldis "Red" Emmerson, family head and CEO, mortgaged all the company's assets in order to purchase 522,000 acres of timberland from the Santa Fe–Southern Pacific Railroad. The following year Sierra Pacific acquired 27,500 acres near Truckee and another 10,000 acres in Tuolumne County. Additional acquisitions in the nineties from Georgia Pacific and Louisiana Pacific, including 127,000 acres and the two plants remaining open at Martell, plus 38,000 acres northeast of Sacramento, made Sierra Pacific by 1997 the largest private landowner in the United States.[17]

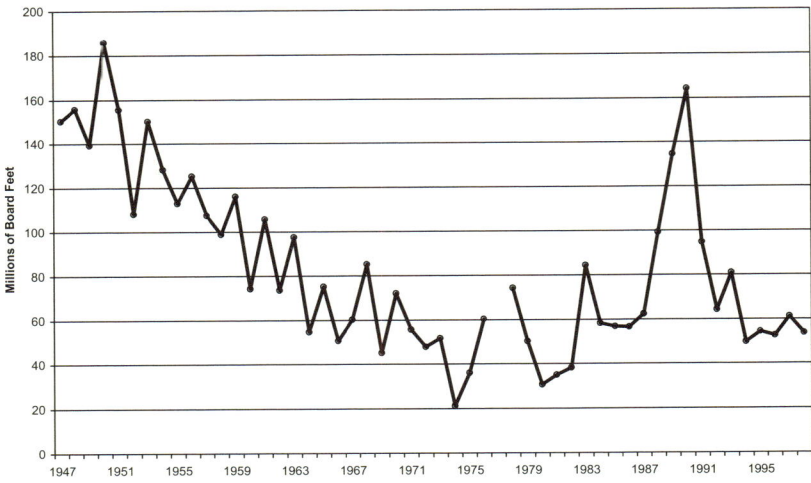

Calaveras Timber Production, 1947–Present. (Authors' database)

While these changes reshaped the corporate landscape, the U.S. Forest Service came under increasing pressure to reduce the timber harvest. Growing concern for spotted-owl habitat in the Pacific Northwest, plus a rising chorus of opposition to clear-cutting, led forestry officials at both the state and the national levels to rethink traditional timber management policies. After a three-year jump in timber harvest during the drought years of the late eighties and early nineties, in 1991 the Forest Service took steps to reduce timber production while scientists studied the impact of logging on the northern California spotted owl and other threatened or endangered species. In Calaveras the logging decline was dramatic. From a 1990 peak production of 164.3 million board feet, production dropped by 70 percent over the next four years. By 1995 only 2.4 percent of the state's timber harvest was coming from Calaveras County.[18]

Though timber production in Calaveras since the mid-nineties has remained far below historic levels, as the century closed the logging industry seemed poised for a comeback. Despite the environmental community's strong opposition to clear-cutting, Forest Service studies have been unable to document any adverse effects of that logging method on endangered or threatened wildlife populations. In response to pressure both from industry and from Congress during a sustained economic growth period, the U.S. Forest Service has proposed several plans to increase logging in the national forests of northern California. However, implementation has been delayed pending full public review and possible court action.[19]

Logging on private lands, in contrast, took a giant leap forward early in 2000, when Sierra Pacific Industries secured permits to clear-cut more than 900 acres of its timberlands in the central Sierra as part of a long-range policy to harvest most of the mature timber on its 1.5 million acres over the next twenty years. Before the logging trucks began rolling through the little town of Arnold, however, a surprising surge of opposition arose from local business leaders, the timber industry's traditional allies. Arguing that clear-cutting harms the watershed above Arnold, defaces the landscape, and threatens tourism, they echoed the sentiments of a coalition of business representatives in Nevada City, who earlier threatened a lawsuit to stop the Forest Service from implementing a plan to triple logging levels in three northern California national forests. Stung by the criticism, Sierra Pacific defended its logging practices but finally agreed to scale back its clear-cutting plans. The political landscape had clearly shifted in the froth of these policy battles, but even if logging accelerates in Calaveras the economic impact on the county will remain small.[20]

The post–World War I era began with two railroad companies and three lines operating in Calaveras County. By the end of the thirties, only one line served the county's northwest corner. This decline was due primarily to the burgeoning truck and passenger car traffic that placed an increasing strain on railroad revenue and forced several cutbacks during the era. Railroading in Calaveras might then have ended altogether had it not been for the rising cement traffic on the Southern Pacific's extension from Valley Springs to Kentucky House, constructed in 1925, and the increasing lumber traffic on the verge of World War II. However, by that time passenger and express traffic was only a memory. Passenger service to Valley Springs ended in 1932.[21] The Milton station agent left before 1933, and a few years later Southern Pacific officials abandoned the line and ripped up the tracks. Service to Angels Camp over the Sierra Railway ended in 1935, and construction crews removed the tracks not long after. Today almost all traces of the routes to Milton and Angels have disappeared.

The surviving rail line into the county, terminating at the Kentucky House cement plant at San Andreas, increased its daily haul substantially when that plant converted from fuel oil to coal in 1977, requiring a third of a ton of Utah coal for each ton of cement produced. With the closure of the cement plant in 1982 all traffic on the line above Lockeford ceased, and the rails were pulled in the late 1990s. Calaveras County is now without any rail connections to the outside world, and local officials are divided whether to purchase the county's last rail corridor, now owned by the Union Pacific Railroad, or to lose this "viable asset" forever.[22]

## A New Gold Boom and Bust

Like Lazarus, the California gold industry came back to life in the late 1970s and early 1980s. The landmark event that sparked the revival was the 1971 change in international monetary policy that allowed the price of gold to float on the open market. Freed from federal regulation for the first time in U.S. history, gold now became simply a commodity, subject to all the whims of the world marketplace. Riding in on a tidal wave of inflation following the oil embargo of 1973 and the subsequent energy crisis, gold prices skyrocketed. For a brief, panicky moment early in 1980, when inflation was running rampant, gold topped $800 an ounce. By the early eighties, however, after oil prices collapsed and the economic hard-liners took over during the Reagan administration, the market price had settled back to a level more closely reflecting international supply and demand, and has continued to the present in a trading range of $250 to $350. But the prospects of a sustained market and a handsome profit margin were enough to revitalize dozens of gold mining projects in California and beyond.[23]

Calaveras County, astride the heart of the Mother Lode, rode through this inflationary cycle with increasing excitement and rising expectations. Goldbugs resurfaced by the score. Mining was a major topic of conversation up and down the Sierra foothills. Typical of the popular sentiment was a comment by Ralph L. Reynolds, a Grass Valley mining broker: "The prospectors of 1849 just scratched the surface. The major underground deposits have hardly been touched." He thought gold mining "had a very rosy future."[24] Everyone did not share such optimism, however. A few small mines, like the Blazing Star in the West Point district, reopened quickly, but production was limited and results disappointing. The failures reinforced the views of most technical experts, who believe that, except for some surface deposits that might be mined profitably by open-pit methods, essentially all of the significant deeper Mother Lode ore bodies have already been worked out.

Those same experts were no doubt quite surprised in the mid-eighties, when Carson Hill reopened after a forty-three-year interlude. The mine had lain undisturbed and seemingly abandoned since the 1942 fire. All the shops and equipment had been torn down and sold during the war, leaving only the surface scars. On the south side of the hill, along the Stanislaus and its local tributaries, the rising waters of New Melones reservoir buried the visible dumps and pits after the dam's completion in 1979. The most visible surface feature on the north side was the old Morgan glory hole, a large, unbenched conical gash that could be seen for miles along Highway 49 south of Angels Camp.

Long before this latest revival the old Carson Hill mine had changed owners. In 1952 the Stevenot family acquired all rights to the property previously held by Anglo American, the corporate investment firm that had abandoned earlier plans to reopen Carson Hill. Fred Stevenot, a San Francisco banker and grandson of the first man to consolidate Carson Hill mines, wrote to his brother Archie soon after the deal, explaining what the acquisition meant to the family: "At least the old family mines will be coming back in the Stevenot name. I guess that's plain sentiment. Yet the thought pleases me, and yourself as well. Without a change in conditions, it is just a big unproductive investment."[25]

Both Fred and Archie Stevenot died before those conditions changed, but Barden Stevenot, Fred's grandson, took over the family interests in the 1960s and was in position to negotiate when Cyprus Minerals, Inc., a Los Angeles firm that had reopened and worked ancient copper deposits on the island in the Mediterranean until after World War II, tested the Calaveras site with an extensive program of diamond drilling in 1975. Barden Stevenot, with a background of business experience as a real estate developer and chief operating officer of Kirkwood Meadows ski resort, helped engineer the exploration of Carson Hill for Cyprus Exploration Company, a subsidiary of Cyprus Minerals. That firm's extensive diamond drilling developed an ore body showing values up to .05 ounces per ton, but in the late 1970s that was not enough to work profitably under existing technology and market conditions. Nearly ten years later Grandview Resources, a Canadian syndicate that had used heap leaching successfully elsewhere, leased the mine and began active planning for full-scale operations.[26]

By 1985 the procedure for developing an open-pit mine had changed considerably. Over the previous two decades the environmental movement had caught the attention of the American public. Concerns over groundwater contamination, air pollution, surface noise, hazardous-waste disposal, wildlife protection, and other matters historically ignored or at least glossed over by miners and their supporters now became public issues that attracted the attention of governmental entities at every level. They responded by imposing a series of laws and regulations, climaxed by California's Mine Reclamation Act of 1975, which dramatically altered the circumstances for doing business. After the 1970s corporate mining interests faced an expensive and time-consuming regulatory process that greatly increased the initial capital investment. The

A sign of the times in modern mining: mitigation requirements to protect and restore wildlife habitats. (Authors' collection)

new regulations require extensive studies embodied in environmental impact reports, applications for mining permits, detailed mining reclamation plans, and numerous public hearings. The latter have often been stormy, as residents with no close ties to the mining industry, especially newcomers from urban industrial metropolises seeking the peace and quiet of country life, raise the familiar cry, "not in my backyard." Calaveras County and other rural provinces along the Mother Lode have frequently encountered the "nimby" syndrome in recent years. But county supervisors and older residents, more responsive to economic than aesthetic arguments, have generally not been very sympathetic to the newcomers. The result is often a court battle, as the "nimbys" reinforce their determination by filing lawsuits in efforts to block or delay new mining ventures, or in some cases to force compliance with permit conditions. A case in point was the Alto mining project near Lake Tulloch. It was delayed in the early permitting stage by public outcry from residents of Copper Cove, a nearby subdivision, who sued on grounds that the county supervisors, who had approved the project, violated state law by not requiring a comprehensive environmental impact report.[27]

At Carson Hill, Grandview spent four hundred thousand dollars acquiring the proper permits before shallow-pit operations could begin. Barden

Stevenot was once again a key player in the development. His "long suit," as he explained it, was his experience in acquiring the necessary permits from various federal, state, and local agencies. Working with regional consultants and appropriate officials, as well as with financial backers from Vancouver, the Grandview management completed the start-up work within three years. In April 1986 the company started full-scale mining, and by December the first gold values had been recovered from the dilute cyanide solution continuously spraying over piles of crushed ore on the leach pads. Within two years the operators were producing 150 ounces per day from 7,000 tons of ore. A virtual mountain of rock had been eaten away by huge shovels in eleven-yard bites, and the old Melones, Morgan, and Reserve glory holes became one giant open pit three hundred feet deeper. For every ton of ore mined, 3 tons of country rock had to be moved. This stripping ratio, plus the low gold recovery rate of 75 percent or less, meant that despite the estimated 15 million tons of ore reserves, operating costs came dangerously close to the profit margin.[28]

Heap leaching next to the primary water supply for downstream farmers in the Central Valley was a red flag to environmentalists, but proactive mining interests and county economic boosters drowned out the voices of concern during the permitting stage of the Carson Hill project. In 1989 a spillway coupling failed on one of the leach pads, dumping ninety thousand gallons of cyanide-laden wastewater. Before it could be contained, some ran down into New Melones reservoir. The amount was small enough to be quickly diffused, however. Tests by water-quality officials showed "no measurable amount of cyanide in the reservoir and no threat to public health," as local newspapers reported. Without fish kills or other visible evidence of a degraded water supply, and reassured by statements that cyanide was biodegradable and quickly broke down into harmless chemical compounds, the public shrugged its shoulders and mining continued unabated.[29]

In 1990 the Western Mining Company, a large Australian concern with considerable working capital and a huge acquisitions budget, paid seventy-five million dollars to buy out the Carson Hill operation.[30] It was bad timing, for the Gulf War and an international recession depressed gold prices worldwide. Full-scale mining proceeded for more than a year after the buyout, but the price of gold continued to weaken and the gold recoveries from heap leaching continued to decline as deeper mining entered fresher sulfide ore. No longer profitable, the mine closed in 1991 and began a long reclamation regime. This last period of activity produced about 100,000 ounces, making Carson Hill's lifetime gold production just over 1.5 million ounces.

In the Madam Felix–Hodson district the revived interest in gold mining after World War II brought Fresno gravel operator Charles Stewart, senior partner of Stewart and Nuss, to Calaveras County. In 1947 his firm bought out Frank Tower and merged the Royal with their own Mountain King holdings. Tower ended his Calaveras mining career with a flourish at the Butcher Shop mine in Hodson, the site of a small but spectacular high-grade pocket. Stewart's consolidation was followed by a series of reexaminations and tests by the American Smelting and Refining Company (ASARCO), New Jersey Zinc, and Yuba mining companies. In 1970, Stewart sold his interests to Lawrence Monte Verda's nephew Bernard, who encouraged Homestake to enter the district. That prominent company test drilled once more, then bowed out.

Even though the economic climate was not yet right for the resumption of gold mining, in 1974 Frank Adams, a San Franciscan long associated with the mining industry, formed Mother Lode Gold Mines, Consolidated, and proceeded to acquire promising gold properties. His ventures took him to the Madam Felix district by 1977, when he made a deal with Monte Verda to renew the test drilling and mineral beneficiation testing. After obtaining the preliminary mining permits the company sought a partner with the capability and finances needed to take over the operation. By 1984 it had signed up Nerco Minerals, a division of the Pacific Power and Light Company, but after a vigorous two-year program of testing and feasibility studies, Nerco also decided not to proceed.

The Mother Lode company's efforts finally succeeded in 1986, when Calaveras mining interests learned that the Royal and Mountain King properties had been taken over by Meridian Minerals, a subsidiary of the Burlington Northern Railroad. Meridian had previously made a deal with the Felix Mining Company, owners of the adjoining Gold Knoll mine, so with the new acquisition the Meridian company had control of the entire district. Meridian moved decisively in that year to finish the test drilling that had been conducted intermittently by several companies since World War II. After completing extensive geological and mineralogical studies, the new operator developed a comprehensive mining plan and proceeded to acquire the remaining permits. Meridian then designed a mill and all the complex components of a large, modern open-pit mining operation, including tailings ponds, mine waste dumps, access and haulage roads, and water supply. The nature of the mineralization required a conventional milling procedure with extremely fine grinding to liberate the gold values. By February 1989 the company had completed its construction program and mine development, and was in full production. The first gold bullion was shipped the following month.

In 1903, J. C. Kemp van Ee had built the Royal mine into the most impressive and up-to-date operation along the Mother Lode. Eighty-six years later the Meridian company matched and then exceeded that effort by once

The Royal Mountain King open pit, 1991. (Authors' collection)

more transforming the Madam Felix–Hodson district into a showpiece of the Sierra gold country. Exploration work at the Royal Mountain King "proved up" some 9 million tons of low-grade ore running an average 0.07 ounces of gold per ton. The stripping ratio was quite high, however, requiring 7 tons of waste to be removed for every ton of ore recovered. Meridian nevertheless moved into full production with one of the best-managed and most technologically advanced mining operations ever seen on the Mother Lode.[31]

Despite the efficiency of the operation the recession of the early 1990s and changing market conditions cut short the Royal Mountain King's revival. Partway through its scheduled mining and milling program, Meridian was taken over by the FMC Gold Company. The Meridian project had been planned when gold prices appeared to be holding around $450 an ounce. By the early nineties, however, gold dropped below $350, and the new owners faced the necessity of curtailing operations to produce the maximum cash flow in the shortest time interval. In 1994 they reluctantly shut down the mine and began the reclamation work required by their permits. The exemplary mining operation at Royal Mountain King had recovered nearly 7 million tons of ore in less than six years, producing more than 318,000 ounces of gold as well as 335,000 ounces of silver. These figures brought the Madam Felix district total to over 500,000 ounces of the yellow metal. The milling process at Royal Mountain King recovered some 0.058 ounces per ton, for a recovery rate of 77 percent, an excellent figure for such fine-grained low-grade sulfide ore. However, more than 50 million tons of waste had to be removed to obtain such results. Meridian had an operating profit of some $22 million, but even after salvage and sale of equipment, only part of the original capital investment was recovered.[32] As we have seen before, high production was no guarantee of financial success in Calaveras mining history.

## Other Extractive Industries

For a brief period in the seventies and eighties an important new mineral revived Calaveras mining prospects. The large asbestos deposit southeast of Copperopolis, near the Stanislaus River, has been known, prospected, and promoted since the early years of the century. Not until the 1960s was it taken seriously, when the Jefferson Lake Sulphur Company decided to open it up. After a decade of unsuccessful attempts to make it profitable, Gordon Coats, a businessman with varied experience in several countries, set up the Calaveras Asbestos Company to take over the operation. In a short time this mine, with a state-of-the-art mill, became the largest asbestos producer in the United States and the second-largest industrial employer in the county. By the mid-eighties, however, Calaveras Asbestos faced serious depletion of the ore reserve previously blocked out, and could develop additional ore only by undertaking an uneconomical stripping project. Also troubling was the growing national concern over studies linking asbestos particulates and lung cancer. Although Calaveras Asbestos sold only about 20 percent of its output in the domestic market, the health issue and the new state and federal restrictions on asbestos use in the United States were important factors in closing the mine. But the primary reason was economic, not environmental. As was the case with so many other mines, this very successful operation closed because the ore body was exhausted. Later the site was converted to a toxic dump primarily for asbestos waste materials, but not without controversy, however. County supervisors were slapped with several lawsuits for approving the project.[33]

Calaveras has long been known for its soapstone and low-grade talc deposits. In the 1970s, with the rising demand for low-grade industrial talc for roofing products and for similar applications, the talc mining division of Johnson and Johnson opened up their Western Source quarry on Red Hill Road between Carson Hill and Vallecito, and built a small processing plant at Toyon, on the Southern Pacific Railroad line to Kentucky House. This successful little company, however, found that trucking transportation was more economical than rail. Johnson and Johnson then sold it to Cyprus Minerals, earlier active at Carson Hill, which had a number of other talc operations. Eventually, Western Source was taken over by Luzenac, the talc division of Rio Tinto Zinc, one of the largest mining concerns in the world. However, changing economic conditions in the talc industry led to the closure of the quarry and plant late in 1997.

The dam construction boom, started in the twenties with the Pardee, Melones, and Hogan Dams, has continued with Camanche, New Hogan, New Melones, and most recently the North Fork projects. This work has required large, though short-lived, quarry and tunnel operations, and has brought temporary waves of prosperity to the county. Of a more durable nature has been the small but steady development of the sand and gravel and crushed-rock industry, mostly centered in Chili Gulch, in Jenny Lind, and at the Calaveras Cement company's waste dumps near San Andreas and Calaveritas. Demand for these materials doubtless will increase over the long term, and thus Calaveras will continue, on a minor scale, to carry on its mining heritage.

## Recreation and Retirement: New Mother Lode Industries

The loss of railroads and the decline of mining and other resource industries in the eighties and early nineties did not trouble a new wave of Calaveras residents who flocked to the foothills from valley and coastal cities during the same period. New subdivisions such as Rancho Calaveras, Copper Cove, and La Contenta just east of the San Joaquin County line opened in the 1960s but boomed in the early 1980s, attracting working families from Stockton and Modesto and other urban centers in the San Joaquin Valley. As a feature writer for a Stockton newspaper explained, commuters came to Calaveras "to escape the stress of big city living. They look forward to outdoor activities as soon as they step from the home. They are lured by the prospect of smaller schools. Mostly, they come for space. And they are willing to drive the distance to pay for that slice of elbow room." [34]

In the eighties and early nineties Calaveras also benefited from white flight. Coastal California in general, and the Los Angeles Basin in particular, lost thousands of middle-class older whites fleeing the high-crime, high-priced, earthquake-prone urban and suburban landscape. Recession and the transition to a peacetime economy following the end of the cold war accelerated the exodus to less expensive rural communities. [35] It made economic sense to move to new foothill subdivisions where a half-acre parcel with electricity and water cost half the price of a city lot. The result in Calaveras was a population boom that began in the late seventies and continued almost without letup into the late nineties. Between 1979 and 1990 the county's population grew

by more than three-fifths, finally exceeding the Gold Rush figure of 20,183
set in 1852. The peak year was 1987, when the county per capita growth rate
was the highest in the state. The statewide recession following the end of the
cold war and the conversion of the state's huge defense economy slowed foothill growth for a time, but as the economy healed the rapid growth patterns
returned. In 1997 alone 1,200 newcomers came to Calaveras, boosting the
county's annual growth rate to 3.4 percent, more than twice that of neighboring counties.[36]

The new residents have dramatically altered the county's economic outlook in recent years, offsetting to a large extent the loss of jobs and payrolls due
to cutbacks in mining and other traditional industries. Whether still working
or retired, with their payroll checks and bank deposits the newcomers have
infused new life in the county economy. They drive expensive automobiles,
purchase and restore historic properties, open or patronize chic new shops,
play golf, and stimulate the growth of local cultural events. But the downside
has been an offsetting loss of rural values as city dwellers move in and bring
their urban values with them. John Maison, researcher for the California Department of Finance, underscored the basic conflict in a recent interview:
"People want to get away from the congested spaces and into the wide open
spaces of the country," he said. "But they don't want to give up all the city
comforts." "We are just moving the valley up the hill," conceded Ed Brock-
man, president of the Tuolumne County Historical Society.[37]

Tuolumne and neighboring Calaveras are discovering the social costs of
urbanization, which has strained services and forced unwelcome changes in
lifestyles and the cost of living. County officials, hard-hit by semistatic revenues in the wake of Proposition 13 and subsequent reductions in state funding, have increased development fees to pay for fire protection, roads, and
sewers. They also have responded to the rising concern over environmental
impacts by tightening restrictions, and in some cases imposing moratoriums,
on new septic systems. The result has been an increase in the cost and difficulty of moving to higher ground, equalizing to some extent the attractions
that once drew valley residents to the foothills. Yet the urban flight still continues, with Calaveras County in the forefront of foothill counties undergoing an economic and social metamorphosis.

This process of foothill urbanization has not been kind to the mining industry. Expatriate urbanites from smoggy coastal and valley towns, and retirees looking for peace and quiet in a rural environment, are not inclined to
appreciate mining activity in their newly acquired backyards. When the Royal
Mountain King sought permission to reopen in 1988, for example, nearby
property owners in the newly established Diamond XX subdivision protested

out of fear that "their well water will be drained, the dust will choke them in the sweltering summer months, and vibrations from blasting will crack and ruin their homes." One homeowner who moved to the area in 1976 said: "I've seen the gold veins in the mine shafts. I know what they're going after. I just don't want to be a victim."[38] Royal Mountain King mining advocates persisted in the face of such fears and ultimately secured the necessary permits, and after mining ceased the company even did more to reclaim the site than the permits required. In other cases adverse public reaction has been enough to scare away potential mining investors. The open clash of interests makes it unlikely that the industry will ever again be universally welcomed in Calaveras County.

## Mining Families and Fading Memories

For deep-rooted Calaveras families, mining no longer dominates but still influences family occupations and opinions. It also brings back memories for some retired family miners, though time is taking its toll on those who personally witnessed or worked in Calaveras mines. Howard Tower, whose uncle Frank was one of the shrewdest mining men in the region, vividly recalls the sparkling fingers of crystallized gold in the Butcher Shop mine. Working carefully by hand, Frank's men broke up this high-grade enrichment on the Royal vein—quite unusual in an otherwise low-grade deposit—and carried it out in powder boxes. Neighbor to the Towers in Salt Spring Valley are the McCarty heirs who still hold land and mineral rights to property Thomas McCarty acquired in the 1860s. Ronald McCarty runs an equipment company in Copperopolis but does contract work for the Royal Mountain King crew closing down the operation at Hodson.[39]

In Angels Camp, Tony Dutil, though long retired, has strong recollections of mining in the thirties and forties, and still prospects occasionally on family property near Copperopolis. In Milton, Glen Nevens remembers the dredges his father worked on in the Jenny Lind area, and the four-by-four-foot prospect hole he later dug for another dredge company. At seventeen feet below the surface he could still throw out cobbles without hoisting. In Sonora, retired school superintendent Ted Bird, at ninety, can describe in detail the Christmas meals he enjoyed every January 7 as a kid visiting Serbian Orthodox mining families at Melones, where his father worked. Charlie Stone at Copperopolis recalls riding the skip at the Royal mine on a weekend outing in the thirties while he was in high school. In Stockton, Vickie Saunders fondly remembers her "great adventure" as a child in Calaveras at the beginning of World War II, when her father invested in mines near Paloma and

The collapsing roof of the Calaveras Central hoist house symbolizes the fading memories of Mother Lode mining. (Authors' collection)

at Quail Hill. While her mother fretted, she accompanied her father overland and underground, learned to pan, shoot a gun, and ride a horse, and thoroughly enjoyed the outdoor life. On the family ranch on Bald Hill, Dick Rolleri remembers the full stock of blacksmith tools left behind in the machine shop of the Calaveras Central mine when his family acquired the property in the early 1960s. The double-drum hoist still remains there today, intact but rusting away under the crumbling roof of the hoist house.[40]

Among the mining families of Calaveras, no one personifies the regional mining heritage better than Barden Stevenot. A fifth-generation Calaveras mine promoter and developer, his personal career in mining started after college and a navy stint with a suction-dredge operation on the Stanislaus River in the 1960s, right below Melones. Confronted by a fixed gold price and minimal returns, Stevenot gained nothing but experience from that endeavor. For a decade he shifted from mining to ski resort management and real estate development at Kirkwood Meadows in the high Sierra on the eastern border of Calaveras, yet all the while retaining mining interests as manager of the Stevenot properties on Carson Hill. In the 1970s, on historic ranch property above Murphys, he started the Stevenot Winery, soon one of the largest and most successful wineries in the entire Mother Lode. But always "fascinated by mining," in the 1980s he turned attention to the district his great-great-grandfather had first examined in 1850. First with Cyprus Mines, then with Grandview, Stevenot was instrumental in bringing mining back to Calaveras, possibly for the last great hurrah. Motivated by a traditional sense of the importance of jobs and development to the public well-being, Stevenot is also a long-standing environmentalist and preservationist, with a record of support for protecting fragile landscapes and historic properties from reckless or environmentally degrading development.[41]

Unlike many newcomers to the Mother Lode, Stevenot recognizes the importance of mining to Calaveras. "Mining opened up California," he says emphatically; "that was the bottom line." Yet he also recognizes the reality of modern mining. His exploratory diamond drilling along the Mother Lode from Angels Camp to Carson Hill in the early 1990s has convinced him that in the global context mining has little future in Calaveras. The county's ore bodies, in his opinion, are insufficient to sustain a significant program of development. "When you can mine as easy in Indonesia as in the U.S.," he says, there is little chance that major developers will return to the Mother Lode. Besides, he asserts, the current regulatory regime and the "social impact" of modern mining in a changing and proactive environmental era make it highly unlikely the industry will ever revive in the California foothills. For Stevenot, mining in Calaveras still has an important heritage, but it is "just part of the tapestry" of modern life in the Mother Lode. As a real estate developer he has consolidated properties and "built over the residue of old mines" where historic surface structures are not involved. Thus, the county's visible mines have faded as the landscape changes, but the legacy lives on in popular culture and in tourist attractions that annually bring thousands of visitors to the Mother Lode.[42]

These are fleeting images, a collage of memory fragments. The mining birthright of Calaveras is much richer and deeper than a few scattered recollections, but the direct ties are unraveling with the march of time.

## The Mining Heritage of Calaveras

Mining has dimmed to a faint fraction of its former radiance in Calaveras County. Nearly all the headframes and hoist houses have disappeared, the adits caved in, and the boardinghouses torn down or converted. Public officials continually warn of open shafts lurking in the bushes, a peril to man or beast, but the most visible ones have long been covered, fenced, or filled in.[43] Most of the Gold Rush camps faded from memory long ago, and even the open pits and surface plants of more recent operations are falling under the onslaught of natural or human forces of change. The Calaveras Cement

plant's limestone pit at Kentucky House is now an artificial lake; the man-
made moraines that represent the huge waste dumps of the Royal Mountain King mine are today green with the vegetation planted by the mitigation
crew left behind to clean up the site as required by law. Nothing remains of
the great mines at Angels Camp but a few rusting pieces of metal in Utica
Park and two or three unmarked and fenced-off shaft collars. The unnatural
angles of the Carson Hill skyline will take longer to erode, but casual visitors
probably do not even glance that way on their drive south along Highway
49 toward the Stevenot Bridge that spans the Stanislaus above New Melones
reservoir. Few who cross that bridge today would know that a once-important
mining town lies deep beneath the water below.

The miners, too, have faded from the memory of most Mother Lode residents. A few were visible during the last "rush" of the 1980s, but modern
mines use more machines than men. By 1989 mining employment in California had dropped to only 0.6 percent of the workforce, compared with the
national average of 1.6 percent. Those that were left had few illusions about
the industry, and neither did the public. As John Leshy has noted, no modern
American could "imagine a grimy, tired miner reaching for a Marlboro" like
the once fashionable billboard cowboy.[44] But it is not necessary to see active
miners or operating mines today, or even to spot the disappearing traces, to
understand the economic importance of mining to Calaveras. A quick glance
through the volumes of claim notices in the Calaveras County Archives is
enough to grasp the historic dimensions of the industry. William B. Clark
and Philip A. Lydon's county report in 1962 listed more than nine hundred
patented or located gold claims alone in Calaveras, not counting scores of
older claims absorbed by consolidation. This book has covered only the "big"
producers, those with a total gold output of forty thousand ounces or more.
But to keep Calaveras gold mining in perspective one must recognize that the
hundreds of small mining operations in Calaveras history also contributed
significantly to the county economy despite their limited production in the
aggregate—perhaps as little as 10 percent of the county's total gold output.
Only a few mines were really profitable, and those for only brief periods of
operation. Without doubt, mine operators in Calaveras spent more money
than they took out. On the other hand, leasing or selling mine property was
profitable for some speculators and owners whether or not the mine ever came
into production.

Regardless of production or profit figures, the mines had a significant economic multiplier effect. The Gold Rush stimulated overland and foreign commerce, food production, manufacturing and distribution, transportation near
and far, and wholesale and retail trade. Thousands who came for mining

shifted to farming, ranching, blacksmithing, merchandising, and dozens of other ways to earn a living. Vast amounts of new wealth transformed people and institutions, both at home and abroad. The global impact, in the words of economic historian Gerald Nash, was a "veritable economic revolution in the state, the nation, and the world."[45] Once started, the revolution could not be reversed after the Gold Rush ended in the mid-1850s.

Seen in microcosm the economic multiplier altered patterns of growth and development up and down the Mother Lode long after the Gold Rush ended. The wealth produced by mine workers also multiplied the economic power of their communities. As a San Andreas mining journal reported in the 1960s, one state survey "found the average miner's family to be 3 1/2 persons, and for every man engaged in mining and allied industries, 2 1/3 jobs were created in service industries. Each miner supported 12 local persons including merchants, mechanics, doctors, lawyers, and other professional trades and service people."[46]

The validity of these estimates cannot be tested empirically, but county tax records can provide some basis for measuring the economic impact of mining on the Calaveras economy. Since its inception in 1850, Calaveras has taxed mine-related activities to help pay annual expenses. Across the United States the primary source of county income has come from property tax revenue, but during the Gold Rush that source was limited in the mining West. Most mines were on public lands, not subject to state or local property taxes. Those obligations fell primarily on ranchers, farmers, and other private property owners who complained bitterly against discrimination. Mining advocates resisted any efforts to tax mining property or production, however, and were powerful enough in the Gold Rush era to protect the principle of "free mining." The net effect was to give miners nearly free and unregulated access to public land and resources at least until the 1860s.[47]

They did not escape other forms of taxation, however. To raise money for schools, public roads, bridges, and other services, western communities relied on licensing fees and taxes on personal property as well as real estate taxes on private property. Of course, those burdens also fell on nonminers, but where mining predominated miners and those who depended on them paid a significant share of poll taxes, liquor licenses, excise taxes, and other fees. They also began to pay taxes on mining property after 1866, when federal mining codes provided a procedure for privatizing public mining lands, thus making them subject to state and local property taxes.

In the mid-nineteenth century an important source of county revenue was the notorious foreign miners' tax, started in 1850 and imposed on Chinese and other minority immigrants (and some Native Americans, as we have seen in

TABLE 8.1  A Century of Mining Tax Revenue, Angels City and Townsip

| Revenue | 1860 | 1880 | 1900 | 1920 | 1940 | 1960 |
|---|---|---|---|---|---|---|
| Total tax revenue | $10,484 | $6,892 | $36,832 | $76,598 | $53,264 | $88,456 |
| Property taxes paid by mineowners | $1,270 | $2,514 | $21,484 | $42,623 | $14,450 | $15,467 |
| Nonmine-related general assessments | $9,214 | $4,378 | $15,348 | $33,975 | $38,814 | $72,989 |
| Total population | 2,366 | 1,381 | 4,258 | 3,165 | 3,447 | 4,941 |
| Number of miners* | 1,348 | 299 | 655 | 472 | 400 | 35 |
| Per capita general assessments | $3.89 | $3.17 | $3.60 | $10.73 | $11.26 | $14.77 |
| Other taxes paid by mining families† | $5,243 | $948 | $2,358 | $5,065 | $4,504 | $517 |
| Total mine-related tax revenue† | $6,513 | $3,462 | $23,842 | $47,688 | $18,954 | $15,984 |
| Ratio of mine-related taxes to total tax | 62.1% | 50.2% | 63.8% | 62.3% | 35.6% | 18.1% |

*estimates 1940, 1960
†estimates 1860–1960

chapter 3) until it was finally declared unconstitutional in the 1870s. In 1860, Calaveras earned $23,929 from this source, or 25.7 percent of its total tax revenue. After a decade of declining surface placers and anti-Chinese agitation the same tax in 1870 raised only $3,139, or 4.2 percent of county revenue. Other states adopted similar measures and evidently raised more income, mostly from Chinese placer miners.[48]

The economic impact of mining-related taxation, defined as taxes paid both by owners of mining property and by miners and their families, is suggested by looking at data from Angels Township, one of nine within Calaveras County. At one time Angels covered a sizable portion of southeast Calaveras, including Angels Camp, Copperopolis, and Carson Hill. The one hundred–year table shown here, prepared from tax assessment rolls and county population data at twenty-year intervals, shows the extent to which mines and mining families contributed to county revenue generated within Angels Township (and the incorporated City of Angels after 1910). It represents two kinds of revenue: taxes paid on mining-related personal property and all other assessments, including licensing fees, school and road taxes, and levies on real estate. In 1860–1861, for example, the township raised $1,270, or 12.1 percent of its total revenue, by assessing stamp mills, arrastras, machine shops, waterwheels, ditches, and other types of mining equipment. The remaining $9,214, or 87.9 percent, came from general assessments on real estate, licensing fees, and special taxes for various goods and services. Assuming mining families were representative of the general township population, their per capita share of these general assessments for 1860–1861 was $3.89, or $5,243. Thus, mining-related taxation for 1860–1861 in Angels Township totaled $6,513 ($1,270 + $5,243), or 62 percent of the township revenue base.

As mining expanded and contracted so did mining-related tax revenue. By 1880, with the contraction of mining after the Civil War, miners and their families within Angels Township were contributing only 50 percent of school taxes and other fees. During the golden years of lode mining more than 63 percent of the taxes paid within Angels Township came from its mines and mining families. The ratio of mine-related taxation to general revenue dropped precipitously after 1920, reflecting both the decline of the number of miners and large-scale mining operations and the rise of nonmining-related general tax assessments in the twentieth century.

These are suggestive statistics at best, and do not adequately reflect the breadth and depth of mining's economic impact in Calaveras. Better data, for example, might enable us to determine how much money generated by mining activity was spent locally, or the percentage of investment capital local mining ventures attracted to Calaveras. Even the gross amount of bullion produced can only be estimated, given the vague and often misleading data available, especially for nineteenth-century mines. The "backward and forward linkages," to use economists' jargon, between mining and ancillary industries like farming, lumbering, and railroading cannot be statistically demonstrated, though there is no denying the direct connections, as we have shown in chapter 4 and elsewhere.[49] These subsidiary activities, at least from the Gold Rush through World War I, depended on and existed primarily for the mining business. However fragmentary, the evidentiary record is still indicative. For well over a century, despite intermittent growth and long periods of decline, the mining industry made substantial contributions to Calaveras County's economic development.

Development can also produce negative results, in both human and environmental terms. Unfathomable greed, disorderly growth, price inflation and deflation, egregious waste, dislocated populations, and disturbed ecosystems characterized the rush for riches. Early mining and milling methods polluted air and water. A new mid-nineteenth-century "plunge and grab" attitude, transformed into Euro-American aggression and ethnocentrism, harmed minority populations and made mockery of American democratic idealism.[50] Boom and bust cycles spotted the West with ghost towns and abandoned equipment. Industrial mining exploited workers and accelerated the pace and scope of environmental harm. In the twentieth century, the rise of powerful unions and environmental legislation mitigated some of the most troubling aspects of the mining heritage, but scars remain in degraded landscapes, isolated and economically distressed rural communities, and displaced minorities still fighting to right past wrongs.

The mining history of Calaveras reflects some of these negative consequences, but the county's favorable geographic location and the diversity of its population and resources insulated it from the worst problems attributed to extractive industrial growth and change. Other small western communities have not been so fortunate. An example is Coos Bay, Oregon. Once the "lumber capital of the world," this lumber-dependent community in the 1980s suffered from a severe recession in the forest products industry, accompanied by mill closures in the Far West and expansion of Canadian lumber sales to the United States. The results, as William Robbins has described in a seminal study, were visibly illustrated in closed buildings, high unemployment, economic doldrums, and the depressed spirit of the people who live there.[51]

In contrast to Coos Bay, Calaveras and other Mother Lode counties, though heavily influenced by extractive industry, escaped the single-industry incubus that often characterized remote lumber-dependent towns. Industrial mining came to the Sierra foothills only after a long formative period of diversified, if inchoate, economic activity. By the time surface placer mining ended, most Mother Lode towns had metamorphosed into trade and supply centers for farmers, water developers, quartz mine operators, artisans, ranchers, and lumbermen. Though Mother Lode counties lost population after the Gold Rush, a diversified core remained, sufficient to survive hard times and resilient enough to prosper during the golden years of mining between the 1890s and World War I. In the meantime, foothill and valley were linked by an ever expanding infrastructure of railroads, highways, telegraph lines, power poles, pipelines, and ditches. Though the rail system deteriorated, the automobile age breathed new life in old Mother Lode communities, providing vital transportation links between rural upland communities and mushrooming population centers in the Central Valley and the San Francisco Bay Area. The linkage became ever more important after World War II, bringing tourists and retirees to the foothills, and providing commuter routes to industrial centers in the valley. Though hurt by the decline of logging in the 1980s, many Mother Lode working families had occupational alternatives not available to residents of more remote lumber towns in the Pacific Northwest.

If the Calaveras economy survived the decline of mining and logging, the region's air and water resources did not escape the degradation that has dogged old mining sites in Montana, Colorado, Idaho, and other "hard places" in the West. Miners everywhere exacted a heavy toll on nearby vegetation, streambeds, topsoil, and water quality, but the long-term effects have only recently come under intensive scrutiny. In 1972 the California Bureau of Mines and Geology, in a study of potential mine pollution, identified sixty-seven mines in Calaveras County. Only the copper and asbestos mines around Copperopolis, however, were noted as potential problem sites. The information was based largely on data gathered in the late 1950s and early 1960s, before the rising eco-tide.[52]

More recently, acid mine drainage (AMD) and heavy metals have become the chief culprits in the West's environmental wars. In addition to liquid mercury, dissolved zinc, copper, arsenic, cadmium, and other mine-waste contaminants may enter groundwater and surface drainage, degrade water quality, and kill freshwater organisms. Acidity increases in the presence of certain bacteria, making a potent brew that can dissolve a variety of heavy metals often found in sulfidic ore bodies. At pH values below minus 1.0, and iron concentrations of 141 grams per liter, the Iron Mountain mine in Shasta County, California, has the most potent acid mine drainage known in the United States.[53]

In Calaveras County public concern in recent years has focused on AMD and heavy metal contaminants from the abandoned Penn mine in the old

Campo Seco copper district. Generated by the oxidation of iron sulfides in the pyritic wallrocks exposed in the ten miles of underground workings as well as in hundreds of tons of excavated debris left on the 140-acre site, AMD has been spilling into the Mokelumne River since the mine opened during the Civil War. For more than a half century the river diluted and washed pollutants downstream, out of sight and mind of most Californians. But explosive coastal growth in the Roaring Twenties, both in the Los Angeles Basin and in the San Francisco Bay Area, placed increasing demands on all freshwater supplies throughout the state. While Los Angeles looked to Owens Valley and the Colorado River for additional sources of water, coastal Californians in the North set their sights on Sierran streams. Six years after San Francisco completed Hetch Hetchy Dam on the Tuolumne River, the East Bay Municipal Utility District (MUD) built Pardee Dam on the Mokelumne a mile and a half above the Penn mine. Upstream diversion, however, had downstream consequences that became increasingly apparent on the Mokelumne after World War II.[54]

In 1963, East Bay MUD, accepting responsibility to recharge downstream aquifers and regulate flow on the lower Mokelumne, constructed Camanche Dam eight miles below the Penn mine. Coming at a time of increased environmental and political activism, Camanche construction, instead of solving regional water problems, led to decades of controversy. Surface runoff from the Penn mine now drained into Camanche Reservoir instead of flowing to the sea. Increasingly stringent federal and state water-quality standards imposed ever greater obligations on Mokelumne water managers. In 1978, to contain the acid buildup at the mine itself, the utility district built an impound dam and evaporation ponds, but in wet years the polluted water spilled over, dumping lethal doses of concentrated AMD and dissolved metals directly into the reservoir and killing thousands of fish. Neutralizing the acid by adding lime helped only for short periods. After years of debate and perfunctory cleanup efforts, a lawsuit in the early 1990s brought by a coalition of environmentalists and sport fishermen against East Bay MUD and the regional Water Quality Control Board led to a massive restoration project. Begun in 1997 and completed three years later, the ten million–dollar program has eliminated the containment ponds, covered the acidic rocks with topsoil and grass, and substantially reduced copper and zinc discharges. It was a costly, but instructive, lesson in the environmental consequences of unregulated mining. Out of this experience and dozens more like it throughout the mining West have come bonding requirements and other reclamation rules to hold mining companies accountable for the cleanup costs. In Calaveras the contrast between the Penn mine, abandoned in the 1960s with taxpayers picking up the reclamation tab, and the Royal Mountain King mine, closed in the 1990s but with company funds financing the reclamation plan, illustrates the regional impact of the new regulatory regime.[55]

If mining's lingering economic and environmental accounts have been partially audited and adjusted, one outstanding debt remains unresolved. Those who suffered the tragic human consequences of territorial aggression and de facto warfare on minority populations were never adequately compensated, though some of the costs of land lost by a few Native tribes in the West were addressed years ago in a series of claims brought before special federal courts established for the purpose. Most California Indians lost their tribal identities during the Gold Rush and its aftermath, however, and the thought of reparations to families whose ancestors were victimized decades earlier is not a compelling idea to modern Americans. Historian Patricia Limerick recently noted the paradox of Americans seemingly willing to accept responsibility for restoring the environment but not for restoring people damaged by past development.[56]

Though mining is gone from Calaveras, the mineral legacy still persists not only in economic and environmental terms but also in the vestigial elements of mining culture that can be found nearly everywhere in the foothills. Memories of active mining and miners are fading, but the legacy is obvious in the museums, historical societies, and restored buildings dedicated to the preservation of cultural resources and artifacts of an earlier era. Indirectly, the mining past also still influences the present, whether deliberately or consciously. The towns and trade centers that supplied the camps are spotted with structures or facades that date from early mining days, though often radically altered from the original. Gold Rush nomenclature persists in the names of commercial buildings, schools, sports teams, newspapers, public parks, annual parades and ceremonies, and dozens of regional advertisements. The private homes of newcomers are often decorated with mining artifacts or facsimiles. There is also a level of discourse, both public and private, that can arguably be traced to the Gold Rush psychology of individual entrepreneurship and the desire for quick wealth.[57] Mining, therefore, is more than a matter of historic machines and methods of extraction, more than rocks and people in mortal combat. Mining was essential to the rapid rise of Calaveras as a modern geopolitical unit, just as it was to so many other counties in the American West. It shaped the economy, determined the social structure, provided technology and capital for industrial development, molded the culture, and, despite the changing values and lifestyles of newcomers, still permeates the fabric of life in the Mother Lode.

APPENDIX

Calaveras County–Selected-Population Data, 1850–1940

| Census | 1850 | 1852 | 1860 | 1870 | 1880 | 1890 | 1900 | 1910 | 1920 | 1930 | 1940 |
|---|---|---|---|---|---|---|---|---|---|---|---|
| Total (State) | 93,000 | — | 380,000 | 560,000 | 865,000 | 1,213,000 | 1,485,000 | 2,478,000 | 3,427,000 | 5,677,000 | 6,907,000 |
| Miners (State)** | 57,979 | — | 82,573 | 36,339 | 37,147 | 21,310 | 12,964 | 6,622 | 4,889 | 54,730 | 21,137 |
| % Miners To State | 62.3 | | 21.8 | 6.5 | 4.3 | 1.8 | 0.8 | 0.3 | 0.1 | 1.0 | 0.3 |
| Total (County) | 16,884 | 20,183 | 16,299 | 8,895 | 9,094 | 8,702 | 11,200 | 9,171 | 6,183 | 6,008 | 8,221 |
| Miners (County)*** | (90.6%)* | — | 1,026 | 1,249 | 2,172 | 776 | 1,198 | 1,378 | 821 | 614 | 825 |
| % Miners To County | 90.6 | | 6.3 | 14.0 | 23.9 | 8.7 | 10.7 | 15.0 | 13.3 | 10.2 | 10.0 |
| Male | 16,617 | 18,679 | 13,698 | 6,246 | 5,988 | 5,455 | 4,552 | 5,452 | 3,642 | 3,497 | 4,671 |
| Female | 267 | 1,504 | 2,601 | 2,649 | 3,106 | 3,247 | 4,438 | 3,719 | 2,541 | 2,511 | 3,550 |
| Foreign Born | 5,855 | 10,735 | 9,359 | 4,218 | 3,349 | 2,332 | 2,375 | 1,813 | 1,035 | 659 | 657 |
| % Foreign To County | 34.7 | 53.2 | 57.4 | 47.4 | 36.8 | 26.3 | 21.2 | 19.8 | 16.7 | 10.8 | 7.9 |
| Chinese | (.5%)* | 1,441 | 3,657 | 1,441 | 1,037 | 333 | 237 | 10 | 11 | — | — |
| Black | 84 | 169 | 95 | 31 | 56 | 77 | 69 | 17 | 11 | 8 | 9 |
| American Indian | 3,000 | 1,982 | 1 | 18 | 169 | 77 | 100 | 161 | 65 | 125 | — |
| Mexican/Latino | (13.4%)* | — | (5.8%)* | 255 | 174 | 183 | 213 | 58 | 71 | 8 | 42 |
| Irish | — | — | — | 466 | 384 | 286 | 4 | 87 | 25 | 2 | 21 |
| Italian | — | — | — | 446 | 391 | 363 | 82 | 634 | 349 | 205 | 160 |

| | | | | | | | | | | | |
|---|---|---|---|---|---|---|---|---|---|---|---|
| German | — | — | — | 409 | 294 | 314 | 3 | 179 | 73 | 47 | 56 |
| French | — | — | — | 356 | 252 | 192 | 3 | 59 | 35 | 21 | 18 |
| U.K. (England, Scotland, Wales) | — | — | — | 337 | 311 | 306 | 127 | 173 | 94 | 79 | 68 |
| Canadian | — | — | — | 70 | 122 | 96 | 179 | 86 | 92 | 12 | 74 |
| Austrian | — | — | — | — | — | 37 | 233 | 280 | — | 24 | 8 |
| Finnish | — | — | — | — | — | — | 323 | 3 | 4 | 6 | 8 |
| Yugoslav | — | — | — | — | — | — | — | — | 57 | 44 | 47 |
| Hungarian | — | — | — | — | — | — | 513 | — | 6 | 2 | 4 |
| Other NW Europe | — | — | — | 122 | 37 | 159 | 47 | 159 | 126 | 104 | 116 |
| Other SE Europe | — | — | — | — | — | 29 | 20 | 11 | 40 | 57 | 26 |

*Figures in parentheses are estimated percentages of total population, based on census random sampling.

— = no data available.

**1880 data an estimate from 1882; 1890 census figures from 1889, when underground miners alone were estimated to number 5,522, with surface labor at 6,231; 1910 data include 1909 gold and silver miners only; 1920, 1950, 1960, and 1970 state figures include metals mining only; 1930, 1980, and 1990 state figures include petroleum and nonmetals mining; 1940 state figure excludes petroleum and coal mining.

***1890 data a estimate from 1893–94 State Mineralogist figures; 1900 data an estimate from 1895–1896 state mineralogist figures.

Sources: U.S. Bureau of the Census. Census publications and returns for 1850–1990; U.S. Bureau of Mines, Information Circulars, Technical Papers, 1911–1925; California Bureau of Mines, various publications, 1890–1970; Study #3: Historical, Demographic, Economic, and Social Data: The United States, 1790–1970, Inter-university Consortium for Political and Social Research (ICPSR), University of Michigan.

INTRODUCTION

1. U.S. Census Bureau at http://www.census.gov/population/estimates/metro-city /scful/SC98F-CA-DR.txt; *California Statistical Abstract,* table B4 (Sacramento: California Department of Finance, 2001), online at http://www.dof.ca.gov/html/fs_data/ stat-abs/toc.htm.

2. Eugene L. Conrotto, *Miwok Means People: The Life and Fate of the Native Inhabitants of the California Gold Rush Country* (Fresno: Valley Publishers, 1973), 8–9.

3. William Cronon, *Changes in the Land: Indians, Colonists, and the Ecology of New England* (New York: Hill and Wang, 1983), 9–14, 48–51; Cronon, "Under an Open Sky: Rethinking America's Western Past," in *Kennecott Journey: The Paths Out of Town,* ed. W. Cronon, G. Miles, and J. Gitlin (New York: W. W. Norton, 1992), 28–51; Lary M. Dilsaver and William C. Tweed, *Challenge of the Big Trees of Sequoia and Kings Canyon National Parks* (Three Rivers, Calif.: Sequoia Natural History Association, 1990), 22–24; M. Kat Anderson, Michael G. Barbour, and Valerie Whitworth, "A World of Balance and Plenty: Land, Plants, Animals, and Humans in a Pre-European California," *California History* 76 (summer and fall 1997): 14–16, 33–39.

4. Some early accounts described the "Mother Vein" as stretching all the way from Mariposa to Grass Valley, thus helping inflate the image of Mother Lode wealth. See *Calaveras County Illustrated and Described, Showing Its Advantages for Homes* (1885; Fresno: Valley Publishers, 1976), 24.

5. The lack of accurate statistics makes it impossible to determine precisely how much of Calaveras County's gold production came from lode mining as opposed to placer. Rough estimates indicate that about half the county's total production, or about 4.5 million ounces since 1848, came from hardrock mines. The bulk of the county's post–Gold Rush production was of lode origin, and since 1896 only one-fifth of the gold recovered has been placer. This overall county ratio of lode-to-placer roughly matches the state figures. Today it is estimated that about 60 percent of the 113 million ounces of gold produced in California to date has come from lode deposits. P. Joralemon, "California's Foothill Gold Belt," *Mining Engineering* 39 (July 1987): 489.

6. See the following modern books, for example: Susan Lee Johnson, *Roaring Camp: The Social World of the California Gold Rush* (New York: W. W. Norton, 2000); Malcolm J. Rohrbough, *Days of Gold: The California Gold Rush and the American Nation* (Berkeley and Los Angeles: University of California Press, 1997); and James J. Rawls and Richard J. Orsi, eds., *A Golden State: Mining and Economic Development in Gold Rush California* (Berkeley and Los Angeles: University of California Press, 1999). See also Mary Hill, *Gold: The California Story* (Berkeley and Los Angeles: University of California Press, 1999). For a more traditional and beautifully illustrated social

and economic history of the Gold Rush era, see Peter J. Blodgett, *Land of Golden Dreams: California in the Gold Rush Decade, 1848–1858* (San Marino: Huntington Library, 1999).

7. Data on nationwide county-population growth rates can be found on the Website for the U.S. Bureau of the Census, http://www.census.gov. For more demographic data on Calaveras County, consult app. 3.

8. For the most useful publications that deal extensively with Calaveras, see Edna Buckbee, *Pioneer Days of Angel's Camp* (Angel's Camp: Calaveras Californian, 1932); "Calaveras County," *State Resources: An Illustrated Monthly Publication* (Mar. 1890): 75–104; James H. Carson, *Recollections of the California Mines* (1852; reprint, Oakland: Biobooks, 1950); Joseph Giovinco, *The Ethnic Dimension of Calaveras County History* ([San Andreas]: Calaveras Heritage Council, [1977]); C. E. Julihn and F. W. Horton, *. . . Mines of the Southern Mother Lode Region, Part 1: Calaveras County,* U.S. Bureau of Mines Bulletin 413, Mineral Industries Survey of the United States (Washington, D.C.: GPO, 1938); Mark B. Kerr, ed., *Mining Resources of Calaveras County, California* ([San Andreas]: Calaveras County Exhibit, 1898); James Gary Maniery, *Six Mile and Murphys Rancherias: An Ethnohistorical and Archaeological Study of the Two Central Sierra Miwok Village Sites,* San Diego Museum Papers, no. 22 (San Diego: San Diego Museum of Man, 1987); William Perkins, *Three Years in California: William Perkins' Journal of Life at Sonora, 1849–1852,* ed. Dale L. Morgan and James R. Scobie (Berkeley and Los Angeles: University of California Press, 1964); Rhoda Stone and Charles A. Stone, *The Tools Are on the Bar: The History of Copperopolis, Calaveras County, California* (Copperopolis: privately printed, 1991); Harvey Wood, *Personal Recollections* (1896; reprint, Angels Camp: Mountain Echo Job Printing Office, 1955); and three books by Richard Coke Wood, *Murphys, Queen of the Sierra* (Angels Camp: Calaveras Californian, 1948); *Tales of Old Calaveras* ([Angels Camp: Calaveras Californian, 1949]); and *Calaveras, the Land of Skulls* (Sonora: Mother Lode Press, 1955). See also Mark Twain, *The Celebrated Jumping Frog of Calaveras County, and Other Sketches* (New York: C. H. Webb, 1867; reprint, New York: Oxford University Press, 1996).

9. For contemporary observations, see, for example, John S. Hittell, *The Resources of California, Comprising Agriculture, Mining, Geography, Climate, Commerce, &c., and the Past and Future Development of the State,* 2d ed. (San Francisco: A. Roman, 1866), 238–40, 431–61; and Hubert Howe Bancroft, *History of California, 1848–1859,* vol. 23 of *The Works of Hubert Howe Bancroft,* facsimile ed. (1888; reprint, Santa Barbara: Wallace Hebberd, 1970), 786–87. For modern scholarship, see Rodman W. Paul, *California Gold: The Beginning of Mining in the Far West* (Cambridge: Harvard University Press, 1947; reprint, Lincoln: University of Nebraska Press, 1967); and John Walton Caughey, *Gold Is the Cornerstone* (Berkeley and Los Angeles: University of California Press, 1948), esp. 291–99.

10. William R. Freudenburg and Scott Frickel, "Digging Deeper: Mining-Dependent Regions in Historical Perspective," *Rural Sociology* 59 (summer 1994): 266–67.

11. Richard V. Francaviglia, *Hard Places: Reading the Landscape of America's His-*

*toric Mining Districts* (Iowa City: University of Iowa Press, 1991); Kevin Starr, "The Gold Rush and the California Dream," *California History* 77 (spring 1998): 60.

12. For an Australian parallel to our study, see C. R. Doran, "The Minerals Sector and the Australian Economy: An Historical Perspective on Mining and Economic Change," in *Special Study No. 6,* ed. L. H. Cook and M. G. Porter (Sydney: George Allen and Unwin Australia, 1984), 37–51. Gerald D. Nash provides a broad overview in "A Veritable Revolution: The Global Economic Significance of the California Gold Rush," in *Golden State,* ed. Rawls and Orsi, 276–92.

13. Richard White, "The Gold Rush: Consequences and Contingencies," *California History* 77 (spring 1998): 46.

I: THE EARLY PLACER ERA

1. Owen C. Coy, *Guide to the County Archives of California* (Sacramento: California State Printing Office, 1919), 124–25.

2. George P. Hammond, *The Weber Era in Stockton History* (Berkeley: Friends of the Bancroft Library, 1982), 27–74, 87–88.

3. Ronald H. Limbaugh, "Making Old Tools Work Better: Culture, Pragmatic Adaptation, and Innovation in Gold-Rush Technology," in *Golden State,* ed. Rawls and Orsi, 27–29.

4. Her nickname was Ellen. See G. P. Hammond, *Weber Era,* 99–104; and Earl F. Schmidt, *Who Were the Murphys?* (Murphys: Mooney Flat Ventures, 1989), 3–9.

5. John McHenry Hollingsworth, *The Journal of Lieutenant John McHenry Hollingsworth of the First New York Volunteers, Stevenson's Regiment, September 1846–August 1849* (San Francisco: California Historical Society, 1923), v–vii; Bancroft, *Works of Bancroft,* 22:499–518; James D. Hart, *A Companion to California,* new ed. (Berkeley and Los Angeles: University of California Press, 1987), 502.

6. Carson, *Recollections,* 5.

7. James J. Rawls, "Gold Diggers: Indian Miners in the California Gold Rush," *California Historical Quarterly* 55 (spring 1976): 28; A. L. Kroeber, *Handbook of the Indians of California,* Bureau of American Ethnology Bulletin 78 (Washington, D.C.: GPO, 1925), 444, 445, pl. 37; Robert F. Heizer, *Languages, Territories, and Names of California Indian Tribes* (Berkeley and Los Angeles: University of California Press, 1966), 44, end maps.

8. George H. Tinkham, *A History of Stockton* (San Francisco: W. M. Hinton, 1880), 22–23.

9. Henry William Bigler, *Bigler's Chronicle of the West: The Conquest of California, Discovery of Gold, and Mormon Settlement As Reflected in Henry William Bigler's Diaries,* ed. Erwin G. Gudde (1848; reprint, Berkeley and Los Angeles: University of California Press, 1962), 56, 87; George Frederic Parsons, *The Life and Adventures of James W. Marshall, the Discoverer of Gold in California,* introduction and notes by G. Ezra Dane (Sacramento: James W. Marshall and W. Burke, 1870; reprint, San Francisco: George Fields, 1935), 56–57.

10. Henry P. DeGroot, *Recollections of California Mining Life* (San Francisco:

Dewey, 1884), 8–9; Albert L. Hurtado, *Indian Survival on the California Frontier* (New Haven: Yale University Press, 1988), 101–4, 118; on Savage, see James J. Rawls, *Indians of California: The Changing Image* (Norman: University of Oklahoma Press, 1984), 124–25.

11. Edward Gould Buffum, *Six Months in the Gold Mines, From a Journal of Three Years' Residence in Upper and Lower California, 1847–8–9*, ed. John Walton Caughey (Philadelphia: Lee and Blanchard, 1850; [Los Angeles]: Ward Ritchie Press, 1959), 103.

12. Hurtado, *Indian Survival*, 106; G. P. Hammond, *Weber Era*, 58–59; Carson, *Recollections*, 8. An early history of San Jose claims that Indians from the old mission at San Jose also accompanied John M. Murphy to the Weber diggings near Placerville, and that it was Murphy, not Weber, that sent them south, along with some Stanislaus Indians, to prospect for gold along the Stanislaus River. See Frederic Hall, *The History of San Jose and Surroundings With Biographical Sketches of Early Settlers* (San Francisco: A. L. Bancroft, 1871), 192–93. Since Murphy was presumably in partnership with Weber at the time, both might take credit for initiating the expedition. Either John or Dan Murphy—or perhaps both—set up a trading post near the new diggings, and did a "lively business" trading utensils and trinkets for gold dust, which he reportedly took in "by the bushel." See *Sonora Herald,* n.d., reprinted in *San Andreas Independent,* Sept. 18, 1858 (TS), MS 187, in Reatha Parcel Smith Collection of Pacific Coast Newspapers, Holt-Atherton Library, University of the Pacific (hereafter cited as UOPWA).

13. Tinkham, *A History of Stockton,* 74; J. B. to editor, Aug. 15, 1848, in *San Francisco Californian,* Sept. 2, 1848, cited in George P. Hammond and Dale L. Morgan, *Captain Charles M. Weber: Pioneer of the San Joaquin and Founder of Stockton, California* (Berkeley: Friends of the Bancroft Library, 1966), 20; see also R. C. Wood, *Murphys,* 2–3. An unidentified newspaper correspondent who visited the site in December 1848 identified it as "Indian Gulch, which runs down to the river from Carson's Hill" (*Sonora Herald,* quoted in *San Andreas Independent,* Sept. 18, 1858 [TS], Smith Collection).

14. By October, when E. G. Buffum passed through, the average daily production on "Weaver's Creek" was down to an ounce of gold per miner, or about ten dollars per day, barely enough to meet expenses (*Six Months,* 47).

15. Tinkham, *A History of Stockton,* 74–75; G. P. Hammond, *Weber Era,* 93–96.

16. Carson, *Recollections,* 3–5.

17. Ibid., 7–8. William Hance located the quartz vein on what became the Morgan mine, Carson Hill's first lode mine, in October 1850. See W. Turrentine Jackson, *Historical Survey of the New Melones Reservoir Project Area* (Sacramento: Army Corps of Engineers, 1976), 34.

18. R. C. Wood, *Calaveras,* 19.

19. Carson, *Recollections,* 13–15; introduction to Carson, "Early Recollections of the Mines," in *Bright Gem of the Western Seas,* ed. Peter Browning (Lafayette, Calif.: Great West Books, 1991), xiv–xv.

20. W. T. Jackson, *Historical Survey,* 22; Edward C. Leonard, *A Brief History of Angel's Camp* (Murphys: Old Timer's Museum, 1973), 3; *Las Calaveras* 35 (Jan. 1987): 11, 26–27. A local journalist once wrote that John W. Mackay, later a bonanza king of the Comstock, began his memorable career probing placer ground at Angels Creek, but no evidence has been found to support this story (Buckbee, *Pioneer Days,* 3).

21. In 1850, Frederick Gerstaecker wrote that the first discoverers were Mexican who "had dug very deep holes . . . and taken out a large quantity of gold" (*California Gold Mines* [1856 {German ed}, 1942 {translation}; reprint, Oakland: Biobooks, 1946], 64).

22. Carson, *Recollections,* 8; R. C. Wood, *Murphys,* 1–2; Gerstaecker, *California Gold Mines,* 64.

23. William B. Clark and Philip A. Lydon, *Mines and Mineral Resources of Calaveras County, California,* report 2 (San Francisco: California Division of Mines and Geology, 1962), 76.

24. *Calaveras County Illustrated,* 21; William Redmond Ryan, *Personal Adventures in Upper and Lower California, in 1848–9* (1850; reprint, New York: Arno Press, 1973), 2:40; Walter Colton, *Three Years in California* (New York: AS Barnes, 1851; Stanford: Stanford University Press, 1949), 277–78.

25. J. Ross Browne, *Report of J. Ross Browne on the Mineral Resources of the States and Territories West of the Rocky Mountains* (Washington, D.C.: GPO, 1868), 57.

26. Jacob Henry Bachman, "Diary of a Used-Up Miner," *California Historical Society Quarterly* 22 (Mar. 1943): 71; John Woodhouse Audubon, *Audubon's Western Journal, 1849–1850* (1906; reprint, Glorieta, N.M.: Rio Grande Press, 1969), 204–5; Joseph Schafer, ed., *California Letters of Lucius Fairchild,* Wisconsin Historical Publications Collections, no. 31 (Evansville: State Historical Society of Wisconsin, 1931), 90.

27. Bachman, "Diary of a Miner," 71; W. T. Jackson, *Historical Survey,* 86; Gerstaecker, *California Gold Mines,* 64; J. R. Browne, *Report on Mineral Resources,* 57; R. C. Wood, *Calaveras,* 16.

28. R. C. Wood, *Murphys,* 3–7.

29. R. C. Wood, *Calaveras,* 18; William B. Clark, *Gold Districts of California,* Bulletin 193 (San Francisco: California Division of Mines and Geology, 1970), 126.

30. Colton, *Three Years in California,* 254. E. G. Buffum reported that in the same month a slightly larger nugget, the largest discovered in 1848, was found "in a dry ravine near the Stanislaus River" (*Six Months,* 107). J. M. Hutchins said that in 1853 he saw a Calaveras County nugget that weighed twenty-six pounds and was shaped "like the kidney of an ox" (*Calaveras County Illustrated,* 20–21).

31. Bayard Taylor, *Eldorado; or, Adventures in the Path of Empire* (New York: George P. Putnam, 1850; New York: Alfred A. Knopf, 1949), 63.

32. Byron Nathan McKinstry, *The California Gold Rush Overland Diary of Byron N. McKinstry, 1850–1852* (Glendale, Calif.: Arthur H. Clark, 1975), 318–77.

33. R. C. Wood, *Calaveras,* 28–30.

34. *San Andreas Independent,* Sept. 24, 1856, p. 3, col. 2.

35. John Hovey journal (MS, Nov. 7, 1849–Jan. 4, 1850), pp. 70–90, Huntington Library.

36. Schafer, *Letters of Fairchild,* 102–3.

37. *San Andreas Independent,* July 30, 1859, p. 3, col. 1; Emmett P. Joy with Ellen H. Ladd, *Chronicles of San Andreas: One Town That Rose From a Golden Channel* (Murphys: Old Timer's Museum, 1972), 3; R. C. Wood, *Calaveras,* 26.

38. Maria del Carmen Ferreyra and David S. Reher, eds. and trans., *The Gold Rush Diary of Ramon Gil Navarro* (Lincoln: University of Nebraska Press, 2000), 38; William Blake, "The Mechanical Appliances of Mining," in *Statistics of Mines and Mining in the States and Territories West of the Rocky Mountains,* ed. Rossiter W. Raymond, 41st Cong., 2d sess., 1870, pt. 4:480–82 (Washington, D.C.: GPO, 1873); Effie Enfield Johnston, "Wade Johnston Talks to His Daughter," *Las Calaveras* 17 (Apr. 1979): 22.

39. Otis E. Young Jr., *Western Mining: An Informal Account of Precious-Metals Prospecting, Placering, Lode Mining, and Milling on the American Frontier From Spanish Times to 1893* (Norman: University of Oklahoma Press, [1970]), 59.

40. Caughey, *Gold Is the Cornerstone,* 25–26; Otis E. Young Jr., "The Southern Gold Rush: Contributions to California and the West," *Southern California Quarterly* 62 (summer 1980): 138.

41. Theodore H. Hittell, *History of California* (San Francisco: Pacific Press and Occidental Publishing, 1885–1898), 2:687; B. Taylor, *Eldorado,* 65–69, 193; O. E. Young, "Southern Gold Rush," 135–36; Colton, *Three Years in California,* 280–81; Joseph Libbey Folsom, *A Letter of Captain J. L. Folsom Reporting on Conditions in California in 1848* (San Francisco: Grabhorn Press, 1944); Caughey, *Gold Is the Cornerstone,* 25–26.

42. B. Taylor, *Eldorado,* 68.

43. Folsom, *Letter of Folsom,* 13; Colton, *Three Years in California,* 276.

44. Philip R. May, *Origins of Hydraulic Mining in California* (Oakland: Holmes Book, 1970), 25–27.

45. B. Taylor, *Eldorado,* 193; Paul, *California Gold,* 272–77.

46. James Ward, *A History of Gold As a Commodity and As a Measure of Value: Its Fluctuations Both in Ancient and Modern Times, With an Estimate of the Probable Supplies From California and Australia* (London: William S. Orr, [ca. 1852]), 104.

47. The conquest theme is treated extensively in Patricia Nelson Limerick, *The Legacy of Conquest: The Unbroken Past of the American West* (New York: W. W. Norton, 1987). See also Daniel Cornford, "'We All Live More Like Brutes Than Humans': Labor and Capital in the Gold Rush," in *Golden State,* ed. Rawls and Orsi, 81–84.

48. J. R. Browne, *Report on Mineral Resources,* 184–86.

49. David L. Drew, a New England argonaut, described the use of sluice boxes in 1856 at Deadman's Bar near Parrot's Ferry ("Placer Mining on the Stanislaus, 1856," *Las Calaveras* 27 [Oct. 1978–Jan. 1979]: 1–18).

50. Randall E. Rohe, "Chinese River Mining in the West," *Montana* 46 (autumn 1996): 14–29.

51. John Steele, *In Camp and Cabin: Mining Life and Adventure in California During 1850 and Later* (1901; reprint, Chicago: R. R. Donnelley and Sons, 1928), 126–27.

52. Ibid., 140–42; Leonard Noyes, "Reminiscences" (TS, n.d.), p. 86, Calaveras County Historical Society, San Andreas, from original in Peabody Museum, Salem, Mass.; Charles William Churchill, *Fortunes Are for the Few: Letters of a Forty-Niner,* ed. Duane A. Smith and David J. Weber (San Diego: San Diego Historical Society, 1977), 57.

53. Leonard Kip, *California Sketches, With Recollections of the Gold Mines* (1850; reprint, Los Angeles: N. A. Kovach, 1946), 45–47.

54. On the drunken prospector, see James J. Ayers, *Gold and Sunshine: Reminiscences of Early California* (Boston: R. G. Badger, 1922), 71–72; on the greenhorn black miners, see David Clarence Demarest, "California Gold" (MS, ca. 1962), MS 28, box 1, UOPWA.

55. Josiah Foster Flagg, "Diary of a Philadelphia 49er," *Pennsylvania Magazine of History* 70 (Oct. 1946): 403.

56. Ibid., 408–11, 419.

57. See, for example, Joseph Libbey Folsom's comments in *Letter of Folsom,* 11–12.

58. "Old Boone" letter, Aug. 15, 1849, in *California Emigrant Letters,* ed. Walker D. Wyman (New York: Bookman Associates, [1952]), 76–77.

59. The classic paper on the Tertiary channel deposits is Waldemar Lindgren (1860–1939), *The Tertiary Gravels of the Sierra Nevada of California,* U.S. Geological Survey Professional Paper 73 (Washington, D.C.: GPO, 1911). For a recent overview, see Powell Greenland, *Hydraulic Mining in California: A Tarnished Legacy* (Spokane: Arthur H. Clark, 2001), 38–40.

60. *The Miner's Own Book, Containing Correct Illustrations and Descriptions of the Various Modes of California Mining, Including All the Improvements Introduced From the Earliest Day to the Present Time* (San Francisco: Hutchings and Rosenfield, 1858; facsimile ed., San Francisco: Book Club of California, 1949), 9–10.

61. *Daily Alta California,* Dec. 14, 1851, p. 2, col. 3; Clark and Lydon, *Mines and Mineral Resources,* 76.

62. On Chili Gulch, see *Sacramento Daily Union,* Dec. 12, 1860 (TS); on Old Woman's Gulch, see excerpt from *Calaveras Chronicle,* Mar. 30, 1861, in *Sacramento Daily Union,* Apr. 1, 1861 (TS), both in Smith Collection.

63. *Annual Report of the State Mineralogist From June 1, 1880, to December 1, 1880* (Sacramento: J. D. Young, 1881), 202–3. The extent of the mining of Tertiary gravels is well documented in Lindgren, *Tertiary Gravels.*

64. May, *Origins of Hydraulic Mining,* 20.

65. "Hydraulic Method Is Introduced in Mines," *Las Calaveras* 10 (Oct. 1961): 2; Roberta Evelyn Holmes, *The Southern Mines of California: Early Development of the Sonora Mining Region* (San Francisco: Grabhorn Press, 1930), 30.

66. Henry G. Hanks, *Second Annual Report of the State Mineralogist* (San Francisco: California State Mining Bureau, 1882), 66. For an extended overview of hydraulic water supply systems in California, see Greenland, *Hydraulic Mining in California,* 71–112.

67. Greenland, *Hydraulic Mining in California,* 117–27.

68. *Golden Dreams, Poisoned Streams: How Reckless Mining Pollutes America's Waters, and How We Can Stop It* (Washington, D.C.: Mineral Policy Center, 1997), 41, 84, 156–57; Michael Huerlach, James J. Rytuba, and Charles N. Alpers, "Mercury Contamination from Hydraulic Placer-Gold Mining in the Dutch Flat Mining District, California," U.S. Geological Survey Toxic Substances Hydrology Program; Proceedings of the Technical Meeting, Charleston, South Carolina, Mar. 8–12, 1999, vol. 2 of *Contamination of Hydrologic Systems and Related Ecosystems,* Water-Resources Investigation Report 99-4018B (USGS Website: http://toxics.usgs.gov/ubs/wri99-4018/volume2/sectionB/2304_Hunerlach/).

69. Randall E. Rohe, "Hydraulicking in the American West: The Development and Diffusion of a Mining Technique," *Montana* 35, no. 2 (1985): 28–34; Rohe, "Man and the Land: Mining's Impact in the Far West," *Arizona and the West* 28, no. 4 (1986): 316; Greenland, *Hydraulic Mining in California,* 105–6.

70. Unidentified newspaper clipping, Aug. 5, 1865, Calaveras County Scraps, p. 130, Bancroft Library, University of California at Berkeley (hereafter cited as Bancroft Scraps Calaveras).

71. Robert L. Kelley, *Gold vs. Grain: The Hydraulic Mining Controversy in California's Sacramento Valley, a Chapter in the Decline of the Concept of Laissez Faire* (Glendale, Calif.: Arthur H. Clark, 1959), 239–40.

72. Raymond F. Dasmann, "Environmental Changes Before and After the Gold Rush," in *Golden State,* ed. Rawls and Orsi, 121; Stewart L. Udall, *The Quiet Crisis* (New York: Holt, Rinehart, and Winston, 1963), 54; David Stiller, *Wounding the West: Montana, Mining, and the Environment* (Lincoln: University of Nebraska Press, 2000), 178–79.

2: THE BEGINNINGS OF LODE MINING, 1850–1885

1. James Stephens Brown, *California Gold: An Authentic History of the First Find With the Names of Those Interested in the Discovery* (Oakland: Pacific Press, 1894), 8; excerpt from letter of Walter Colton, Aug. 29, 1848, in J. Quinn Thornton, *Oregon and California in 1848* (New York: Harper and Bros., 1849), 291–92; Buffum, *Six Months,* 58–69; Ferreyra and Reher, *Diary of Navarro,* 20–21.

2. David T. Ansted, *The Gold Seeker's Manual* (London: John Van Voorst, 1849), 41–46.

3. Walter Colton, *Three Years in California,* 278–79, 286, 295, 312.

4. Bancroft, *History of California,* 6:377, in vol. 23 of *Works of Bancroft.* Recently, Mary Hill, ignoring the earlier Mariposa claim, wrote that George McKnight in 1850 discovered the first quartz gold in a Grass Valley outcrop (*Gold: The California Story,* 133). Bancroft distinguishes between Mariposa, where the first quartz vein was located,

and Grass Valley, where some claim the first mills were established (415). Caughey, *Gold Is the Cornerstone,* 251.

5. Phil T. Hanna, *Dictionary of California Land Names* (Los Angeles: Auto Club of Southern California, 1946, revised and enlarged 1951), 203. Sheep Ranch is also sometimes spelled "Sheepranch."

6. Ross E. Browne, "The Mother Lode of California," in *California Mines and Minerals* (San Francisco: California Miners Association, 1899), 58–63; Joralemon, "California's Foothill Gold Belt," 489–91.

7. James F. Kemp, *The Ore Deposits of the United States,* rev. ed. (New York: Scientific Publishing, 1895), 288; Joralemon, "California's Foothill Gold Belt," 491; R. E. Browne, "Mother Lode of California," 58–63.

8. W. Turrentine Jackson, "Lewis Richard Price, British Mining Entrepreneur and Traveler in California," *Pacific Historical Review* 29 (Nov. 1960): 331; Samuel Frank Marryat, *Mountains and Molehills; or, Recollections of a Burnt Journal,* ed. Marguerite Eyer Wilbur (London: Longman, Brown, Green, and Longmans, 1855; facsimile ed., Stanford University Press, 1952), 241–43, 272–330.

9. Peter Y. Cool, "Goodness, Gold, and God: The California Mining Career of Peter Y. Cool, 1851–52, a Journal," ed. William A. Clebsch, *Pacific Historian* 10 (summer 1966): 19–42.

10. John Rowe, *The Hard-Rock Men: Cornish Immigrants and the North American Mining Frontier* (Liverpool: University Press, 1974), 109–12; Thomas Allsop, *California and Its Gold Mines* (London: Groombridge and Sons, 1853), 58–60; Ferreyra and Reher, *Diary of Navarro,* 189; Paul, *California Gold,* 130–32; Caughey, *Gold Is the Cornerstone,* 251–56; and Maureen A. Jung, "Capitalism Comes to the Diggings: From Gold-Rush Adventure to Corporate Enterprise," in *Golden State,* ed. Rawls and Orsi, 63–67.

11. Paul, *California Gold,* 165–68; J. Arthur Phillips, *The Mining and Metallurgy of Gold and Silver* (London: E. and F. N. Spon, 1867), 54–55.

12. W. H. Storms, "Report on Mother Lode Mines" (TS, n.d.), MS 197, box 17, Archie D. Stevenot Papers, UOPWA.

13. Willard P. Fuller Jr., "The Gwin Mine at Paloma," *Las Calaveras* 16 ( Jan. 1968): 5.

14. Rockwell D. Hunt, *California's Stately Hall of Fame* (Stockton: College of the Pacific, 1950), 249–53.

15. On purchase, see *San Francisco Post,* Apr. 13, 1877, Bancroft Scraps Calaveras; letter quoted in Lately Thomas, *Between Two Empires* (Boston: Houghton, Mifflin, 1962), 373; W. P. Fuller, "Gwin Mine," 5.

16. Henry G. Hanks, *Sixth Annual Report of the State Mineralogist,* pt. 2 (San Francisco: California Bureau of Mines, 1887), 32; W. H. Storms, "Report on Mother Lode Mines," Stevenot Papers. See also Hanks, *Sixth Annual Report,* 32. Despite the criticism, by the profit-making standards of the day the Gwins did very well.

17. Clark C. Spence, *British Investments in the American Mining Frontier, 1860–1901* (Ithaca: American Historical Association, 1958), 121–38.

18. Estimated total gold production for the two districts, 1848–1995, in ounces: Angels Camp, 1,765,000; Carson Hill, 1,525,000. Compiled from company records, publications of the California Division of Mines, and various historical studies.

19. For a discussion of mining camp laws in the West from a traditional perspective, see Charles Howard Shinn, *Mining Camps: A Study in American Frontier Government* (1884; reprint, New York: Alfred A. Knopf, 1948), 223–46; on 1872 law, see United States Code, title 30, chap. 2, available online at http://www4.law.cornell.edu/uscode/30/ch2.html; Paul, *California Gold,* 210–38; Caughey, *Gold Is the Cornerstone,* 225–33. The apex of a vein, whether vertical or dipping, is the highest part that exists on a lode claim. Generally, it is the surface outcrop, but sometimes the vein apex does not reach the surface. The claim owner or his assigns normally have the right, under the 1872 law, to follow indefinitely any vein that apexes on or within the claim end lines, even if the vein dips under the surface of another claim. Unfortunately, complications are frequent, partly because of the differing shapes, positions, and relative ages of the claims involved, and partly because of variations in the geology of the veins.

20. J. R. Browne, *Report on Mineral Resources,* 59.

21. *San Joaquin Republican,* July 27, 1852, p. 3, col. 1; depositions of W. S. Rowe and Zenas Wheeler, witnesses, in *James G. Fair* v. *Gabriel Stevenot et al.,* California District Court, Fifth Judicial District, Feb. 12, 1864, Calaveras County Archives, San Andreas; W. Turrentine Jackson and Stephen Mikesell, "Mexican Melones," *Las Calaveras* 28 (1979): 5–7.

22. Noyes, "Reminiscences," p. 59, Calaveras County Historical Society; *San Joaquin Republican,* Apr. 17, 1852, p. 2, col. 1.

23. Noyes, "Reminiscences," p. 59, Calaveras County Historical Society; *Daily Alta California,* Dec. 24, 1851, Bancroft Scraps Calaveras.

24. *San Joaquin Republican,* July 10, 1852, p. 2, col. 4.

25. Ibid., Dec. 13, 1851, p. 4, col. 3; *Sonora Herald,* cited in ibid., Dec. 20, 1851, p. 2, col. 5.

26. *San Joaquin Republican,* Dec. 20, 1851, p. 2, col. 5; Noyes, "Reminiscences," p. 59, Calaveras County Historical Society.

27. Noyes, "Reminiscences," pp. 59–60, Calaveras County Historical Society; Alfred A. Lacy, "Reminiscences" (MS, ca. 1924), pp. 8–11, Bancroft Library.

28. J. R. Browne, *Report on Mineral Resources,* 59.

29. Subsequent accounts of the Morgan dispute followed the Browne version. Recently, W. Turrentine Jackson and Stephen Mikesell argued convincingly that Browne had confused Mulligan with James Finnegan, who held a discovery claim to the Morgan property and had been subsequently left out when the Morgan company was formed. It was Finnegan, not Mulligan, say these scholars, that participated in the miners' meetings in the late fall of 1851. Their conclusion: "If Mulligan was there at the time, his presence was so insignificant as to go unnoticed. More likely, the Mulligan story has become a part of local mythology, an unfortunate legacy to Browne's questionable scholarship" ("Mexican Melones," 12). This revisionist account is an im-

provement over earlier versions but not the entire story. By seriously doubting, if not denying, that Mulligan participated in the Carson Hill dispute, they contradict at least two eyewitnesses, Leonard Noyes and Albert A. Lascy, both of whose reminiscences state categorically that Mulligan was there in the spring of 1852. See Noyes, "Reminiscences," pp. 60–61, Calaveras County Historical Society; and Lascy, "Reminiscences," pt. 4, pp. 9–12, Bancroft Library.

30. Noyes, "Reminiscences," pp. 60–61, Calaveras County Historical Society.

31. John David Borthwick, *Three Years in California* (1857; New York: Biobooks, 1948), 264. For a colorful description of placer pocket mining, see Mark Twain, *Roughing It* (New York: New American Library, 1962), 325–26. See also *Stockton Times,* Mar. 23, 1850, p. 4, col. 2; May 25, 1850, p. 4, col. 2; Sept. 7, 1850, p. 2, col. 5; Sept. 14, 1850, p. 3, col. 1; Nov. 30, 1850, p. 2, col. 1; and *San Joaquin Republican,* July 5, 1851, p. 1, col. 5; May 19, 1852, p. 2, col. 4; July 21, 1852, p. 2, col. 4; Aug. 25, 1852, p. 2, col. 5; Jan. 15, 1853, p. 2, col. 7; Feb. 9, 1853, p. 1, col. 7.

32. *Stockton Times,* Oct. 5, 1850, p. 2, col. 1.

33. James A. Smith, "Carson Hill Gold," *Las Calaveras* 8 (July 1960): 4.

34. *Calaveras County Illustrated,* 21. For accounts of fabulous results from Carson pocket mines, see the *Stockton Times,* Mar. 23, 1850, p. 4, col. 2; Sept. 14, 1850, p. 3, col. 1; and *San Joaquin Republican,* July 5, 1851, p. 1, col. 5.

35. William Jeffrey, "History of the Carson Mine, Carson Hill, Calaveras County, California" (ts, ca. 1890), ms 275, Segerstrom Family Collection, UOPWA.

36. Ibid. Other Carson Hill properties were similarly treated. Lower-grade ore was packed in hide sacks and transported on the backs of mules to mule-powered arrastras on Coyote Creek and on the north bank of Carson Creek. See also Edna Bryan Buckbee, *Pioneer Days,* 73–74.

37. Jeffrey, "History of the Carson Mine," Segerstrom Family Collection; J. R. Browne, *Report on Mineral Resources,* 60.

38. Jeffrey, "History of the Carson Mine," Segerstrom Family Collection.

39. Ibid.

40. Ibid.

41. Ibid.

42. Cosmorama letter, Mar. 3, 1863, unidentified newspaper clipping, Bancroft Scraps Calaveras. *Lead,* in this context, is synonymous with *lode* or *lode vein.* In placer mining the term is also used to indicate a pay streak.

43. Hubert Howe Bancroft has Fair at "Tabor" Mountain, doubtless a typographical error. See *Chronicles of the Builders* (San Francisco: History Company, 1892), 4:212–17. Johnston, "Wade Johnston Talks to His Daughter," *Las Calaveras* 8 (Apr. 1970): 25–28.

44. Willard P. Fuller Jr., Judith Cunningham, and Julia Costello, "Carson Hill and Its Gold Quartz Mills," *Las Calaveras* 39 (Apr. 1990): 31–32; Julia Costello and Judith Cunningham, *History of Mining at Carson Hill,* report for the Carson Hill Gold Mining Corporation (privately printed, 1988).

45. W. P. Fuller, Cunningham, and Costello, "Carson Hill," 28–31.

46. Buckbee, *Pioneer Days,* 13–14. Edward C. Leonard considered the story "unfounded" in his "Early Quartz Mining in Angels Camp," *Las Calaveras* 17 (Oct. 1968): 1. U.S. Bureau of the Census, Census of 1880, Calaveras County, seventh township, 27.

47. U.S. Bureau of the Census, Census of 1860, Calaveras County, eighth township, 101.

48. D. C. Demarest, "California Gold," UOPWA. The Demarest manuscript identified this claim as the Baltimore mine, but Edward Leonard found it recorded in 1855 as the Winter brothers' mine ("Early Quartz Mining," 2).

49. Demarest, "California Gold," chap. 5, UOPWA; *A Memorial and Biographical History of the Counties of Merced, Stanislaus, Calaveras, Tuolomne, and Mariposa, California* (Chicago: Lewis Publishing, 1892), 162.

50. Leonard, "Early Quartz Mining," 1, 5.

51. *The Bay of San Francisco: The Metropolis of the Pacific Coast and Its Suburban Cities: A History* (Chicago: Lewis Publishing, 1892), 2:342–43.

52. Edward C. Leonard, "The Mills of Angels," *Las Calaveras* 19 (Jan. 1971): 9; Demarest, "California Gold," chap. 5, pp. 1–2, UOPWA.

53. Bancroft, *Chronicles of the Builders,* 4:217; Leonard, "Early Quartz Mining," 4; U.S. Bureau of the Census, Census of 1860, Calaveras County, eighth township, 52.

54. Cornelius B. Demarest, "Chronicles of Calaveras," part 3, *Las Calaveras* 25 (Apr. 1977): 28.

55. Buckbee, *Pioneer Days,* 37.

56. Leonard, "The Mills of Angels," 9; D. C. Demarest, "California Gold," chap. 6, p. 3, UOPWA.

57. Leonard, "The Mills of Angels," 11.

58. Leonard, "Early Quartz Mining," 4.

59. Leonard, "The Mills of Angels," 9–10.

60. Clark, *Gold Districts of California,* 68, 144; W. A. Wallace, MS dictation, ca. 1887–1889, Bancroft Library; Lev Johnson, "History of the Sheepranch Mine From an Old Record of Lev Johnson Through Courtesy of Clarence L. Feusier," *Las Calaveras* 7 (Oct. 1958): 1–2. The original spelling, *Sheepranch,* has been revised in favor of modern practice.

61. U.S. Bureau of the Census, Census of 1880, Calaveras County, seventh township.

62. See Selim Woodworth Papers, Huntington Library; and Noyes, "Reminiscences," p. 67, Calaveras County Historical Society.

63. J. W. Willard, *Simon Willard and His Clocks* (1911; reprint, Brookline, Mass.: Dover Publishing, 1968), 73.

64. Willard P. Fuller Jr., Judith Marvin, and Julia G. Costello, *Madam Felix's Gold: The Story of the Madam Felix Mining District, Calaveras County, California* (San Andreas: Calaveras County Historical Society and Foothill Resources, 1996), chap. 3.

65. *Stockton Times,* July 20, 1850, p. 3, col. 1; Jeffrey, "History of the Carson Mine," Segerstrom Family Collection; J. R. Browne, *Report on Mineral Resources,* 59.

66. Borthwick, *Three Years in California,* 265, 306–7. On rat-hole mining, consult O. E. Young, *Western Mining,* 79–81.

67. Paul, *California Gold,* 140; O. E. Young, *Western Mining,* 72–79.

68. Bancroft, *History of California,* 6:356–57, in vol. 23 of *Works of Bancroft;* Perkins, *Three Years in California,* 268, 298; Ferreyra and Reher, *Diary of Navarro,* 194–96, 222, 249–50.

69. Cool, "Goodness, Gold, and God," 34–35; U.S. Bureau of the Census, Census of 1852, Calaveras County, schedule 2 (ms, California State Library, Sacramento), 249; William Higby to father, May 2, 1856, ms box 218, Higby Papers, California State Library; *Daily Alta California,* Oct. 12, 1858, p. 1, col. 5; Leonard, "The Mills of Angels," 5. For an excellent modern study of San Francisco's foundries and machine shops, see Lynn R. Bailey, *Supplying the Mining World: The Mining Equipment Manufacturers of San Francisco, 1850–1900* (Tucson: Westernlore Press, 1996). Demarest, "California Gold," chap. 23, uopwa.

70. Perkins, *Three Years in California,* 268.

71. *Mining and Scientific Press* 10 (May 20, 1885): 211; J. D. Whitney, *Geology of California* (Sacramento: California State Legislature, 1865), 1:263; Oliver E. Bowen Jr. and Richard A. Crippen Jr., "California's Mother Lode Highway: Chinese Camp to Mokelumne Hill," *California Geology* 50 (May–June 1997): 90.

72. Julihn and Horton, *Southern Mother Lode Region,* 99.

73. Clark C. Spence, *The Lace-Boot Brigade: Mining Engineers in the American West* (New Haven: Yale University Press, 1970), 236.

74. Laura DeForce Gorden letter, Mar. 26, 1877, quoted in *San Francisco Post,* Apr. 13, 1877, Bancroft Scraps Calaveras.

75. Lewis Atherton, "Structure and Balance in Western Mining History," *Huntington Library Quarterly* 30 (Nov. 1966): 57–58.

76. For a good case study, see Lewis Atherton's seminal essay on Theodore J. Lamoreaux in "The Mining Promoter in the Trans-Mississippi West," *Western Historical Quarterly* 1 (Jan. 1970): 35–50.

77. Albin J. Dahl, "British Investment in California Mining, 1870–1890" (Ph.D. diss., University of California at Berkeley, 1961), 23–25; Richard H. Peterson, *The Bonanza Kings: The Social Origins and Business Behavior of Western Mining Entrepreneurs, 1870–1900* (Lincoln: University of Nebraska Press, 1977), 89–90.

78. Richard H. Peterson, "The Frontier Thesis and Social Mobility on the Mining Frontier," *Pacific Historical Review* 44 (Feb. 1975): 54–67.

79. Dahl, "British Investment," 18–21; Jung, "Capitalism Comes to the Diggings," 54–55.

80. Irving Stone, *The Global Export of Capital From Great Britain, 1865–1914: A Statistical Survey* (London: Macmillan, 1999), 6–8, 19–21, 42–51; Jean-Jacques van Helten, "Mining, Share Manias, and Speculation: British Investment in Overseas Mining, 1880–1913," in *Capitalism in a Mature Economy: Financial Institutions, Capital Exports, and British Industry, 1870–1939,* ed. J.-J. van Helten and Y. Cassas (Hants, England: Edward Elgar, 1990), 159–85.

81. Jung, "Capitalism Comes to the Diggings," 68–73; Dahl, "British Investment," 83–84; Spence, *British Investments,* 86–91.

82. *Stockton Evening Mail,* Mar. 30, 1897, p. 3, col. 3; Frances E. Bishop, "The Mysterious Disappearance of Windsor A. Keefer," *Las Calaveras* 31 (Apr.–July 1983); D. C. Demarest, "California Gold," chap. 32, UOPWA.

83. Thomas A. Rickard, *Retrospect: An Autobiography* (New York: Whittlesey House, 1937), 36–38.

84. L. Atherton, "Mining Promoter," 44.

85. Otis E. Young Jr., "Philipp Deidesheimer, 1832–1916, Engineer of the Comstock," *Historical Society of Southern California Quarterly* 57 (1975): 361–70.

86. L. Atherton, "Structure and Balance," 56–57; Dahl, "British Investment," 51.

87. Joralemon, "California's Foothill Gold Belt," 490.

88. Martin Ridge, "The Legacy of the Gold Rush," *Montana* 49 (autumn 1999): 61.

89. Limerick, *Legacy of Conquest,* 123.

90. Donald Macleod, "Miners, Mining Men, and Mining Reform: Changing the Technology of Nova Scotian Gold Mines and Collieries, 1858 to 1910" (Ph.D. diss., University of Toronto, 1981), 35, 490–93.

91. W. A. Williams to J. W. Stow, Aug. 22, 1865, Stow Papers, Huntington Library.

92. Henry George Jr., *The Life of Henry George* (New York: Doubleday and McClure, 1900), 138–41.

93. L. Helen Lewis, "One Hundred Years in Copperopolis," *Las Calaveras* 9 (Oct. 1960): 2; *San Francisco Bulletin,* Jan. 9, 1862 (TS), Smith Collection; *Sacramento Daily Union,* June 27, 1860 (TS), Smith Collection; Johnston, "Wade Johnston Talks to His Daughter," *Las Calaveras* 20 (Jan. 1972): 9.

94. Printed letter from "M," Apr. 14, 1862, unidentified newspaper clipping, Bancroft Scraps Calaveras.

95. "Copper Discoveries in California," *Mining and Scientific Press* 7 (Aug. 17, 1863): 1.

96. On market price, see *San Francisco Herald,* May 8, 1861 (TS), Smith Collection.

97. The figures for each mine: Union, more than 2,000 tons; Keystone, 650 tons; Hughes (Quail Hill), 150 tons; Napoleon, 75 tons; Copper Hill (Campo Seco), 30 tons; smaller mines, 57 tons. *Mining and Scientific Press* 5 (July 31, 1862): 5.

98. Excerpt from *Daily Alta California,* in *Stockton Daily Independent,* Sept. 30, 1963 (TS), Smith Collection; Jane Meader Nye, "California Connection" (MS, Meader Family Association, Balston Lake, N.Y., 1990).

99. *Stockton Daily Independent,* July 17, 1865 (TS), Smith Collection. For additional information on Copperopolis development, see R. Stone and C. A. Stone, *Tools Are on the Bar.*

100. On Union price, see excerpt from *Daily Alta California,* in *Stockton Daily Independent,* Sept. 30, 1863; *Amador Ledger,* Nov. 14, 1863 (TS), Smith Collection; for quote, see *Stockton Daily Independent,* Apr. 20, 1863 (TS), Smith Collection.

101. *Stockton Daily Independent,* Jan. 12, 1866.

102. Clark and Lydon, *Mines and Mineral Resources,* 24.

103. George Stone, "Reminiscences" (MS, ca. 1930, in possession of Charles Stone, Copperopolis).

104. *Mining and Scientific Press* 8 (Jan. 2, 1864): 10.

105. J. Mosheimer letter, *Mining and Scientific Press* 11 (Sept. 30, 1865): 194.

106. "Copper Smelting at Copperopolis," *Mining and Scientific Press* 10 (Apr. 22, 1865): 248.

107. *Stockton Daily Independent,* Mar. 27, 1865 (TS), Smith Collection.

108. *Mining and Scientific Press* 11 (Nov. 11, 1865): 290. The smelter was named for a New England town southeast of Boston where ore was shipped for smelting.

109. *Stockton Daily Independent,* July 19, 1864 (TS), Smith Collection.

110. J. R. Browne, *Report on Mineral Resources,* 208.

111. *San Francisco Commercial Herald and Market Review,* Jan. 11, 1868 (TS), Smith Collection.

112. *Stockton Daily Independent,* July 15, 1864 (TS), Smith Collection.

113. Excerpt from *San Francisco Bulletin,* in *Commercial Herald and Market Review,* Jan. 18, 1868 (TS), Smith Collection. Even in faraway Massachusetts, the fortunes of the Meader family were seriously affected by the depression in the copper industry. They faced even greater reversals with the crash of the Union Pacific's Credit Mobilier scheme, in which they had heavily invested. Charles Meader later recovered financially and moved to Butte, Montana, where he built one of the first smelters in that important copper district. In the meantime, his cousin Thomas B. Meader remained at Copperopolis to look after the Meader-Ames interests for the next two decades. See Nye, "California Connection."

114. One recent scholar has found this same pattern in other mining communities coming under the influence of bonanza kings such as George Hearst or Alvinza Hayward. Peterson, *Bonanza Kings,* 94–109; see also Spence, *Lace-Boot Brigade,* 77–78, 274–75.

3 : MINING SOCIETY IN THE EARLY YEARS

1. Earl Warren, "California's Coming Centennials," address delivered at State Chamber of Commerce luncheon, St. Francis Hotel, Oct. 11, 1946, *Pony Express* 13 (Feb. 1947): 4–5, 9.

2. S. L. Johnson, *Roaring Camp,* 11–12, 52–53, 185–87.

3. For a recent discussion of the connection between violence and masculine predominance in the mining West, see David T. Courtwright, *Violent Land: Single Men and Social Disorder From the Frontier to the Inner City* (Cambridge: Harvard University Press, 1996), 66–86.

4. The dark side of the Gold Rush legacy was recently illustrated by a confrontational discussion on how miners treated Native Americans, aired on KQED radio (a PBS affiliate), San Francisco, Jan. 23, 1998. See also Hurtado, *Indian Survival.*

5. For a statistical and demographic analysis of the 1850 Calaveras census, see Giovinco, *Ethnic Dimension.*

6. John David Borthwick, *The Gold Hunters: A First-Hand Picture of Life in Cali-*

*fornia Mining Camps in the Early Fifties,* original title *Three Years in California* (1857; reprint, Cleveland: International Fiction Library, 1917), 276–90, 310–15. For Borthwick's background, see R. E. Mather, "Borthwick's California: Gold Rush Panorama," *Californians* 12, no. 1 (1994): 16–25.

7. For a summary of archaeological and ethnohistorical data on the Miwok of Calaveras, see Maniery, *Six Mile and Murphys Rancherias.*

8. Thomas S. Wylly, "'Westward Ho—in '49': Memoirs of Captain Thomas S. Wylly," *Pacific Historian* 22 (summer 1978): 123–24.

9. John De Laittre, "Reminiscences" (TS, Jan. 1910), p. 11, Bancroft Library; Carson, *Recollections,* 59.

10. Anderson, Barbour, and Whitworth, "World of Balance and Plenty," 26.

11. Antonio F. Coronel, *Tales of Mexican California* (1848; reprint, Santa Barbara: Bellerophon Books, 1994), 61–62; S. L. Johnson, *Roaring Camp,* 89–95, 136–37; Hurtado, *Indian Survival,* 104–6, 154–56; Rawls, *Indians of California,* 172–201.

12. David E. Stannard, *American Holocaust: Columbus and the Conquest of the New World* (New York: Oxford University Press, 1992). For a good review of postmodern trends in Native American historiography, consult W. R. Swagerty, ed., *Scholars and the Indian Experience: Critical Reviews of Recent Writings in the Social Sciences* (Bloomington: Indiana University Press, 1984).

13. See, for example, Liping Zhu, *A Chinaman's Chance: The Chinese on the Rocky Mountain Mining Frontier* (Niwot: University Press of Colorado, 1997); Donald L. Fixico, ed., *Rethinking American Indian History* (Albuquerque: University of New Mexico Press, 1997); and Devon A. Mihesuah, ed., *Natives and Academics: Researching and Writing About American Indians* (Lincoln: University of Nebraska Press, 1998).

14. S. L. Johnson, *Roaring Camp,* 218–34; Lascy, "Reminiscences," pt. 1, pp. 7–11, Bancroft Library.

15. John Doble, *John Doble's Journal and Letters From the Mines: Mokelumne Hill, Jackson, Volcano, and San Francisco, 1851–1865,* ed. Charles L. Camp (Denver: Old West Publishing, 1962), 43–44; Pringle Shaw, *Ramblings in California: Containing a Description of the Country, Life at the Mines, State of Society, &c.* (Toronto: James Bain, [1856]), 52; Rawls, *Indians of California,* 115–33, 176–86; Hurtado, *Indian Survival,* 107–11.

16. Ferreyra and Reher, *Diary of Navarro,* 50–51; Hovey journal, Sept.–Nov. 1849, Huntington Library; Ferreyra and Reher, *Diary of Navarro,* 51.

17. De Laittre, "Reminiscences," p. 12, Bancroft Library.

18. William Hanson letter, Jan. 14–15, 1851 (TS), Bancroft Library.

19. James Madison Grover, "Memoirs" (MS, 1905), pp. 105–6, Bancroft Library.

20. *Stockton Times,* Mar. 23, 1850, p. 4, col. 1.

21. J. Doble, *Journal and Letters,* 258; John Wallis diary, 1851–1857, UOPWA.

22. J. Doble, *Journal and Letters,* 45–46. Robert F. Heizer documents the deleterious effects of liquor on California Indians in *The Destruction of California Indians* (Santa Barbara: Peregrine Smith, 1974), 271–78. For a recent examination of the North

American liquor trade and its consequences from the 1830s to the 1870s, consult Margaret A. Kennedy, *The Whiskey Trade of the Northwestern Plains: A Multidisciplinary Study* (New York: Peter Lang, 1997).

23. Cool, "Goodness, Gold, and God," 29, 38.

24. William Ellison, "The Federal Indian Policy in California, 1846–1860," *Mississippi Valley Historical Review* 9 (June 1922): 37–67; Hurtado, *Indian Survival,* 126–48.

25. Helen Giorgi, "Benjamin Franklin Jones and Kin," *Las Calaveras* 32 (Oct. 1983): 1–9; Stephen Farbotnik, "The Canepas of Vallecito," *Las Calaveras* 14 (Apr. 1966): 1–2. A somewhat dated but still reliable study of Native American population decline in California is Shelburne F. Cook, *The Conflict Between the California Indian and White Civilization* (Berkeley and Los Angeles: University of California Press, 1976), 255–361. See also Wilbur R. Jacobs, *The Fatal Confrontation: Historical Studies of American Indians, Environment, and Historians* (Albuquerque: University of New Mexico Press, 1996), 91–106.

26. U.S. Bureau of the Census, Census of 1880, Calaveras County, seventh township.

27. Ibid.

28. Maniery, *Six Mile and Murphys Rancherias,* 9–35.

29. Jean Kirkpatrick, "Sheep Ranch Indians," *Las Calaveras* 2 (Jan. 1953): 3.

30. Rowe, *Hard-Rock Men,* 100–109.

31. Richard E. Lingenfelter, *The Hardrock Miners: A History of the Mining Labor Movement in the American West, 1863–1893* (Berkeley and Los Angeles: University of California Press, 1974), 6–7; Ronald M. James, "Defining the Group: Nineteenth-Century Cornish on the North American Mining Frontier," *Cornish Studies* 2, no. 2 (1994): 32–47; "Miner" letter, Oct. 9, 1871, quoted in Rossiter W. Raymond, "The Burleigh Drill," in *Statistics of Mines and Mining,* chap. 20, p. 488.

32. Johnston, "Wade Johnston Talks to His Daughter," *Las Calaveras* 20 (Jan. 1972): 9–19.

33. Doris M. Wright, "The Making of Cosmopolitan California: An Analysis of Immigration, 1848–1870," *California Historical Society Quarterly* 20 (1941): 65–68; Effie Enfield Johnston, "Wade Johnston Talks to His Daughter," *Las Calaveras* 18 (Apr. 1970): 25–28; Effie Enfield Johnston, "Wade Johnston Talks to His Daughter," *Las Calaveras* 19 (Oct. 1970): 1–6.

34. Effie Enfield Johnston, "Wade Johnston Talks to His Daughter," *Las Calaveras* 19 (July 1971): 25–32; Robert Briggs biographical sketches (ca. 1887), Hubert Howe Bancroft Collection, Bancroft Library.

35. U.S. Bureau of the Census, Census of 1870, Calaveras County, sixth township, 36.

36. Mary Jane McSorley Garamendi, "The McSorleys of Chili Gulch," *Las Calaveras* 40 (Jan. 1991): 13–21.

37. *Las Calaveras* 36 (July 1988): 41–46; Giovinco, *Ethnic Dimension,* 23.

38. Giovinco, *Ethnic Dimension,* 4; Robert E. Levinson, *The Jews in the California*

*Gold Rush* (New York: Ktav Publishing House, 1978), 15; store ledger book photostats, Kathleen Mitchel Collection, Calaveras County Historical Society.

39. A. P. Nasatir, *A French Journalist in the California Gold Rush: The Letters of Etienne Derbec* (Georgetown, Calif.: Talisman Press, 1964), 15–20; S. L. Johnson, *Roaring Camp,* 79–81; Borthwick, *Gold Hunters,* 287; "Rich in History, Fricot City Faces Uncertain Future," *Las Calaveras* 35 (Apr. 1987): 29–30, 37–39.

40. Borthwick, *Gold Hunters,* 342–47; Rufus King Wyllys, "The French of California and Sonora," *Pacific Historical Review* 1 (Sept. 1932): 341–42.

41. Rufus K. Wyllys, *The French in Sonora (1850–1854): The Story of French Adventurers From California Into Mexico,* University of California Publications in History, no. 21 (Berkeley and Los Angeles: University of California Press, 1932), 37–38; S. L. Johnson, *Roaring Camp,* 226.

42. Schafer, *Letters of Fairchild,* 100; Hovey journal, pp. 128–29, Huntington Library.

43. William Hanson letter, Apr. 27, 1851 (TS), Bancroft Library; Hovey journal, pp. 130–39, Huntington Library; *Stockton Times,* Apr. 26, 1851, p. 3, col. 1.

44. Hovey journal, pp. 130–39, Huntington Library.

45. *San Joaquin Republican,* June 30, 1852, p. 3, col. 2.

46. Giovinco, *Ethnic Dimension,* 12–23.

47. Norman Lagomarsino, "The Lagomarsino-Werle Family," *Las Calaveras* 26 (Oct. 1977): 1–11.

48. Olivia Barden Harbinson, "The Rolleri Family," *Las Calaveras* 14 (Apr. 1966): 3–4.

49. Michael S. Coray, "Negro and Mulatto in the Pacific West, 1850–1860: Changing Patterns of Black Population Growth," *Pacific Historian* 29 (winter 1985): 18–27. For the Oregon double ban and its national political consequences, see Elizabeth McLagan, *A Peculiar Paradise: A History of Blacks in Oregon, 1788–1940* (Portland: Georgian Press, 1980), 24–28; and Robert W. Johannsen, *Frontier Politics and the Sectional Conflict: The Pacific Northwest on the Eve of the Civil War* (Seattle: University of Washington Press, 1955), 15–23.

50. Albert S. Broussard, "Slavery in California Revisited: The Fate of a Kentucky Slave in Gold Rush California," *Pacific Historian* 29 (spring 1985): 17–21; S. L. Johnson, *Roaring Camp,* 188–90; *Ex Party Archy,* 9 CA 147 (1858); Ferreyra and Reher, *Diary of Navarro,* 94.

51. Noyes, "Reminiscences," p. 44, Calaveras County Historical Society.

52. Rudolph M. Lapp, *Blacks in Gold Rush California* (New Haven: Yale University Press, 1977), 52–53.

53. Effie Enfield Johnston, "Wade Johnston Talks to His Daughter," *Las Calaveras* 18 (Apr. 1970): 21; R. C. Wood, "Buster's Gold," in *Tales of Old Calaveras,* 5–14.

54. Effie Enfield Johnston, "Wade Johnston Talks to His Daughter," *Las Calaveras* 18 (Apr. 1970): 21–22, 25.

55. Lillian Gorham Murphy, "The Chinese at O'Byrnes Ferry," *Las Calaveras* 11 (July 1963): 2; Noyes, "Reminiscences," p. 52, Calaveras County Historical Society;

Archie D. Stevenot, "Chinese on the Mokelumne River," *Las Calaveras* 12 (Oct. 1963): 7.

56. *San Joaquin Republican,* Nov. 15, 1851, p. 2, col. 3; Giovinco, *Ethnic Dimension, 10.*

57. Borthwick, *Gold Hunters,* 302–3. For a brief survey of Calaveras Chinese districts, see *Las Calaveras* 11 (Apr. 1963): 1–4; (July 1963): 1–4; 12 (Oct. 1963): 1–8; (Jan. 1964): 1–4; and 21 (Jan. 1973): 11–22.

58. *San Joaquin Republican,* June 28, 1851, p. 2, col. 4; William Shaw, *Golden Dreams and Waking Realities: Being the Adventures of a Gold-Seeker in California and the Pacific Islands* (London: Smith, Elder, 1851), 86–87.

59. Ping Chiu, *Chinese Labor in California, 1850–1880: An Economic Study* (Madison: State Historical Society of Wisconsin, 1963), 10–29.

60. Stephen Williams, *The Chinese in the California Mines, 1848–1860* (1930; reprint, Stanford: Stanford University Press, 1971), 38; Randall E. Rohe, "After the Gold Rush: Chinese Mining in the Far West, 1850–1890," *Montana* 32 (autumn 1982): 8–9; Williams, *Chinese in California Mines,* 39–40.

61. Chiu, *Chinese Labor in California,* 11, 20.

62. Giovinco, *Ethnic Dimension,* 40–42; Rossiter W. Raymond, *Mining Statistics West of the Rocky Mountains,* 42d Cong., 1st sess. H. Exec. Doc. 10 (Washington, D.C.: GPO, 1871), 28, 31; Alexander Saxton, *The Indispensable Enemy: Labor and the Anti-Chinese Movement in California* (1971; reprint, Berkeley and Los Angeles: University of California Press, 1995), 58–60. One scholar has concluded that at least in the placer mines of Nevada County, charges that Chinese undercut white wages by working for less "contained little truth" because they worked claims few whites wanted (Ralph Mann, *After the Gold Rush: Society in Grass Valley and Nevada City, California, 1849–1870* [Stanford: Stanford University Press, 1982], 87).

63. Charles Schwoerer, "Chinese in Douglas Flat," *Las Calaveras* 12 (Jan. 1961): 2; Giovinco, *Ethnic Dimension,* 43–44.

64. *Pacific States Illustrated Weekly,* Aug. 20, 1887, p. 5; Saxton, *Indispensable Enemy,* 4; Dorothea J. Theodoratus, comp., *An Ethnographic Study of the New Melones Lake Project* (Sacramento: Army Corps of Engineers, 1976), 23–36.

65. Sister M. Colette Standard, "The Sonora Migration to California, 1848–1856: A Study in Prejudice," *Southern California Quarterly* 58 (fall 1976): 336–38; S. L. Johnson, *Roaring Camp,* 59–61; D. M. Wright, "Cosmopolitan California," 324–26.

66. R. C. Wood, *Calaveras,* 25; Coronel, *Tales of Mexican California,* 53–54.

67. R. C. Wood, *Calaveras,* 4–5, 19.

68. W. T. Jackson and Mikesell, "Mexican Melones," 6; J. R. Browne, *Report on Mineral Resources,* 59; Noyes, "Reminiscences," p. 60, Calaveras County Historical Society.

69. Noyes, "Reminiscences," p. 66, Calaveras County Historical Society; *San Andreas Independent,* Aug. 25, 1860 (TS), MS 187, Smith Collection. These figures are based on contemporary estimates as well as scholarly assessment of population demographics for the time period. See Giovinco, *Ethnic Dimension,* 36.

70. *Stockton Times,* July 20, 1850, p. 3, col. 1.

71. Ferreyra and Reher, *Diary of Navarro,* 36, 85.

72. Ibid., 22–33; *Stockton Times,* Apr. 6, 1850, as quoted in Richard H. Peterson, "The Foreign Miners' Tax of 1850 and Mexicans in California: Exploitation or Expulsion?" *Pacific Historian* 20 (fall 1976): 265–72. In a later essay, Peterson argues that the primary reason for anti-Mexican sentiment in the mines was cultural bias, not economic competition ("Anti-Mexican Nativism in California, 1848–1853: A Study of Cultural Conflict," *Southern California Quarterly* 62 [winter 1980]: 309–27).

73. S. L. Johnson, *Roaring Camp,* 188–89.

74. Edwin A. Bielharz and Carlos U. Lopez, eds. and trans., *We Were 49ers! Chilean Accounts of the California Gold Rush* (Pasadena: Ward Ritchie Press, 1976), xvii; David M. Potter, *The Impending Crisis, 1848–1861* (New York: Harper and Row, 1976), 80–82; Leigh Bristol-Kagan, "Chinese Migration to California, 1851–1882: Selected Industries of Work, the Chinese Institutions, and the Legislative Exclusion of a Temporary Labor Force" (Ph.D. diss., Harvard University, 1982), chaps. 2–3; Shih-Shan Henry Tsai, *The Chinese Experience in America* (Bloomington: Indiana University Press, 1986), 3–7; S. L. Johnson, *Roaring Camp,* 193.

75. Hovey journal, pp. 84–90, Huntington Library.

76. Jay Monaghan, *Chile, Peru, and the California Gold Rush of 1849* (Berkeley and Los Angeles: University of California Press, 1973), 243–44; *Alta California,* Apr. 26, 1849, as quoted in Neal Harlow, *California Conquered: The Annexation of a Mexican Province, 1846–1850* (Berkeley and Los Angeles: University of California Press, 1982), 327; Josiah Royce, *California: From the Conquest in 1846 to the Second Vigilance Committee in San Francisco: A Study of American Character,* new ed. (1886; reprint, New York: Alfred A. Knopf, 1948), 157–59; Harlow, *California Conquered,* 318–27.

77. Shinn, *Mining Camps,* 78–83, 86–87.

78. Ibid., 117–34, 142–45, 155–72; Harlow, *California Conquered,* 331.

79. Shinn, *Mining Camps,* 83–86. See also Oscar T. Shuck, *History of the Bench and Bar of California* (Los Angeles: Commercial Printing House, 1901), xv–xxi.

80. Ferreyra and Reher, *Diary of Navarro,* 47.

81. Hovey journal, pp. 84–90, Huntington Library. Susan Johnson's recent analysis of the incident incorporates viewpoints of various participants, including one Chilean, Ramon Jil Navarro (*Roaring Camp,* 196–208).

82. Ferreyra and Reher, *Diary of Navarro,* 68–70; Hovey journal, Dec. 15, 1849, p. 76, Huntington Library.

83. Hovey journal, pp. 78–79, Huntington Library. See also Ferreyra and Reher, *Diary of Navarro,* 72–74. *Alta California,* Jan. 2, 1850, p. 2, cols. 3–4; Ayers, *Gold and Sunshine,* 50–51.

84. Robert Wilson letter, Dec. 31, 1849, in *Alta California,* Jan. 2, 1850, p. 2, cols. 3–4.

85. Robert Wilson letter, Jan. 3, 1850, in *Alta California,* Jan. 7, 1850, p. 1, col. 3; Ayers, *Gold and Sunshine,* 51–57. Although it disputes Ayers's version of the affair, the best modern account is in Monaghan, *Chile, Peru, and California,* 243–47. For

Chilean versions, see Ferreyra and Reher, *Diary of Navarro,* 68–79; and Bielharz and Lopez, *We Were 49ers!* 101–49.

86. Hovey journal, pp. 78–81, Huntington Library; Ferreyra and Reher, *Diary of Navarro,* 82–84.

87. Hovey journal, p. 82, Huntington Library.

88. Ibid., p. 83; Ferreyra and Reher, *Diary of Navarro,* 84; S. L. Johnson, *Roaring Camp,* 208.

89. Leonard Pitt, *The Decline of the Californios: A Social History of the Spanish-Speaking Californians, 1846–1890* (Berkeley and Los Angeles: University of California Press, 1966), 60–64; Peterson, "Foreign Miners' Tax," 265–72.

90. "The Discoverer of Mosquito Gulch," *Las Calaveras* 5 (Apr. 1957): 3.

91. Effie Enfield Johnston, "Wade Johnston Talks to His Daughter," *Las Calaveras* 18 (Oct. 1969): 8.

92. Ferreyra and Reher, *Diary of Navarro,* 119–23.

93. Pitt, *Decline of the Californios,* 60–67; Richard H. Morefield, "Mexicans in the California Mines, 1848–1853," *California Historical Society Quarterly* 35 (1956): 38–39; Coronel, *Tales of Mexican California,* 65; Edward J. Phillips, "Seeing the Elephant," *Pacific Historian* 18 (spring 1974): 14–15.

94. Pitt, *Decline of the Californios,* 66–67; S. L. Johnson, *Roaring Camp,* 211–15, 247–48.

95. *San Joaquin Republican,* May 19, 1852, p. 2, col. 4.

96. *Stockton Times,* May 25, 1850, p. 4, col. 2; *San Joaquin Republican,* July 24, 1854, p. 2, col. 3.

97. "Harvey Wood of Robinson's Ferry," *Las Calaveras* 29 (Oct. 1980): 1–8.

98. McKinstry, *Gold Rush Diary,* 364; Borthwick, *Gold Hunters,* 297–99.

99. Wallis diary, 1851–1857, MS 131, UOPWA.

100. Noyes, "Reminiscences," p. 60, Calaveras County Historical Society.

101. *San Joaquin Republican,* Sept. 15, 1852, p. 2, col. 4; Oct. 11, 1853, p. 3, col. 4.

102. See Clare V. McKanna Jr., *Homicide, Race, and Justice in the American West, 1880–1920* (Tucson: University of Arizona Press, 1997); Roger D. McGrath, *Gunfighters, Highwayman, and Vigilantes: Violence on the Frontier* (Berkeley and Los Angeles: University of California Press, 1984); Courtwright, *Violent Land,* 81–84; and Martin Ridge, "Disorder, Crime, and Punishment in the California Gold Rush," *Montana* 49 (autumn 1999): 12–27.

103. For an extensive study of the roots of Gold Rush violence, consult Jason R. Beck, "California Gold Rush Violence, 1849–1854: A Psychological Interpretation" (Ph.D. diss., University of Southern California, June 1978).

104. Joseph Pownall, "To California Over Southern Trails: The Joseph Pownall Papers [1850]," ed. Albert Dressler (TS), Huntington Library.

105. *Stockton Times,* July 6, 1850, p. 2, col. 1; *Sonora Herald,* July 13, 1850, pp. 2–3; July 27, 1850, pp. 1–3; William Hanson letter, Jan. 14, 1851 (TS), Bancroft Library.

106. *Calaveras Chronicle,* Sept. 25, 1852, p. 2, col. 3.

107. *Sonora Herald,* July 5, 1851, p. 3; Ferreyra and Reher, *Diary of Navarro,* 201.

108. Borthwick, *Gold Hunters,* 299–302; Alfred R. Doten, *The Journals of Alfred Doten, 1849–1903,* ed. Walter Van Tilburg Clark (Reno: University of Nevada Press, 1973), 1:98–107.

109. James Megeath to Miss M. F. Megeath, May 20, 1853 (TS), Bancroft Library; *Sonora Herald,* July 13, 1850, pp. 2–3; *San Joaquin Republican,* Jan. 29, 1853, as quoted in Frank Latta, *Joaquin Murrieta and His Horse Gangs* (Santa Cruz: Bear State Books, 1980), 35–36. On the absence of guns and security of property before the 1850s, see Rohrbough, *Days of Gold,* 86–90; and Ridge, "Disorder, Crime, and Punishment," 14–17.

110. Wallis diary, Aug. 20, 1852, photocopy of original, UOPWA; William Brown letter, Feb. 14, 1852, 1849–1853, in Brown Papers, Huntington Library.

111. Elias S. Ketchum diaries, 1851, 1853 (MSS), Huntington Library.

112. Doten, *Journals of Alfred Doten,* 140.

113. *San Joaquin Republican,* Jan. 29, 1853, facsimile in Latta, *Joaquin Murrieta,* 37–38; *San Joaquin Republican,* Feb. 14, 1853, facsimile in Latta, *Joaquin Murrieta,* 42.

114. *Calaveras Chronicle,* Jan. 29, 1853, p. 2, col. 2, cited in *San Joaquin Republican,* Feb. 2, 1853. See also Latta, *Joaquin Murrieta,* 40.

115. Wallis diary, 1851–1857, photocopy of original, UOPWA.

116. Mrs. Herbert Grutzmacher, "One Hundred Years in Calaveras," *Las Calaveras* 7 (Apr. 1959): 6; James Alexander Smith, "San Antone Camp," *Las Calaveras* 8 (July 1960): 1.

117. California State Assembly, *Journal* (Vallejo: State Printer, 1853), 344, 574, 657; Harry Love to Governor Bigler, Aug. 4, 1853, reprinted in *Pacific Historian* 16 (winter 1972): 70–71; Latta, *Joaquin Murrieta,* 1–19, 59–70; Raymund F. Wood, "New Light on Joaquin Murrieta," *Pacific Historian* 14 (winter 1970): 54–65; Monaghan, *Chile, Peru, and California,* 214–19; Maria Mondragon, "'The (Safe) White Side of the Line': History and Disguise in John Rollin Ridge's 'The Life and Adventures of Joaquin Murieta: The Celebrated California Bandit,'" *American Transcendental Quarterly* 8 (Sept. 1994): 173–87. For the most scholarly recent account, which relies heavily on Latta's seminal study, see S. L. Johnson, *Roaring Camp,* 25–53.

118. Rohrbough, *Days of Gold,* 75–78, 141–44; Riley Senter, *Crossing the Continent to California Gold Fields* (Exeter, Calif.: Exeter Sun, 1938; reprint, Lemon Grove, Calif.: W. R. Senter, 1938), 19–22.

119. Alonzo W. Merrill to Miss Mary A. Merrill, July 9, 1854, Bancroft Library.

120. George Lundy Hunt diary, 1867 (TS), p. 3, Bancroft Library.

121. Anthony Lorenz, "Scurvy in the Gold Rush," *Journal of the History of Medicine and Allied Sciences* (1957): 473–510; S. L. Johnson, *Roaring Camp,* 108–10, 127–28; Rohrbough, *Days of Gold,* 190–92; Audubon, *Audubon's Western Journal,* 223; Liping Zhu, "No Need to Rush: The Chinese, Placer Mining, and the Western Environment," *Montana* 49 (autumn 1999): 48–51; Patricia Nelson Limerick, "The Gold Rush and the Shaping of the American West," *California History* 77 (spring 1998): 32.

122. Matthew Scott, "Gold Rush Letters," *Pacific Historian* 6 (Nov. 1962): 161–78.

123. Harriet Steele, "Gold Rush Letters Copied From an Old Letter Book," *Pacific Historian* 8 (Feb. 1964): 43–52; Rohrbough, *Days of Gold,* 182–84.

124. George A. Locke to "Dear Father," May 26, 1854, Jonathan F. Locke Papers, Bancroft Library; JoAnn Levy, *They Saw the Elephant: Women in the California Gold Rush* (Hamden, Conn.: Archon Books, 1990), 113; "The Swank Family Had Mining in Its Blood," *Las Calaveras* 36 (Apr. 1988): 29, 32–36.

125. Harrold family, "Letters from 'The Land of Gold,'" ed. Anne Lohrli, *Pacific Historian* 20 (fall 1976): 252–64.

126. Lascy, "Reminiscences," pt. 1, pp. 11–12, Bancroft Library; S. L. Johnson, *Roaring Camp,* 30–31, 119–20, 124–25, 157–59; U.S. Bureau of the Census, Census of 1870, Calaveras County, fourth township.

127. Brian Eugene Roberts, "The California Gold Rush and the Making of the American Middle Class" (Ph.D. diss., Rutgers University, 1995), 1–60; Ryan, *Personal Adventures,* 2:50.

128. Schafer, *Letters of Fairchild,* 95–96; George A. Locke to "My Father," Oct. 15, 1854, Locke Papers.

129. Harrold family, "Letters from 'The Land of Gold,'" 259; T. S. Wylly, "'Westward Ho—in '49,'" 292–95; Doten, *Journals of Alfred Doten,* 117.

130. Lascy, "Reminiscences," pt. 2, pp. 8–9, Bancroft Library; George Adams Locke to "Dear Mother," Jan. 5, 1852, Locke Papers.

131. Louis B. Wright, *Culture on the Moving Frontier* (Bloomington: Indiana University Press, 1955), 131–32; J. S. Hittell, *Resources of California,* 440–41; Mann, *After the Gold Rush,* 34–39; McGrath, *Gunfighters, Highwaymen, and Vigilantes,* 149–64; Limerick, *Legacy of Conquest,* 48–54.

132. Roberts, "Gold Rush and Middle Class," 90–136; Rohrbough, *Days of Gold,* 243–55. See also Malcolm J. Rohrbough, "The California Gold Rush As a National Experience," *California History* 77 (spring 1998): 16–29. Jane Apostol, "Gold Rush Widow," *Pacific Historian* 28 (summer 1984): 49; James H. Rushton, "The Rushton-Norton Letters from the Gold Fields," *Pacific Historian* 7 (May 1963): 76–77.

133. Cool, "Goodness, Gold, and God," 20.

134. Wallis diary, Dec. 21, 1852 (photocopy of original), UOPWA; D. C. Demarest, "California Gold," chap. 28, UOPWA; Giovinco, *Ethnic Dimension,* 47.

135. John E. Fletcher to wife, June 20, 1850, UOPWA; William Brown to Cornelia T. Brown, Dec. 13, 1852, Huntington Library; Francis X. Hall to Alexander Dickinson, Dec. 28, 1853, Bancroft Library; W. H. Newell to Benjamin Harrison, June 15, 1851, Pownall Collection; William Hanson to Alma Morse Hanson, Feb. 21, 1852, Bancroft Library; Schafer, *Letters of Fairchild,* 107.

136. Rohrbough, "Gold Rush As National Experience," 26; Jonathan Frost Locke to Mary Locke, Oct. 18–24, 1852, Bancroft Library.

137. Chloe H. Sprague to James Madison Grover, Mar. 16, 1878, Bancroft Library.

138. Franklin Walker, *San Francisco's Literary Frontier* (New York: Alfred A. Knopf, 1939), 265–65; George Hoeper, "Gold Rush Miners Feared Spectre of Horrendous Min," *Las Calaveras* 36 (Oct. 1987): 7. Townsend's stories entertained a new genera-

tion of readers when they reappeared with new embellishments nearly three decades later in Calaveras newspapers. See R. C. Wood, "The Calaveras Serpent," in *Tales of Old Calaveras,* 30–36.

139. Kevin Starr, *Americans and the California Dream* (New York: Oxford University Press, 1973, 1986), 134; Walker, *San Francisco's Literary Frontier,* 11–15.

140. Walker, *San Francisco's Literary Frontier,* 64–69.

141. Roy F. Hudson, "Roaring Camp Revisited," *Pacific Historian* 5 (May 1961): 69–76; Hudson, "From Poker Flat to Sandy Bar," *Pacific Historian* 6 (Aug. 1962): 129–37; Mark Twain to William Dean Howells, Apr. 15, 1879, quoted in Margaret Duckett, *Mark Twain and Bret Harte* (Norman: University of Oklahoma Press, 1964), 190.

142. Frederick Anderson, Michael B. Frank, and Kenneth M. Sanderson, eds., *Mark Twain's Notebooks and Journals* (Berkeley and Los Angeles: University of California Press, 1975), 1:66–80; Duckett, *Twain and Harte,* 20–22; William R. Gillis, *Memories of Mark Twain and Steve Gillis,* 2d ed. (Sonora: Banner, 1929), 49–50; Walker, *San Francisco's Literary Frontier,* 193–94.

143. Anderson, Frank, and Sanderson, *Twain's Notebooks and Journals,* 1:81–82.

144. R. C. Wood, *Tales of Old Calaveras,* 37–46.

145. Ibid., 49–57; Ralph W. Dexter, "Historical Aspects of the Calaveras Skull Controversy," *American Antiquity* 51, no. 2 (1986): 365–69; R. E. Taylor, Louis A. Payen, and Peter J. Slota Jr., "The Age of the Calaveras Skull: Dating the 'Piltdown Man' of the New World," *American Antiquity* 57, no. 2 (1992): 269–75. Harte's lampoon, "To the Pliocene Skull, a Geological Address," appeared in the *Californian* (July 28, 1366). See also J. D. Whitney, "The Auriferous Gravels of the Sierra Nevada," *Memoranda of the Museum of Comparative Zoology* 6 (1879); and Lindgren, *Tertiary Gravels.* Scribner's "discovery" may have inspired a similar prank on the Comstock a few years later. Dan DeQuille's tongue-in-cheek account appears in *The Big Bonanza* (1876; New York: Alfred A. Knopf, 1947), 392–93.

146. Ridge, "Disorder, Crime, and Punishment," 20–27; *Daily Alta California,* July 17, 1851, p. 2, col. 4; Thomas Jefferson Matteson, *Diary* (San Andreas: Calaveras County Historical Society, 1954), 25; Noyes, "Reminiscences," p. 66, Calaveras County Historical Society.

147. Royce, *California,* 214–19, 295–96.

148. Quoted in Raymond, *Statistics of Mines and Mining,* 69.

4: ANCILLARY INDUSTRIES

1. For a colorful, if somewhat dated, account of California in the 1840s, see Ray Allen Billington's classic, *The Far Western Frontier, 1830–1860* (New York: Harper and Row, 1956), 1–13.

2. Lawrence James Jelinek, "'Property of Every Kind': Ranching and Farming During the Gold-Rush Era," *California History* 77 (winter 1998–1999): 233–37, 242–43. For a comprehensive and authoritative study of the wheat trade, see Kenneth Aubrey Smith, "California: The Wheat Decades" (Ph.D. diss., University of Southern California, 1969).

3. *Daily Alta California,* June 18, 1853, pp. 131–32, Bancroft Scraps Calaveras; Jelinek, "'Property of Every Kind,'" 237–41; Nash, "Veritable Revolution," 278–79.

4. For data on California grain production by county, 1852–1859, see *Transactions of the California State Agricultural Society, During the Year 1859* (Sacramento: State Printer, 1860), tables 1–14, pp. 325–34.

5. W. P. Fuller, Marvin, and Costello, *Madam Felix's Gold,* 11–13.

6. Lary M. Dilsaver, "From Boom to Bust: Post–Gold Rush Patterns of Adjustment in a California Mining Region" (Ph.D. diss., Louisiana State University, 1982), 1:155–57; Paul, *California Gold,* 243–47.

7. "Sheriff Joseph Zwinge, a Respected Lawman," *Las Calaveras* 36 (July 1988): 41–46; Rose Hogarth Fletcher, "Alexander Love Family," *Las Calaveras* 15 (July 1966): 5; Farbotnik, "The Canepas of Vallecito," 1–2; Ted Bird, interview by authors, Sonora, Calif., May 7, 1998.

8. *Calaveras County Illustrated,* 45. Numbers compiled from *Study No. 3: Historical, Demographic, Economic, and Social Data: The United States, 1790–1970.* Census data tabulated and distributed by the Inter-university Consortium for Political and Social Research, Ann Arbor, Mich., 1992.

9. See, for example, *San Joaquin Republican,* Dec. 13, 1851, p. 4, col. 3.

10. *Stockton Times,* Sept. 21, 1850, p. 2, col. 4; *San Joaquin Republican,* Dec. 13, 1851, p. 4, col. 3; Sept. 25, 1852, p. 2, col. 3; Dec. 1, 1852, p. 2, cols. 6–7.

11. *Stockton Times,* Mar. 16, 1850, p. 1, col. 3.

12. *San Joaquin Republican,* Apr. 21, 1852, p. 2, col. 4; excerpts from *Calaveras Chronicle,* in *San Joaquin Republican,* July 28, 1852, p. 2, col. 6.

13. *Calaveras Chronicle,* reprinted in *San Joaquin Republican,* Jan. 15, 1853, p. 2, col. 7.

14. Gordon R. Miller, "Shaping California's Water Law," *Southern California Historical Quarterly* 55 (spring 1973): 9–42.

15. *Stockton Times,* Mar. 12, 1851, p. 2, cols. 1–2; *San Joaquin Republican,* Jan. 7, 1852, p. 2, col. 4. For the drainage problems of Murphys Flat, see chap. 1.

16. Noyes, "Reminiscences," p. 82, Calaveras County Historical Society; W. T. Jackson, *Historical Survey,* 84–85. Jackson contends that the Bedrock Flume, mentioned in chap. 1, was necessary because the Union Water Company "brought too much water into Murphys" (85–86). A more likely cause was the cycle of wet weather in the late 1850s.

17. W. P. Fuller, Marvin, and Costello, *Madam Felix's Gold,* 19; Ephriam Cutting, "The Union Water Company in 1884," *Las Calaveras* 8 (Oct. 1959): 2–4; J. A. Smith, "The Salt Spring Valley Reservoir," *Las Calaveras* 8 (Apr. 1960): 1–2.

18. *Daily Alta California,* June 23, 1853; J. A. Smith, "The Mokelumne Hill Ditch," *Las Calaveras* 8 (Oct. 1959): 1; Borthwick, *Gold Hunters,* 286–87; *San Joaquin Republican,* Dec. 18, 1852, p. 2, cols. 3–4; June 4, 1853, p. 2, col. 5.

19. R. C. Wood, *Calaveras,* 38; Borthwick, *Three Years in California,* 247.

20. Bancroft, *History of California,* 6:375, in vol. 23 of *Works of Bancroft;* Douglas

R. Littlefield, "Water Rights During the California Gold Rush: Conflicts Over Economic Points of View," *Western Historical Quarterly* 14 (Oct. 1983): 415–34.

21. *San Andreas Independent,* June 27, 1857, p. 2, col. 2; Effie Enfield Johnston, "Wade Johnston Talks to His Daughter," *Las Calaveras* 19 (Oct. 1970): 1–2.

22. Samuel Linus Prindle, Records and Papers, 1858–1889, Bancroft Library; J. A. Smith, "The Mokelumne Hill Ditch," 1–2.

23. Cutting, "Union Water Company," 2–4; Frances E. Bishop, "Gold Rush Water: The Union Water Company of Murphys" (TS, ca. 1987), pp. 44–48, Calaveras County Archives.

24. Bishop, "Gold Rush Water," 49–53, Calaveras County Archives.

25. Cornelius B. Demarest, "Chronicles of Calaveras," *Las Calaveras* 25 (Oct. 1976): 6.

26. Charles R. Joy, "Calaveras County Goes Electric," *Las Calaveras* 23 (Apr.–July 1975): 29–31; Duane Oneto, telephone interview by authors, July 7, 1998.

27. Oneto, interview.

28. A. C. W. Bethel, "The Golden Skein: California's Gold-Rush Transportation Network," in *Golden State,* ed. Rawls and Orsi, 263.

29. *Stockton Times,* Mar. 16, 1850, p. 1, col. 3; *Stockton Independent,* Feb. 26, 1896, p. 3, col. 5; B. Taylor, *Eldorado,* 76; Oscar O. Winther, *Express and Stagecoach Days in California From the Gold Rush to the Civil War* (Stanford: Stanford University Press, 1936), 9–10.

30. Holmes, *Southern Mines of California,* 7; Bethel, "Golden Skein," 256–58.

31. Giorgi, "Jones and Kin," 1–9; Bethel, "Golden Skein," 258–59; *San Joaquin Republican,* Dec. 17, 1851, p. 4, col. 2.

32. Audubon, *Audubon's Western Journal,* 203; George R. Underhill, *Voyage to California and Return: George R. Underhill, 1849–1852* (Manhasset, N.Y.: Underhill Society of America, 1980), 33–40.

33. Harvey M. Seaman, "Letter Written by Harvey M. Seaman From Stockton," *Pacific Historian* 21 (spring 1977): 27; De Laittre, "Reminiscences," p. 1, Bancroft Library; *San Joaquin Republican,* Dec. 17, 1851, p. 4, col. 2.

34. *San Joaquin Republican,* Dec. 17, 1851, p. 4, col. 2.

35. Ibid., Dec. 31, 1851, p. 2, col. 1.

36. Flagg, "Philadelphia 49er," 401.

37. Oscar O. Winther, *Via Western Express and Stagecoach* (Stanford: Stanford University Press, 1945), 10.

38. Larry Bradfield, "The Elephant at the Depot Door" (term paper, University of the Pacific, 1979), 10–12.

39. Ibid., 13–17; *San Andreas Register,* Dec. 7, 1867 (TS), Smith Collection.

40. "Boston" letter to editor, May 1, 1867, *Stockton Daily Independent,* June 7, 1867 (TS), ibid.

41. Bradfield, "Elephant," 18–19; George H. Tinkham, *California, Men, and Events: Time, 1769–1890,* 2d ed. (Stockton: Record Publishing, 1915), 287.

42. Leonard Noyes to brother, May 2, Sept. 24, 1856, "Reminiscences," pp. 61, 65, Calaveras County Historical Society.

43. Bradfield, "Elephant," 22–24.

44. Ibid., 24–35; J. A. Smith, "Stockton and Copperopolis Railroad," *Las Calaveras* 2 (Apr. 1954): 3–4.

45. Bradfield, "Elephant," 35–39.

46. California State Railroad Commission, "Ninth Annual Report" (1887), in *Appendix to Journals of the Senate and Assembly, 28th Session,* vol. 2 (1889), pp. 243–44, California State Library.

47. Ward McAfee, "Local Interests and Railroad Regulation in Nineteenth-Century California" (Ph.D. diss., Stanford University, 1966), 111–12.

48. James C. Wagers, "The San Joaquin and Sierra Nevada Railroad," *San Joaquin Historian* 11 (July–Sept. 1975): 82–83.

49. David F. Myrick, *Railroads of Nevada and Eastern California* (Berkeley: Howell-North Books, 1962–1963), 1:210–11; *Memorial and Genealogical History of Representative Citizens of Northern California* (Chicago: Standard Genealogical Publishing, 1901), 146; *San Francisco Call,* Apr. 24, 1900, p. 9, col. 5.

50. Wagers, "San Joaquin and Sierra Nevada Railroad," 82.

51. Ibid., 83–86; Donald Ray Floyd, "The Role of Narrow-Gauge Railroads in California's Transportation Network" (master's thesis, University of California at Berkeley, 1968), 67–68.

52. Philip Charles Habib, "Some Economic Aspects of the California Lumber Industry and Their Relation to Forest Use" (Ph.D. diss., University of California at Berkeley, [1952]), 65–66.

53. Ruby E. Taylor, "Glencoe," *Las Calaveras* 6 (July 1958): 3; Arthur R. Wilson, *My Life and Boyhood Days in West Point, 1881–1959* ([West Point?], Calif.: privately printed, 1959), 21.

54. William H. Storms, *Timbering and Mining: A Treatise on Practical American Methods* (New York: McGraw-Hill, 1909), 1, 5.

55. Richard M. MacKinnon, "The Historical Geography of Settlement in the Foothills of Tuolumne County, California" (Ph.D. diss., University of California at Berkeley, 1967), 101; Ruth Lemue, "A Pioneer Daughter Remembers Girlhood in the Mother Lode," *Las Calaveras* 35 (July 1987): 45–49.

56. Kerr, *Mining Resources,* 58–59.

57. *Annual Report From June 1, to December 1, 1880,* 202–3; George W. Stewart, "Early Government Attempts at Forest Conservation," *Sierra Club Bulletin* 16 (Feb. 1931): 19.

58. *Daily Alta California,* June 23, 1853, p. 1, col. 6; excerpt from *Report of Civil Engineers of Calaveras County,* in *Daily Alta California,* Dec. 16, 1856, p. 1, col. 2.

59. *San Joaquin Republican,* June 4, 1853, p. 2, col. 5.

60. Grover, "Memoirs," pp. 86–90, Bancroft Library.

61. R. C. Wood, *Calaveras,* 26, 30, 37.

62. Frances E. Bishop, "McKay's Clipper Mill," *Las Calaveras* 24 (Jan. 1976): 14–16.

63. On Kimball and Cutting, see *Daily Alta California,* May 29, 1862, p. 1, col. 2; on Hanford's mill, see excerpt from *Report of Civil Engineers,* in *Daily Alta California,* Dec. 16, 1856, p. 1, col. 2; on Fisher's mills, see *Las Calaveras* 3 (Jan. 1955): 3; excerpt from *Stockton Daily Independent,* in *San Joaquin Republican,* June 10, 1860 (TS), MS 187, box 4, Smith Collection.

64. *Sacramento Daily Union,* Apr. 5, 1861, Jan. 20, 1862 (TSS), MS 187, Smith Collection; *San Andreas Register,* July 2, 6, 1864 (TSS), ibid.; *Daily Alta California,* Oct. 12, 1868, p. 1, col. 5; D. C. Demarest, "California Gold," chap. 12, UOPWA.

65. *Transactions of the California State Agricultural Society During the Year 1877* (Sacramento: F. P. Thompson, 1878), 146; Emanuel Fritz, *The Development of Industrial Forestry in California* (Seattle: University of Washington Press, 1960), 9.

66. Rodney S. Ellsworth, "Discovery of the Big Trees of California, 1833–1852" (Ph.D. diss., University of California at Berkeley, 1933), 114–16; Francis P. Farquhar, *History of the Sierra Nevada* (Berkeley and Los Angeles: University of California Press, 1965), 41–42, 83–88.

67. Farquhar, *History of the Sierra Nevada,* 84–86; Hank Johnston, *They Felled the Redwoods: A Saga of Flumes and Rails in the High Sierra* (Los Angeles: Trans-Anglo Books, [1966]), 17–18; *San Joaquin Republican,* June 22, 1854, p. 2, col. 1.

68. J. A. Smith, "Calaveras Grove of Big Trees," *Las Calaveras* 1 (Oct. 1953): 2–3; H. Johnston, *They Felled the Redwoods,* 17.

69. Dilsaver and Tweed, *Challenge of the Big Trees,* 56.

70. D. C. Demarest, "California Gold," chap. 15, UOPWA; U. S. Bureau of the Census, Census of 1880, Calaveras County, seventh township, 41.

71. John Muir, *Our National Parks* (1901; reprint, Madison: University of Wisconsin Press, 1981), 356.

72. Stewart, "Early Government Attempts," 20–21; William S. Brown and S. B. Show, *California Rural Land Use and Management: A History of the Use and Occupancy of Rural Lands in California* ([Washington, D.C.]: Department of Agriculture, Forest Service, California Region, 1944), 1:208–13.

73. *American Eagle* 2 (Jan. 1945): 1, in Calaveras County Historical Society.

74. Bailey, *Supplying the Mining World,* 8–29.

75. Ibid., 15–26.

76. *San Andreas Independent,* Mar. 31, 1860, as quoted in Richard F. Epstein, *Old Mok (Mokelumne Hill): The Story of a Gold Camp* (San Francisco: Dry Bones Press, 1995), 73; Spence, *Lace-Boot Brigade,* chap. 3; Rossiter W. Raymond, "Condition of the Mining Industry—California," in *Statistics of Mines and Mining,* 19; "Anglo-Columbian" letter to editor, Apr. 29, 1879, *Mining and Scientific Press* 38 (May 3, 1879): 285.

77. D. C. Demarest, "California Gold," chap. 24, UOPWA.

78. Ibid., chap. 23.

79. Ibid., chap. 30.

5 : PREPARING FOR MODERN MINING

1. One prominent recent example was the Silicon Valley dot.com boom, touted as "California's second Gold Rush," which ended with a spectacular "dot.bomb" crash in 2000. *Stockton Record,* May 22, 2000, p. F1. For other examples, see *Los Angeles Times,* Record ed., Aug. 22, 1999, p. 17; Sept. 26, 1999, p. 17; Mar. 24, 2000, p. 1; May 22, 2000, p. 6; July 6, 2000, p. 42; Sept. 27, 2000, p. 9; Apr. 28, 2001, p. D11; and May 29, 2001, p. D5.

2. Dilsaver, "From Boom to Bust," 1:192–98; *Stockton Independent,* Mar. 20, 1863, p. 2, col. 3.

3. U.S. decennial census records remain the primary source for calculating the number of miners in Calaveras, but those records are inconsistent at best and inaccurate at worst. They may not adequately represent the actual number of miners employed in any given year or active mining cycle. In the case of missing or obviously erroneous census data we have supplied estimates.

4. For an analysis of the economic impact of this transitional period on California mining, see Paul, *California Gold,* 171–87.

5. Greenland, *Hydraulic Mining in California,* 264.

6. J. J. Crawford, *Thirteenth Annual Report of the State Mineralogist* (Sacramento: California Mining Bureau, 1896), 14.

7. Calaveras Public Utility District Records, Calaveras County Archives.

8. Clark C. Spence, "The Golden Age of Dredging: The Development of an Industry and Its Environmental Impact," *Western Historical Quarterly* 11 (Oct. 1980): 404.

9. Lewis E. Aubury, *Gold Dredging in California,* California State Mining Bureau Bulletin 57 (Sacramento: W. W. Shannon, 1910), 205–7.

10. Willard P. Fuller Jr., "Calaveras Gold Dredges," *Las Calaveras* 33 (July 1985): 33–34.

11. Julihn and Horton, *Southern Mother Lode Region,* 14; W. P. Fuller, "Calaveras Gold Dredges," 27.

12. Paul, *California Gold,* 154; Aubury, *Gold Dredging in California,* 102–3.

13. O. E. Young, *Western Mining,* 132–36.

14. For a good technical overview of mining innovations, see O. E. Young, *Western Mining.*

15. Saxton, *Indispensable Enemy,* 58–60.

16. O. E. Young, *Western Mining,* 212–14.

17. W. C. Ralston, "Notes on the Cost of Tunneling at the Melones Mine, Calaveras County, Calif.," *Transactions of the American Institute of Mining Engineers* 28 (1898): 547–53.

18. O. E. Young, *Western Mining,* 205–9; U.S. Public Health Service quoted in J. C. Foster, "Western Miners and Silicosis: 'The Scourge of the Underground Toiler,' 1890–1943," *Industrial Labor Relations Review* 37 (Apr. 1984): 372.

19. D. C. Demarest, "California Gold," chap. 50, UOPWA.

20. Ibid., chaps. 8, 24.

21. Robert E. Kendall, "Deep Enough: The Pitfalls and Perils of Deep Mining on the Comstock," *Nevada Historical Society Quarterly* 39 (fall 1996): 216–31; D. C. Demarest, "California Gold," chap. 24, pp. 10–12, UOPWA.

22. For a discussion of impulse-wheel development, see W. A. Doble, "The Tangential Water-Wheel," *Transactions of the American Institute of Mining Engineers* 29 (1900): 852–93.

23. Dahl, "British Investment," 209–13; "Electricity Comes to Calaveras County," *Las Calaveras* 23 (Apr.–July 1975): 30.

24. Eliot Lord, *Comstock Mining and Miners* (Washington, D.C.: GPO, 1883; reprint, San Diego: Howell-North Books, 1959, 1980), 89–90; O. E. Young, "Philipp Deidesheimer," 361–69; Buckbee, *Pioneer Days,* 40–41.

25. Watson Parker, *Gold in the Black Hills* (Norman: University of Oklahoma Press, 1966), 189; Vernon H. Jensen, *Heritage of Conflict: Labor Relations in the Nonferrous Metals Industry Up to 1930* (Ithaca: Cornell University Press, 1950), 8–9. For further discussion of labor-saving technology and its impact on the mining industry, see Harry Braverman, *Labor and Monopoly Capital: The Degradation of Work in the Twentieth Century* (New York: Monthly Review Press, 1974), 85–100, 426–47; Lingenfelter, *Hardrock Miners,* 83–84; Richard Edwards, *Contested Terrain: The Transformation of the Workplace in the Twentieth Century* (New York: Basic Books, 1979), 115–26; Larry Lankton, *Cradle to Grave: Life, Work, and Death at the Lake Superior Copper Mines* (New York: Oxford University Press, 1991), 86–92; Larry Lankton and Jack K. Martin, "Technological Advance, Organizational Structure, and Underground Fatalities in the Upper Michigan Copper Mines, 1860–1929," *Technology and Culture* 28 (Jan. 1987): 42–66; Jeremy Mouat and Logan Hovis, "Miners, Engineers, and the Transformation of Work in the Western Mining Industry, 1880–1930," *Technology and Culture* 37 (July 1996): 429–56; and David Noble, *America by Design: Science, Technology, and the Rise of Corporate Capitalism* (New York: Alfred A. Knopf, 1977), 50–83.

26. Mouat and Hovis, "Miners, Engineers, and Work," 429.

27. Ruth Harper Lemue, "The John Peirano Family," *Las Calaveras* 14 (Jan. 1966): 3–5; Mouat and Hovis, "Miners, Engineers, and Work," 429–56.

28. Spence, *Lace-Boot Brigade,* 142.

29. D. C. Demarest, "California Gold," chap. 36, UOPWA.

30. O. E. Young, *Western Mining,* 136.

31. Ibid., 274.

32. Leonard, "The Mills of Angels," 1.

33. Raymond, "Mining Industry—California," 22–23; Gray Brechin, *Imperial San Francisco: Urban Power, Earthly Ruin* (Berkeley and Los Angeles: University of California Press, 1999), 250–51; Michelle I. Hornberger, Samuel N. Luoma, Alexander van Geen, Christopher Fuller, and Roberto Anima, "Historical Trends of Metals in the Sediments of San Francisco Bay, California," *Marine Chemistry* 64 (1999): 39–55; Peter I. Ritson, Robin M. Bouse, A. Russell Flegal, and Samuel N. Louma, "Stable

Lead Isotopic Analyses of Historic and Contemporary Lead Contamination of San Francisco Bay Estuary," *Marine Chemistry* 64 (1999): 71–83.

34. Clarence A. Logan, *The Mother Lode Gold Belt of California,* California Division of Mines Bulletin 108 (San Francisco: California Division of Mines, 1935), 193.

35. On Bodie, see Spence, *Lace-Boot Brigade,* 241; Dahl, "British Investment," 208.

36. Logan, *Mother Lode Gold Belt,* 193.

37. A. Scheidel, *The Cyanide Process: Its Practical Application and Economical Results,* California Mining Bureau Bulletin 5 (Sacramento: State Printing Office, 1894), 79, 88–95.

38. For a recent study of the Australian contribution to flotation, see Jeremy Mouat, "The Development of the Flotation Process: Technological Change and the Genesis of Modern Mining, 1898–1911," *Australian Economic History Review* 36 (Mar. 1996): 3–31. O. E. Young, *Western Mining,* 231–32.

39. Mouat, "Flotation Process," 3–31; Edward H. Nutter to William J. Loring, Nov. 1, 1916, Pacific Coast Gold Mining Corporation records, box 4; W. J. Loring to W. B. Devereux Jr., Feb. 4, 1920, Pacific Coast Gold Mining Corporation records, box 5; Loring to Oolf Wenstrom, Aug. 30, 1924; "Annual Report of the Carson Hill Gold Mining Company for the Year Ending December 31, 1919" (n.p., 1920), all in Segerstrom Family Collection.

40. *Las Calaveras* 3 (Apr. 1955): 2.

41. *Sacramento Daily Union,* Feb. 4, 1888, p. 1, col. 1; Feb. 11, 1888, p. 3, cols. 1–2; Bishop, "McKay's Clipper Mill," 13–23.

42. D. C. Demarest, "California Gold," chap. 12, UOPWA.

43. Bishop, "McKay's Clipper Mill," 17; *Las Calaveras* 3 (Apr. 1955): 2–3.

44. Carolyn B. Bryan, "The Calaveras Timber Story" (MS, 1965), Calaveras County Historical Society.

45. W. S. Brown and Show, *California Rural Land Use,* 1:209.

46. George H. Tinkham, *History of Stanislaus County, California: With Biographical Sketches of the Leading Men and Women of the County Who Have Been Identified With Its Growth and Development From the Early Days to the Present* (Los Angeles: Historic Record, 1921), 972; Bryan, "The Calaveras Timber Story," p. 10, Calaveras County Historical Society.

47. Bryan, "The Calaveras Timber Story," p. 10, Calaveras County Historical Society.

48. *San Francisco Chronicle,* May 29, 1932, p. 6, col. 4; D. C. Demarest, "California Gold," chap. 27, UOPWA; Michael B. Arkin and Franklin T. Laskin, *From the Depth of the Mines Came the Law* (Clovis, Calif.: Word Dancer Press, 2000), 49–50.

49. Bryan, "The Calaveras Timber Story," pp. 10–11, Calaveras County Historical Society.

50. Jack R. Wagner, *Short Line Junction* (Fresno: Academy Library Guild, 1956), 119–23.

51. Dorothy N. Deane, *Sierra Railway* (San Francisco: Howell-North Books, 1960), 8–9.

52. Poniatowski married Elizabeth Sperry, niece of James Sperry and sister-in-law of William H. Crocker. Memorandum, R. C. Ashworth to R. Saclise, Oct. 19, 1914, Public Utility Commission Records, case 910, California State Archives; Deane, *Sierra Railway*, 79–83.

53. D. C. Demarest, "California Gold," chap. 3, UOPWA.

54. John Hayes Hammond, "Changes of Fifty Years in Mining Engineering," *Mining and Metallurgy* 9 (Oct. 1928): 439–41.

55. Data showing the decline of both state and county miners are listed in the appendix. See also the chart on p. 173.

56. Lingenfelter, *Hardrock Miners*, 19. See the appendix for statistical details of Calaveras demographic changes, 1850–1940.

57. Grover, "Memoirs," pp. 106–7, Bancroft Library.

58. George Hoeper, "Chris Porovich, Master Miner," *Las Calaveras* 41 (Apr. 1992): 31–36; *Sonora Times,* Nov. 23, 1908, scrapbook clipping, Pacific Coast Gold Mining Corporation records, MS 275, Segerstrom Family Collection; Effie Enfield Johnston, "Wade Johnston Talks to His Daughter," *Las Calaveras* 20 (Jan. 1972): 9–19; Sheryl Waller, "A Sheep Ranch Native Remembers the Old Days," *Las Calaveras* 44 (July 1996): 50–52. For examples of the type and frequency of Mother Lode mine accidents, see *Sonora Banner,* July 10, 1908; *Sonora Democrat,* Aug. 29, 1908; and *Sonora Times,* Aug. 19, Nov. 23, 1908, all in Pacific Coast Gold Mining Corporation Scrapbook, Segerstrom Family Collection.

59. California Special Committee on the Argonaut Mine Disaster Report (Nov. 1922), unpaged photocopy of TS (Sacramento, 1922), California State Archives. For a list of national mine tragedies, see U.S. Department of Labor Mine Safety and Health Administration, "Historical Data on Mine Disasters in the United States" (2003), at http://www.msha.gov/MSHAINFO/FACTSHET/MSHAFCT8.HTM.

60. Alan Derickson, *Workers' Health, Workers' Democracy: The Western Miners' Struggle, 1891–1925* (Ithaca: Cornell University Press, 1988), 36–38; *Mother Lode Banner,* June 5, 1903, p. 1, col. 7.

61. Albert H. Fay, *Metal-Mine Accidents in the United States During the Year 1915,* U.S. Bureau of Mines Technical Paper 168 (Washington, D.C.: GPO, 1917), 7–47; William W. Adams, *Metal-Mine Accidents in the United States During the Calendar Year 1920,* U.S. Bureau of Mines Technical Paper 299 (Washington, D.C.: GPO, 1922), 13–26.

62. Derickson, *Workers' Health,* 28–54; Foster, "Miners and Silicosis," 371–85.

63. Mark Wyman, *Hard Rock Epic: Western Miners and the Industrial Revolution, 1860–1910* (Berkeley and Los Angeles: University of California Press, 1979), 120–45; Limerick, *Legacy of Conquest,* 108–11.

64. M. Wyman, *Hard Rock Epic,* 58–60, 134–35, 156–200, 203–25; Derickson, *Workers' Health,* 63–120, 155–80.

65. M. Wyman, *Hard Rock Epic,* 28–29.

66. Lingenfelter, *Hardrock Miners,* 32–65; Sally Zanjani, *Goldfield: The Last Gold Rush on the Western Frontier* (Athens, Ohio: Swallow Press, 1992), 42–43.

67. Charles F. Jackson, "Operating the Small Mine," *Engineering and Mining Journal* 141 (Aug. 1940): 42; Loring to James L. Brooks, May 1, 1916, MS 275, Segerstrom Family Collection.

68. Drew, "Placer Mining," 1–18.

6: LODE MINING IN THE GOLDEN YEARS

1. Julihn and Horton, *Southern Mother Lode Region,* 12–17. See chart 6.1.

2. Joralemon, "California's Foothill Gold Belt," 491–92.

3. For a concise discussion of late-nineteenth-century monetary issues, see the classic text by Samuel Eliot Morison and Henry Steele Commager, *The Growth of the American Republic,* 4th ed. (New York: Oxford University Press, 1950), 2:242–49.

4. Designation of Person of Whom Process May Be Served, signed by G. W. Huddleston, president, Sultana Mining Company, Sept. 12, 1902, State Incorporation Records, California State Archives; Leonard, "Early Quartz Mining," 2; D. C. Demarest, "California Gold," chap. 5, UOPWA.

5. Clark and Lydon, *Mines and Mineral Resources,* 40; D. C. Demarest, "California Gold," chap. 2, UOPWA.

6. On being Comstock's nephew, see Leonard, "The Mills of Angels," 10; *Mining and Scientific Press* 59 (Dec. 21, 1889): 468.

7. D. C. Demarest, "California Gold," chap. 6, UOPWA; Clark and Lydon, *Mines and Mineral Resources,* 40; Leonard, "Early Quartz Mining," 2.

8. Proposition, Duncan MacVichie to J. V. Coleman for 75 percent of Angels Mine, Oct. 16, 1915, Segerstrom Family Collection.

9. Leonard, "The Mills of Angels," 9; *Stockton Weekly Mail,* Oct. 24, 1896, p. 14, col. 5.

10. *Stockton Evening Mail,* Oct. 2, 1897, p. 1, col. 1; *Lightner Mining Company* v. *Alvinza Hayward, C. D. Lane, and Hobart Estate Company, Partners in Utica Mining Company,* Calaveras County Superior Court, July 10, 1902; plaintiff exhibit no. 2, *Lightner* v. *Lane,* Nov. 22, 1904, Calaveras County Superior Court; docket book, Calaveras County Superior Court (Dec. 19, 1904), 2:62, Calaveras County Archives; Arkin and Laskin, *Depth of the Mines,* 45.

11. Chalmers's daughter Harriet married a mining engineer and traveled the world, eventually becoming a nationally known photo essayist (Durlynn Anema, *Harriet Chalmers Adams: Explorer and Adventurer* [Greensboro, N.C.: Morgan Reynolds, 1997]).

12. A. M. Hunt and Wynn Meredith, "Electric Power: Its Generation and Utilization for Mining Work on the Pacific Coast," in *California Mines and Minerals,* 85; and H. W. H. Penniman, "Mines of Calaveras County," in ibid., 338.

13. D. C. Demarest, "California Gold," chap. 7, UOPWA.

14. Ibid.

15. *Stockton Independent,* Oct. 31, 1905, p. 5, cols. 1–2.

16. Ibid.

17. Ibid., Nov. 2, 1905, p. 3, col. 1.

18. D. C. Demarest, "California Gold," chap. 7, UOPWA; Leonard, "Early Quartz Mining," 2; Clark and Lydon, *Mines and Mineral Resources,* 60–61.

19. Thomas A. Rickard, *Interviews With Mining Engineers* (San Francisco: Mining and Scientific Press, 1922), 290; Lord, *Comstock Mining and Miners,* 283–88; Grant H. Smith, "The History of the Comstock Lode, 1850–1920," *University of Nevada Bulletin* 37 (July 1, 1943): 128–34; Thomas A. Rickard, "The Re-opening of Old Mines Along the Mother Lode, California," *Mining and Scientific Press* 112 (June 24, 1916): 935–39; *San Francisco Call,* Feb. 15, 1904, p. 4, cols. 2–3; *Calaveras Prospect,* Feb. 20, 1904, p. 1, col. 5; Rickard, "Re-opening of Old Mines," 938–99.

20. D. C. Demarest, "California Gold," chap. 24, pp. 14–15, UOPWA.

21. W. E. Robinson to Charles D. Lane, Apr. 12, 1885, Selkirk-McCauley Letters, Calaveras County Archives; Buckbee, *Pioneer Days,* 38–39. See also D. C. Demarest, "California Gold," chap. 8, UOPWA. Spiritualism also influenced another mining venture at Angels Camp, according to D. C. Demarest. He said that a St. Louis medium opened the Ghost mine east of the Utica, expecting to find a continuation of the Utica's bonanza ore body. He sank a thirteen hundred–foot shaft but found nothing. See ibid., chap. 2.

22. Edward C. Leonard, "Deep Gold Mining in Angels Camp," *Las Calaveras* 20 (July 1972): 42; Francis E. Bishop, "The Story of the Union Water Company," ed. W. Scott Perry (TS, ca. 1980), p. 112, Calaveras County Historical Society.

23. D. C. Demarest, "California Gold," chap. 40, UOPWA.

24. *San Francisco Call,* Feb. 15, 1904, p. 4, cols. 2–3; Peterson, *Bonanza Kings,* 44–45; D. C. Demarest, "California Gold," chap. 37, UOPWA.

25. Logan, *Mother Lode Gold Belt,* 150.

26. D. C. Demarest, "California Gold," chap. 37, UOPWA; M. Wyman, *Hard Rock Epic,* 117.

27. Storms, *Timbering and Mining,* 205–6.

28. Ibid., 195–96.

29. Buckbee, *Pioneer Days,* 40–41.

30. On Comstock union, see Lingenfelter, *Hardrock Miners,* 31; on Homestake union, see Jensen, *Heritage of Conflict,* 249; Lingenfelter, *Hardrock Miners,* 90–103. Vernon Jensen's classic history of labor violence contends that labor relations also changed with the introduction of bulk mining and hauling techniques such as shrinkage stoping, glory holing, sublevel caving, and electric tramming. Such processes, he says, relied more on machinery than on men, and tended to lower the skill level of those underground workers who remained. "The engineer stepped in more prominently and the skilled miner faded out. In his place came an unskilled, or, at best, a semi-skilled miner" (*Heritage of Conflict,* 8–9). Others challenge this conclusion, insisting that the level of technical skills required to mine underground today is higher than ever before. For a recent study of modern mining from the miner's point of view, see Jerry Dolph, *Fire in the Hole: The Untold Story of Hardrock Miners* (Pullman: Washington State University Press, 1994).

31. George Hoeper, "Ernie Vogliotti Looks Back on 100 Years," *Las Calaveras* 40

(Oct. 1990): 1–4; *Mining and Scientific Press* 12 (Mar. 10, 1866): 147; *Copperopolis Courier,* Jan. 27, 1866; R. Stone and C. A. Stone, *Tools Are on the Bar,* 17; D. C. Demarest, "California Gold," chap. 8, UOPWA; Leonard, "Deep Gold Mining," 45.

32. *Mining and Scientific Press* 68 (Feb. 24, 1894): 115; (May 19, 1894): 317; (May 26, 1894): 333; *Calaveras Prospect,* Mar. 17, 1894, p. 4; Apr. 14, 1894, p. 4; May 19, 1894, p. 4; *Stockton Evening Mail,* May 14, 1894, p. 1, col. 3; Leonard, "Deep Gold Mining," 45–46.

33. E. B. Matthews to Utica Mine owners, Apr. 25, 1898, MS 70, Richard Coke Wood Papers, UOPWA.

34. Jensen, *Heritage of Conflict,* 96–117; Leonard, "Deep Gold Mining," 46.

35. *Stockton Evening Mail,* July 22, 1895, p. 1, cols. 1–2; July 23, 1895, p. 8, col. 1; Buckbee, *Pioneer Days,* 46; on Cross shaft, see Leonard, "Early Quartz Mining," 4; on Angels residents, see D. C. Demarest, "California Gold," chap. 8, UOPWA; J. J. Crawford, *Twelfth Annual Report of the State Mineralogist* (San Francisco: California Mining Bureau, 1894), 122.

36. On Utica-Stickle production, see Leonard, "Early Quartz Mining," 4; on Gold Cliff and Madison, see Edward C. Leonard, "Notebook" (MS, ca. 1968), pp. 25, 48, Calaveras County Historical Society. The Utica-Stickle unit never made a profit after 1904. On the other hand, the Gold Cliff showed a steady profit up to its closure in 1920, more than offsetting the Utica-Stickle losses.

37. *San Francisco Call,* Mar. 15, 1904, p. 9, col. 7; D. C. Demarest, "California Gold," chap. 8, UOPWA.

38. *Stockton Evening Mail,* Mar. 23, 1894, p. 1, col. 7.

39. Ibid., Apr. 30, 1894, p. 1, col. 6.

40. *Mountain Echo* quoted in *Mining and Scientific Press* 68 (May 19, 1894): 317; *Calaveras Prospect,* May 19, 1894, p. 4; *Amador Record,* May 17, 1894, quoted in *Mining and Scientific Press* 68 (May 26, 1894): 333.

41. George Hoeper, "The Calaveras Hotel Was Home to Many," *Las Calaveras* 45 (Jan. 1997): 17–21.

42. George Hoeper, "Guy Castle, a Lifetime of Mining Memories," *Las Calaveras* 43 (Oct. 1994): 2–7.

43. Lemue, "Pioneer Daughter," 45–49.

44. Giovinco, *Ethnic Dimension,* 39–50; Edward C. Leonard, "John Chinaman in Angels Camp," *Las Calaveras* 12 (Oct. 1963): 1–3; Mary Jane Garamendi, "The Chinese in Mokelumne Hill," *Las Calaveras* 12 (Oct. 1963): 4–5; Arlene Westenrider, "The Sam Choy Building," *Las Calaveras* 43 (Oct. 1993).

45. U.S. Bureau of the Census, Census of 1880, Calaveras County, seventh township. See also Shirley Ewart, "Cornish Miners in Grass Valley: The Letters of John Coad," *Pacific Historian* 25 (winter 1981): 38–45.

46. Farbotnik, "The Canepas of Vallecito," 1–2; James Rule, correspondence to the *Virginia Enterprise,* excerpted in *Mining and Scientific Press* 41 (Oct. 14, 1882): 246.

47. Michael Begovich, "Serbianism: How It Motivated the Serbian Pioneers of the California Mother Lode," *East European Quarterly* 23 (June 1989): 202, 207.

48. Ibid., 200–207; Giovinco, *Ethnic Dimension,* 11.

49. Lingenfelter, *Hardrock Miners,* 6–18; William Serrin, *Homestead: The Glory and Tragedy of an American Steel Town* (New York: Random House, 1992), 67–73; *Mining and Scientific Press* 41 (July 3, 1880): 8; W. S. Reid quoted in Derickson, *Workers' Health,* 82–83.

50. H. W. H. Penniman, "Mines of Calaveras County," in *California Mines and Minerals,* 336; *Mining and Scientific Press* 86 (May 30, 1903): 346.

51. *Placerville Nuggett,* May 18, 1903, p. 4, col. 3; Joseph R. Conlin, *Big Bill Haywood and the Radical Union Movement* (Syracuse: Syracuse University Press, 1969), 48–51.

52. *Mining and Scientific Press* 86 (Mar. 21, 1903): 178; *Mother Lode Banner,* Apr. 3, 1903, p. 1, col. 6; Apr. 24, 1903, p. 1, col. 2.

53. Katherine G. Aiken, "'It May Be Too Soon to Crow': Bunker Hill and Sullivan Company Efforts to Defeat the Miners' Union, 1890–1900," *Western Historical Quarterly* 24 (Aug. 1993): 313–20. For the Cripple Creek war, see Frank Waters, *Midas of the Rockies: The Story of Stratton and Cripple Creek,* 3d ed. (Chicago, 1937; reprint, Athens, Ohio: Swallow Press, 1989), 142–47; Richard H. Peterson, "Conflict and Consensus: Labor Relations in Western Mining," *Journal of the West* 12 (Jan. 1973): 1–17; M. Wyman, *Hard Rock Epic,* 203–25.

54. *Mother Lode Banner,* Mar. 27, 1903, p. 3, col. 3; Apr. 3, 1903, p. 1, col. 6; Apr. 24, 1903, p. 1, col. 2; *Georgetown Gazette,* Apr. 16, 1903, p. 2, col. 3.

55. *Calaveras Prospect,* Apr. 18, 1903, p. 1, col. 3; p. 1, col. 5; May 2, 1903, p. 1, col. 4.

56. *Mother Lode Banner,* May 8, 1903, p. 1, col. 3; *Georgetown Gazette,* Apr. 30, 1903, p. 3, col. 2; Elizabeth Jameson, *All That Glitters: Class, Conflict, and Community in Cripple Creek* (Urbana: University of Illinois Press, 1998), 233–47.

57. M. Wyman, *Hard Rock Epic,* 226–43; Derickson, *Workers' Health,* 189–203; *Stockton Record,* Sept. 25, 1922, p. 4, col. 6.

58. W. P. Fuller, "Gwin Mine," 7, 11; Jeffrey Schweitzer, "The Gwin Mine, Lower Rich Gulch, Calaveras County, California," Oct. 4, 1933, MS 174, Jeffrey Schweitzer Collection, UOPWA.

59. D. C. Demarest, "California Gold," chap. 18, UOPWA; W. P. Fuller, "Gwin Mine," 12; Schweitzer, "Gwin Mine," Schweitzer Collection.

60. Ella M. Hiatt and Willard P. Fuller Jr., "The Royal Consolidated Mine," *Las Calaveras* 16 (July 1968): 17; G. O. Argall Jr., "Royal/Mountain King Mine Brings New Technology to the Mother Lode," *Engineering and Mining Journal* (Oct. 1988): 40; Frank T. Gilbert, *History of San Joaquin County, California, With Illustrations Descriptive of Its Scenery, Residences, Public Buildings, Fine Blocks, and Manufactories, From Original Sketches by Artists of the Highest Ability* (Oakland: Thompson and West, 1879), 94.

61. For a description of Bodie's saloons, see McGrath, *Gunfighters, Highwaymen, and Vigilantes,* 111–14.

62. W. P. Fuller, Marvin, and Costello, *Madam Felix's Gold,* 61–62. After the Tioga

project van Ee researched and promoted a "roll film" camera, which George Eastman eventually developed and marketed as the Kodak (62).

63. D. C. Demarest, "California Gold," chap. 19, UOPWA; *San Francisco Call,* Nov. 11, 1905, p. 4, col. 6; Hiatt and W. P. Fuller, "The Royal Consolidated Mine," 18.

64. *Calaveras Prospect,* July 25, 1903, p. 1, cols. 1–3.

65. Phil Bradley, interview by Bancroft Library Oral History Project, 1988, p. 47.

66. D. C. Demarest, "California Gold," chap. 19, UOPWA.

67. *San Francisco Call,* Nov. 19, 1905, p. 49, col. 6; *Calaveras Prospect,* Dec. 19, 1903, p. 1, col. 5, p. 7, col. 3; Jan. 23, 1904, p. 1, col. 3; Jan. 30, 1904, p. 1, col. 3.

68. Herbert Hoover to Theodore Hoover, June 26, 1905, Pre-Commerce Correspondence, box 7, Hoover Presidential Library, West Branch, Iowa.

69. On Sheep Ranch, see Clark and Lydon, *Mines and Mineral Resources,* 69–70; on George Hearst, see Rodman Paul, *Mining Frontiers of the Far West, 1848–1880* (New York: Holt, Rinehart, 1963), 62–63; Gertrude Atherton, *Golden Gate Country,* 186, as quoted in W. A. Swanberg, *Citizen Hearst* (New York: Scribner's, 1961), 15; Parker, *Gold in the Black Hills,* 196.

70. D. C. Demarest, "California Gold," chap. 9, UOPWA; *Calaveras Prospect,* Dec. 3, 1904, p. 1, col. 4; Effie Enfield Johnston, "Wade Johnston Talks to His Daughter," *Las Calaveras* 21 (Oct. 1972): 6–7; Emil Guidicy, interview by authors, Angels Camp, May 18, 1937; Rhoda Dunlap, "Sheepranch," *Las Calaveras* 7 (Jan. 1959): 3; Thomas A. Rickard, *History of American Mining* (New York: McGraw-Hill, 1932), 214–15.

71. Clark and Lydon, *Mines and Mineral Resources,* 69–70.

72. E. H. Schaeffle letter, *Mining and Scientific Press* 59 (Dec. 14, 1889): 448; Clark and Lydon, *Mines and Mineral Resources,* 45–47; D. C. Demarest, "California Gold," chap. 13, UOPWA.

73. Dahl, "British Investment," 201–2; *Eleventh Annual Report of the State Mineralogist* (San Francisco: California Mining Bureau, 1892), 169–70; Crawford, *Thirteenth Annual Report,* 99.

74. Articles of Incorporation, Melones Consolidated Mining Company, Feb. 27, 1875, Incorporation Records, California State Archives; *California Blue Book* (Sacramento: Secretary of State, 1903), 288; *San Francisco Chronicle,* Jan. 23, 1924, p. 8, col. 1; *National Cyclopedia of American Biography,* s.v. "Deveraux, Walter B."; D. C. Demarest, "California Gold," chap. 39, UOPWA.

75. Ralston, "Cost of Tunneling," 547–53.

76. D. C. Demarest, "California Gold," chap. 13, UOPWA; Julihn and Horton, *Southern Mother Lode Region,* 104–6.

77. D. C. Demarest, "California Gold," chap. 13, UOPWA.

78. Julihn and Horton, *Southern Mother Lode Region,* 105–6; Dahl, "British Investment," 211–12.

79. Julihn and Horton, *Southern Mother Lode Region,* 105–6.

80. Jack R. Wagner, *Gold Mines of California* (Berkeley: Howell-North Books, 1970), 69; D. C. Demarest, "California Gold," chap. 13, UOPWA; Oscar H. Hershey,

"Report on Carson Hill Mines," Feb. 5, 1931, in John A. Burgess, "Report on Carson Hill Gold Mining Corporation," Oct. 13, 1932, pp. 15–28, Segerstrom Family Collection.

81. Clark and Lydon, *Mines and Mineral Resources,* 24.

82. D. C. Demarest, "California Gold," chap. 19, UOPWA; Francis Howard Riggs, "A Home in 'Copper,'" *Las Calaveras* 27 (July 1979): 27–33.

83. Riggs, "A Home in 'Copper,'" 27–29.

84. *Stockton Evening Mail,* Mar. 24, 1894, p. 1, col. 7; excerpt from *Calaveras Citizen,* in *San Francisco Chronicle,* July 20, 1908, Pacific Coast Gold Mining Corporation scrapbook, box 1, Segerstrom Family Collection.

85. Richard Coke Wood, "The Tanner Mine As Told by Amon Tanner to Coke Wood," *Las Calaveras* 11 (Jan. 1963): 1–2; on G.A.R. mine, see *Stockton Weekly Mail,* June 27, 1896, p. 9, col. 1; [Willard P. Fuller Jr.], "The Alta Mine," *Las Calaveras* 31 (Oct. 1982): 7–10. The mine is commonly referred to as "Alto," though sometimes listed as "Alta."

86. Derickson, *Workers' Health,* 61–85, 155–73, 189–201; M. Wyman, *Hard Rock Epic,* 114–15; U.S. Department of the Interior, Controller General, *After Years of Effort, Accident Rates Are Still Unacceptably High in Mines Covered by the Federal Metal and Nonmetallic Mine Safety Act* [Washington, D.C.: GPO, 1977].

## 7: THE UNSTABLE TWENTIES AND THIRTIES

1. A. E. Rau Roesler, "Mining on the Mother Lode of California," *Mining and Scientific Press* 122 (June 11, 1921): 807–9.

2. William J. Loring, "The McFadden Bill," *Mining and Scientific Press* 122 (Feb. 19, 1921): 251–52.

3. Richard B. Stockton, "Populations of Calaveras County," *Las Calaveras* 22 (Jan. 1974): table 1, p. 13.

4. D. C. Demarest, "California Gold," chap. 4, UOPWA.

5. Clark and Lydon, *Mines and Mineral Resources,* 74.

6. "Mine Operators of Note: William J. Loring," *Engineering and Mining Journal* 112 (Oct. 15, 1921): 617.

7. Spence, *Lace-Boot Brigade,* 366; Walter R. Crane, *Ore Mining Methods, Comprising Descriptions of Methods of Support in Extraction of Ore, Detailed Descriptions of Methods of Stoping and Mining in Narrow and Wide Veins and Bedded and Massive Deposits Including Stull and Square-Set Mining, Filling and Caving Methods, Open-Cut Work, and a Discussion of Costs of Stoping* (New York: J. Wiley and Sons, 1910), 3.

8. William J. Loring, "Mill Practice of the Utica Mills," *Transactions of the American Institute of Mining Engineers* 28 (1898): 553–65.

9. Spence, *Lace-Boot Brigade,* 298; Ronald H. Limbaugh, "Making the Most of Experience: The Career of William J. Loring, Nevada Mining Engineer," *Mining History Journal* 1 (1994): 9–13.

10. Rickard, *Interviews With Mining Engineers,* 275–91; biographical sketch for *Who's Who in America* (n.d.), MS 197, Stevenot Papers.

11. Excerpt from speech of William J. Loring, *Mining and Scientific Press* 115 (Jan. 10, 1914): 109–10; William J. Loring, "Re-opening of the Plymouth Mine and the Results," *Mining and Scientific Press* 121 (Nov. 27, 1920): 771–72; Logan, *Mother Lode Gold Belt,* 78.

12. Loring did not resign formally from the BMC until 1920, when his son took over his share of the partnership. Clipping from the *Mining Journal* (Oct. 30, 1932), MS 275, Segerstrom Family Collection.

13. Frank J. Solinsky to Segerstrom, Aug. 25, 1916; Loring to Segerstrom, Sept. 8, 1916; supplemental agreement between C. H. Segerstrom and E. J. Segerstrom, and J. F. Thompson and Dutch Consolidated Gold Mining Company, Apr. 5, 1917; Archie D. Stevenot to Segerstrom, May 23, 1931, all in Segerstrom Family Collection. The Dutch-Sweeney shut down in 1920 after extensive underground work failed to find profitable ore. Adolph Knopf, *The Mother Lode System of California,* United States Geological Survey Professional Paper 157 (Washington, D.C.: GPO, 1929), 78.

14. George J. Young, "Gold Mining at Carson Hill, California," *Engineering and Mining Journal* 112 (Nov. 5, 1921): 725–29; Julihn and Horton, *Southern Mother Lode Region,* 106.

15. William J. Loring, "Annual Report of the Carson Hill Gold Mines, Inc., 1919," Feb. 1, 1920, pp. 30–32; Loring to Edward A. Clark, July 22, 1921, both in Segerstrom Family Collection.

16. Loring to Clark, June 18, 1921, enclosed in Loring to Segerstrom, July 9, 1921, ibid. Loring blamed much of the financial problems on his board of directors, who, in his opinion, did not sell enough treasury shares to pay operating expenses. He complained to Segerstrom: "Had this been an English company instead of a blooming American company the shares would have been $30.00, everybody would have had regular dividends and everybody would have been happy." What he neglected to mention is what Segerstrom must have clearly understood: Mining stock of any kind was a drag on the market in 1921.

17. F. J. Solinsky to Segerstrom, Mar. 27, Apr. 4, 1922; William J. Loring, "Annual Report of the Carson Hill Gold Mining Company for the Year Ending December 31, 1921" (n.p., 1922), both in ibid.

18. "Annual Report of the Carson Hill Gold Mining Company for the Year Ending December 31, 1919" (n.p., 1920); Olof Wenstrom to R. C. Eisenhauer, May 11, 1924; Loring to Oolf Wenstrom, Aug. 30, 1924; R. C. Eisenhauer to Loring, Sept. 8, 1924; Minute Book, Pacific Tungsten Company, pp. 7–121; William J. Loring, "Preliminary Report on the Properties of the Pacific Tungsten Company, Inc.," Pershing County, Nev., Nov. 4, 1919; receipt, Segerstrom, First National Bank of Sonora, to Loring, Feb. 27, 1924; Loring to Segerstrom, Nov. 17, Nov. 20, 1924, all in ibid.

19. Loring to Segerstrom, Nov. 2, 1927, ibid.

20. Loring to Archie Stevenot, Apr. 19, 1952, MS 197, Stevenot Papers. For a fuller account of Loring's latter years, see Limbaugh, "Making the Most of Experience," 9–13.

21. Hamilton, Beauchamp, and Woodworth, Consulting Metallurgical Engineers,

to Segerstrom, Mar. 12, 1926; M. H. Manuel, responses to questionnaire, Carson Hill Mining Corporation (ca. 1930), both in Segerstrom Family Collection; Logan, *Mother Lode Gold Belt,* 131; M. H. Manual, "History of Properties Owned and Controlled" (ca. 1930), Carson Hill Mining Corporation, Segerstrom Family Collection.

22. Arthur M. Schlesinger Jr., *The Coming of the New Deal* (Boston: Houghton-Mifflin, 1959), 199–203, 234–46; Loring, "The McFadden Bill," 251–52; J. F. Telfer to Segerstrom, Mar. 12, 1933, Segerstrom Family Collection.

23. W. P. Fuller, Marvin, and Costello, *Madam Felix's Gold,* chap. 3, pp. 52, 87, 93–95; Howard Tower, interview by authors, Salt Spring Valley, May 11, 1998.

24. Tony Dutil, interview by authors, Angels Camp, July 16, 1997.

25. Jeffrey Schweitzer, "Once in a Thousand Times" (1933), MS 174, Schweitzer Collection.

26. *California Mining Journal* 6 (Dec. 1936): 4; (Aug. 1937): 29; 7 (Feb. 1938): 29; 8 (Aug. 1939): 29; (May 1939): 28; 9 (June 1940): 9; 10 (Jan. 1941): 11; (May 1941): 31.

27. *Mining and Industrial News* 10 (May 15, 1942): 3.

28. Segerstrom to Daniel G. Wing, July 14, 1927, enclosed with Lawrence Monte Verda to Segerstrom, May 5, 1927, Segerstrom Family Collection.

29. D. C. Demarest, "California Gold," chap. 26, UOPWA.

30. Draft proposal attached to Monte Verda to Segerstrom, May 5, 1927; Segerstrom to Wing, July 14, 1927, both in Segerstrom Family Collection. Segerstrom's personal stake is reported in "Minutes of Special Meeting of Directors of Carson Hill Gold Mining Corporation," Mar. 18, 1933, ibid. William H. Coolidge and James W. Hazen to Monte Verda, June 5, 1930, ibid.

31. Prospectus, Carson Hill Gold Mining Corporation (1932, 1933), ibid.

32. Segerstrom to R. M. Wright, Dec. 16, 1931, ibid.

33. John A. Burgess to E. H. McMurray, Feb. 3, 1932; Burgess, preliminary report on Carson Hill, Apr. 17, 1932, both in ibid.

34. Oscar H. Hershey, "Report on Carson Hill Mines," Feb. 5, 1931, in Burgess, "Report on Carson Hill Gold Mining Corporation," Oct. 13, 1932, pp. 15–28, ibid.

35. Segerstrom to David Maltman, Feb. 8, 1932, ibid.

36. C. A. Beinhorn to Segerstrom, May 22, 1932; E. H. McMurray to Carson Hill Gold Mining Corporation, May 24, 1932, both in ibid.

37. Edwin A. McKanna to Segerstrom, Oct. 12, 1932; financial statement, Carson Hill Gold Mining Corporation, Jan. 1, 1933, both in ibid.

38. Carson Hill Gold Mining Corporation (ca. Mar. 18, 1933); Segerstrom to W. A. Sutherland, May 27, 1933; Sol Grossbard to Sutherland, June 22, 1933; Segerstrom to John B. Canada, Jan. 9, 1937, all in ibid.

39. Grossbard to Sutherland, June 22, 1933, ibid.

40. Segerstrom to Grossbard, July 11, 1933; Representative Harry L. Englebright to "Friend," Aug. 29, 1933, both in ibid.

41. John A. Burgess, "Report on the Carson Hill Gold Mines at Melones and Carson Hill, Calaveras County, California," Jan. 14, 1933, ibid.

42. *Stockton Record,* Mar. 21, 1934, p. 1, col. 1.

43. Burgess to Segerstrom, Nov. 12, 1932, Segerstrom Family Collection.

44. Milo Bird, *Melones Memories: Recollections of a Mother Lode Gold Mining Camp by a Latter-Day Tom Sawyer* (Sonora: Tuolumne County Historical Society, 1985), 32–39.

45. Edward B. Ransehousen to Segerstrom, Nov. 4, 1935, Segerstrom Family Collection.

46. *Mining and Industrial News* 9 (Dec. 15, 1941): 1; (May 15, 1941): 1.

47. Charles H. Segerstrom, "Progress Report: Carson Hill Gold Mining Corp., March 1, 1935," Segerstrom Family Collection.

48. Report of the annual meeting, Carson Hill Gold Mining Corporation, Feb. 1, 1936, in *California Mining Journal* 5 (Feb. 1936): 16; (May 1936): 10; 6 (Sept. 1936): 18; and (Dec. 1936): 20.

49. Segerstrom to Grossbard, Sept. 1, 1936, Segerstrom Family Collection.

50. Walter Lyman Brown to Rogers, Mayer, and Ball, May 17, 1935; "Minutes of Special Meeting of Directors of Carson Hill Gold Mining Corporation," May 26, 1937, both in ibid. For an example of Segerstrom's concern for stockholders, see Segerstrom to Grossbard, June 18, 1937, in ibid.

51. Grossbard to Segerstrom, Dec. 17, 1938, ibid.

52. List of stockholders, Carson Hill Gold Mining Corporation, Sonora Office, June 14, 1937, ibid.

53. Annual Report for 1941, Carson Hill Gold Mining Corporation, in *Mining and Industrial News* 10 (Feb. 15, 1942): 16; *Mining and Industrial News* 10 ( June 15, 1942): 11; *California Mining Journal* 13 (Oct. 1943): 23.

54. John A. Burgess, "Mining on Carson Hill," in *Geological Guidebook Along Highway 49,* California Division of Mines Bulletin 141 (San Francisco: California Division of Mines, 1948), 90.

55. Warren Ronald Blomquist, "Gold Dredging in California" (master's thesis, University of the Pacific, 1973), 83–85; W. P. Fuller, "Calaveras Gold Dredges," 26–31; on the Milton Gold Dredging Company, see Julien and Horton, *Mines of the Southern Mother Lode,* 82–87; Glenn Bump, interview by authors, Stockton, 1979.

56. Julihn and Horton, *Southern Mother Lode Region,* 75–82.

57. Ibid., 33–75.

58. B. Franklin Serlis, *The Story of the Calaveras Central Gold Mining Co., Ltd: Its History, Development, and Outlook* (San Francisco: Serlis, Coplin, [193_?]); Julihn and Horton, *Southern Mother Lode Region,* 42–58.

59. California Special Committee on the Argonaut Mine Disaster Report (Nov. 1922), California State Archives; "The Stone Family," *Las Calaveras* 15 ( Jan. 1967): 10–11; Charles Stone Jr., interview by authors, Copperopolis, Apr. 26, 1998.

60. Julihn and Horton, *Southern Mother Lode Region,* 17, 20–21, 70–71; Garamendi, "McSorleys of Chili Gulch," 13–21.

61. George Hoeper, "The Dosters Made It Through Hard Times," *Las Calaveras*

44 (Jan. 1996): 13–16; Jesse L. Coffey with George Hoeper, *Bacon and Beans From a Gold Pan* (Garden City, N.Y.: Doubleday, 1972), 38–39.

62. Coffey, *Bacon and Beans*, 16–19.

63. Milo Bird, "Charlie Allen's Gopher Hole," *Las Calaveras* 33 (Apr. 1985): 20–22.

64. Hoeper, "Chris Porovich, Master Miner," 31–36.

65. O. E. Young, *Western Mining*, 225–27; Hoeper, "Chris Porovich, Master Miner," 31–36.

66. Hoeper, "Guy Castle," 2–7; Hoeper, "Chris Porovich, Master Miner," 31–36.

67. Hoeper, "Guy Castle," 2–7.

8 : A HALF CENTURY OF CHANGE

1. *Mining and Industrial News*, 10 (Nov. 15, 1942): 1, 4; 14 (July 1945): 16. Not all remained closed for the duration, however. In Grass Valley, for example, early in 1943 the Idaho-Maryland and the Empire were permitted to reopen on a limited basis, mining just enough ore "to meet maintenance costs." The modification came after WPB decided to "give consideration to appeals from authorized producers who are faced with drastic losses in property under wartime restrictions on manpower and material aid." Such relief was available only in special circumstances, however, and did not alter the situation at Carson Hill or at other closed Calaveras County properties. *Mining and Industrial News* 12 (Jan. 15, 1943): 3.

2. Clark and Lydon, *Mines and Mineral Resources*, 29.

3. Dutil, interview.

4. Charles Stone, interview by authors, Copperopolis, July 10, 1997.

5. George Huyl, "Ore Deposits of Copperopolis, California," in *Copper in California*, Bulletin 144 (San Francisco: California Division of Mines, 1948), 93.

6. Harry Sears: "The Conspiracy Against Gold," *Commercial and Financial Chronicle* 182 (July–Sept. 1955): 64; *Calaveras Californian*, June 15, 1961, p. 10, col. 4.

7. C. Stone, interview.

8. Blomquist, "Gold Dredging in California," 132–33.

9. Arnold R. Ross, *Twenty-five Years of Building the West* (San Francisco: Calaveras Cement, 1950).

10. F. F. Momyer, article on V. F. Pickering for *National Cyclopedia of American Biography* (Feb. 28, 1949), in Pickering Lumber Company file, Sonora Public Library.

11. *San Francisco Chronicle*, Apr. 22, 1930, p. 22, col. 3; Bryan, "The Calaveras Timber Story," Calaveras County Historical Society.

12. *American Eagle* 15 (Aug. 1958): 1–9.

13. Habib, "Some Economic Aspects," 100–101.

14. *California Statistical Abstract, 1970* (Sacramento: State Department of Finance, 1970), 195, 196.

15. Ibid., 196.

16. Habib, "Some Economic Aspects," 9–10.

17. *San Francisco Chronicle*, Oct. 13, 1987, p. C1, col. 1; Dec. 27, 1996, p. D1, col. 1;

*Forest Industries North American Factbook* (San Francisco: Miller Freeman Publications, 1988); *California Journal* 27 (June 1996): 40–42; *Forbes* 160 (Oct. 13, 1997): 122–28.

18. *Los Angeles Times,* July 10, 1991, p. A3; Jan. 14, 1993, p. A3; Feb. 7, 1995, p. A1.

19. Ibid., Aug. 20, 1996, p. A3; Aug. 24, 1996, p. A17; July 10, 1997, p. A3; Aug. 21, 1999, p. A20; *San Francisco Chronicle,* May 3, 2000, p. A4.

20. *Wall Street Journal,* Aug. 12, 1999, p. A6; *San Diego Union-Tribune,* June 19, 2000, p. A3; *San Francisco Chronicle,* June 26, 2000, p. A1; July 26, 2000, p. A3; *Los Angeles Times,* July 7, 2000, p. A1; Aug. 23, 2000, p. A3.

21. Wagers, "San Joaquin and Sierra Nevada Railroad," 88.

22. *Stockton Record,* June 23, 1998, p. B3.

23. G. O. Argall Jr., "The New California Gold Rush," *Engineering and Mining Journal* 188 (Dec. 1987): 30–37.

24. *Stockton Record,* Oct. 18, 1986, p. B1, col. 2.

25. Fred Stevenot to Archie Stevenot, Dec. 6, 1952, MS 197, box 29, Stevenot Papers.

26. Barden Stevenot, interview by authors, Murphys, June 18, 1998; Mike Kizer, Grandview geologist, interview by authors, Carson Hill, Oct. 30, 1987. John Burgess, the son of John Burgess Sr., the general manager at Carson Hill in the 1930s, was the last manager on the island of Cyprus for Cyprus Mines, later Cyprus Minerals. At one time the Cyprus company also controlled the Western Source Company, the talc producer in Calaveras County.

27. *Stockton Record,* Sept. 8, 1989, p. B2, col. 3.

28. B. Stevenot, interview; Kizer, interview.

29. *Stockton Record,* May 18, 1989, p. B1, col. 2.

30. B. Stevenot, interview.

31. Argall, "Royal/Mountain King Mine," 40–42.

32. W. P. Fuller, Marvin, and Costello, *Madam Felix's Gold,* 117–18, 121.

33. *Stockton Record,* Feb. 2, 1986, p. 1, cols. 2–4; Dec. 9, 1987, p. B6, col. 4; Mar. 24, 1989, p. B2, col. 5.

34. *Stockton Record,* Apr. 20, 1987, p. B1, col. 2.

35. Walter Nugent, *Into the West: The Story of Its People* (New York: Alfred A. Knopf, 1999), 363–69.

36. *Stockton Record,* Jan. 11, 1988, p. A1, col. 2; Apr. 1, 1990, p. A1, col. 2; California State Department of Finance population estimates, as quoted in ibid., May 13, 1998, p. B6.

37. *Stockton Record,* Jan. 11, 1988, p. A1, col. 2; Nov. 29, 1987, p. B1, col. 1.

38. Ibid., Apr. 10, 1988, p. B5, col. 2.

39. Tower, interview; Ella (McCarty) Hiatt, telephone interview by authors, May 7, 1998.

40. Dutil, interview; Glen Nevens, telephone interview by authors, May 13, 1998; Bird, interview; C. Stone, interview; Vickie Hunter Saunders, "Daddy and Mamma, the Gold Bug and Me: From Beverly Hills to the Mother Lode, 1939–1942" (MS,

Stockton, 1996); Vickie Hunter Saunders, interview by authors, Stockton, May 4, 1998; Dick Rolleri, interview and site visit by authors, Calaveras Central Mine, Altaville, May 11, 1998.

41. Judith Marvin, "Grand Marshall, Barden Stevenot" (ts, Murphys, July 1993), Stevenot Family Papers, private collection, Murphys.

42. B. Stevenot, interview.

43. For example of warning, see *Stockton Record,* June 29, 1986, p. B1, cols. 2–5.

44. Katherine M. Albert, William B. Hull, and Daniel M. Sprague, *The Dynamic West, a Region in Transition: A Guide for State Policy Makers on the Top Ten Trends Transforming the West* ([Sacramento?]: Council of State Governments, Western Region, 1989), 23–25; John Leshy, "Mining's Diminished Future," in *Reopening the Western Frontier,* ed. Ed Marston (Washington, D.C.: Island Press, 1989), 242–47. Limerick makes the same point in "Gold Rush and Shaping the West," 34.

45. Nash, "Veritable Revolution," 276–92.

46. Ray B. Jepperson, "Mining Made America Great," *American Gold News* 32 (Oct. 1964): 4.

47. Donald J. Pisani, "'I Am Resolved Not to Interfere, but Permit All to Work Freely': The Gold Rush and American Resource Law," in *Golden State,* ed. Rawls and Orsi, 128–31.

48. Despite discriminatory tax measures, one recent study argues that Chinese in Idaho, at least, made a good living at mining, and did not find the tax burden excessive, perhaps because it was not vigorously enforced (Zhu, *Chinaman's Chance,* 47–48, 51–53, 133–41).

49. For a discussion of linkages and their importance in assessing the connection between mining and regional economic development, see Freudenburg and Frickel, "Digging Deeper," 266–88; and David St. Clair, "The Gold Rush and the Beginnings of California Industry," in *Golden State,* ed. Rawls and Orsi, 185–208.

50. For a discussion of Gold Rush attitudes and their impact on minorities, see Elliott West, "Golden Dreams: Colorado, California, and the Reimagining of America," *Montana* 99 (autumn 1999): 2–11.

51. William Robbins, *Hard Times in Paradise: Coos Bay, Oregon, 1850–1986* (Seattle: University of Washington Press, 1988), 5–8, 167–68.

52. For further reading on the environmental problems attributed to western mining, consult Stiller, *Wounding the West;* and Duane Smith, *Mining America: The Industry and the Environment* (Lawrence: University Press of Kansas, 1987). *Principal Areas of Mine Pollution,* California Water Resources Control Board, Basin Planning Study, Basin Planning Area 5C, San Joaquin (Sacramento: California Division of Mines and Geology, 1972).

53. Darrell Kirk Nordstrom, "Chemical Modeling of Acid Mine Waters in the Western United States," U.S. Geological Survey, *Toxic Substances Hydrology Program: Proceedings of the Technical Meeting, Monterey, California, March 11–15, 1991,* Water-Resources Investigations Report 91-4034, ed. G. E. Mallard and D. A. Aronson (Washington, D.C.: U.S. Geological Survey, 1991), 534–38.

54. For the origins and early development of the East Bay Municipal Utility District, see John Wesley Noble, *Its Name Was* M.U.D.: *The Story of Water As It Has Affected the Urban Complex on the Eastern Shore of San Francisco Bay, With Emphasis on the Creation of the East Bay Municipal Utility District* (Oakland: [East Bay Municipal Utility District], 1970).

55. Scott M. Hamlin and Charles N. Alpers, "Hydrogeology and Geochemistry of Acid Mine Drainage in Ground Water in the Vicinity of Penn Mine and Camanche Reservoir, Calaveras County, California: Second-Year Summary, 1992–93," U.S. Geological Survey Water-Resources Investigations Report 96-4257 (ca.water.usgs.gov/rep/penn/sum.html); Stiller, *Wounding the West,* 92–97; Francis P. Garland, "Penn Mine's Revival," *Stockton Record,* Sept. 30, 2000, p. A1.

56. For literature on claims brought by various tribes against the United States, consult the docket books of the Indian Claims Commission and other records of the Indian Claims Section of the U.S. Department of Justice. Limerick, "Gold Rush and Shaping the West," 39–40.

57. Whether modern California attitudes can be traced back to the Gold Rush is still a matter of debate among historians. During the 1998 sesquicentennial, for example, Kevin Starr drew a connection between Gold Rush individualism and the entrepreneurial spirit of Silicon Valley, but Richard White was skeptical of claims based on "contingency," or trying to associate present events with past events, especially over long periods of time. See Starr, "California Dream," 42–54; and White, "Consequences and Contingencies," 56–67.

*Adit:* A nearly horizontal passage from the surface by which a mine is entered; commonly called a tunnel.

*Amalgamation:* The process of collecting free gold and silver from ores and concentrates by combination with mercury (quicksilver). The resulting amalgam is then heated to vaporize the mercury, leaving a gold or silver "sponge" that is remelted, poured into doré bars, and shipped to a smelter for further refinement.

*Apex:* The highest exposed point of a vein, either on the surface or underground.

*Arrastra:* A circular rock-lined pit in which broken ore is pulverized by dragging stones over it. The dragstones are attached to horizontal supports fastened to a central shaft, and powered by animals or water.

*Batea:* A shallow wooden bowl used for panning, brought to California by Hispanic miners.

*Breast:* A horizontal stope in a drift mine from which gold-bearing gravel is mined.

*Cage:* A partially enclosed platform or vehicle that can be raised and lowered in a vertical shaft for transporting miners, loaded or empty mine cars, and some supplies. Vertical wooden guides on each side of the shaft keep the cage properly aligned when in motion.

*Cobble:* A rounded piece of rock larger than a pebble but smaller than a boulder, in streambed gravels.

*Collar:* The horizontal timbering or other reinforcing at the top of a shaft at the surface. Used for both vertical and inclined shafts and sometimes tunnels. Commonly used in reference to the opening of a mine.

*Country rock:* The general mass of adjacent rock as distinguished from a dike or from the mineralized vein or lode.

*Coyoting:* Mining in irregular and generally small openings, somewhat resembling the holes of coyotes, foxes, or gophers.

*Cretaceous:* The geologic time period directly before the Tertiary, ranging from 65 million to 130 million years ago.

*Crosscut:* A horizontal opening on a level, cutting across an ore body or geological formation.

*Dike:* A tabular and thin body of igneous rock that has been injected, while still molten, into a fissure or crack in older rock and then cooled and solidified.

*Dip:* The angle at which beds or strata, veins, dikes, and so on are inclined from the horizontal; to slope downward from the horizontal. Dip is perpendicular to the strike.

*Dog:* A mechanical device, usually affixed to timbers or logs, to aid in handling. A safety dog on a mine cage is designed to grip wooden shaft rails or guides in an emergency, thus stopping further descent.

*Dressing ore:* Any method of picking, sorting, washing, or other processing of crushed and ground ores to produce concentrates for further reduction to metals or mineral products.

*Drift:* A horizontal passage underground, following a vein or other structure, as distinguished from a crosscut, which intersects such structures at an angle.

*Eocene:* The geologic time epoch in the lower (older) part of the Tertiary period.

*Extralateral right:* The right of a lode claim owner or operator to mine a vein downward beyond the vertical projection of the sidelines if the apex of the vein is on the claim. In all cases, however, the owner or operator must stay within the projected end lines of the claim.

*Flotation:* A method of mineral separation in which a froth created in water by a variety of reagents floats some finely ground minerals and causes other minerals to sink.

*Footwall:* The wall or rock on the lower side of an inclined vein.

*Free-milling:* Gold-bearing ore that can be milled by crushing and grinding. The liberated gold is then recovered by simple gravity separation from gangue or waste.

*Gallows frame:* An older name for headframe.

*Gangue:* The nonmetalliferous or nonvaluable metalliferous minerals in ore.

*Glory hole:* A funnel-shaped surface opening or pit that is connected with underground workings, enabling ore or waste broken from the pit walls to fall by gravity into ore chutes or passes, then loaded onto cars or conveyors and trammed or hoisted up to the surface.

*Gob:* Waste or low-grade ore that is stored underground in old, inactive stopes to prevent caving or to save the cost of removal.

*Hanging wall:* The wall or rock on the upper side of an inclined vein.

*Hardrock:* A term used for solid, unweathered rock formations, as opposed to unconsolidated clay, sand, or gravel. It is also applied to the mining of quartz veins.

*Headframe:* The superstructure above the collar of a mine shaft that supports the sheave wheels, hoisting cable, and other equipment for hoisting and handling ore and waste.

*High grade:* Ore that is substantially higher in value than average ore. The term is also applied to specimen ore that has been stolen.

*Inclined shaft:* A sloping shaft as opposed to a vertical shaft.

*Lead:* Pronounced as the present tense of the verb, a synonym for *vein, lode,* or sometimes for *gravel channel.*

*Level:* A drift, crosscut, or other mine working at a certain elevation and generally identified by a number representing the distance down from the collar of the shaft.

*Lode:* A fissure in country rock containing valuable mineral-bearing material; essentially a synonym for *vein.* The term was used in claim designations to differentiate them from placer claims.

*Middlings:* An intermediate product of concentration that has to be upgraded by further treatment.

*Millhead values:* The amounts of valuable metals contained in the ore delivered to the mill for processing.

*Miocene:* The geologic time epoch in the middle of the Tertiary period.

*Muck:* The general term for any material underground that has been blasted or broken. Mucking is the act of loading this material, either manually or by mechanical means, into mine cars for removal.

*Ore:* Mineral material containing sufficient values to be mined at a profit, or hope of profit.

*Ore shoot:* An occurrence of ore within a vein or other structure that has well-defined boundaries; a more specific term than *ore body* or *deposit.*

*Overburden:* Worthless surface material overlying a body of useful or valuable mineral.

*Placer:* A deposit of alluvial material such as gravel, sand, and silt containing particles of valuable minerals that can be recovered by relatively simple washing or other methods.

*Pliocene:* The geologic time epoch in the upper (younger) part of the Tertiary period.

*Quartz:* A silicate, one of the commonest minerals, generally white, very hard, and the principal mineral of a quartz vein.

*Raise:* An opening driven upward from a mine level working such as a drift or crosscut.

*Rake:* The direction of an ore shoot within the plane of a vein, described with relationship to horizontal.

*Reduction; reduction works:* The process of extracting metals from their ores; a processing plant for that purpose.

*Refractory ores:* Ores in which the valuable metals are not free-milling, but are instead intimately mixed or otherwise combined with other minerals, and must be treated with more complex processes involving heat or chemicals or both.

*Secondary enrichment:* When a copper deposit outcrops and is subject to weathering, some of the copper is dissolved by surface waters. As these waters percolate down into the deposit to the water table, copper is reprecipitated, adding to the primary copper minerals, thus enriching or raising the content of copper in the primary deposit. In contrast, gold, being quite insoluble, is enriched only from the weathering and removal of other minerals.

*Shaft:* A mine opening from the surface, with a relatively small cross-section, either vertical or inclined, used for access and for servicing and operating a mine.

*Shrinkage stoping:* A method of stoping in steep veins where miners drill and blast at the top of the stope, working off previously broken ore, which then drops through ore passes and chutes to haulage levels below. On each shift enough ore is drawn out of the stope to give the miners working room.

*Single- and double-jacking:* A manual technique for drilling blast holes in rock by striking a chisel-shaped bit with a hammer. A single-jack driller used a small hammer alone; double-jacking required one miner with a large hammer to hit the drill and a second miner to hold and twist it after each blow.

*Skip:* A vehicle designed for ore and waste haulage in mine shafts, either vertical or inclined, as well as for other uses such as bailing water, transporting supplies, and lowering and hoisting miners in inclined shafts. In vertical shafts skips ran on wooden or metal guides, but in inclined shafts they were equipped with wheels and ran on rails.

*Slimes:* Mineral material ground too finely to respond to normal treatment by water and gravity.

*Stope:* An underground excavation from which ore is being or has been extracted; to remove ore from underground. Stopes are served by the level workings directly under them.

*Strike:* The direction or bearing of a horizontal line in the plane of an inclined geological structure such as a vein, fault, bed, or other structural feature. Strike is perpendicular to the dip.

*Tailings:* The waste material or gravel piles left after crushing and processing ores, or working placer deposits. Untreated waste material discarded from a mining operation, however, is called *mine waste,* and is deposited in waste dumps.

*Tertiary:* The geologic time period after the Cretaceous period and before the Quaternary period, ranging from 3 million to 65 million years ago.

*Throughput:* The amount of material that can be processed in a plant during a given time interval.

*Tunnel:* Technically, an underground passage open at both ends; also used colloquially as a synonym for *adit.*

*Vanner:* A mechanical device used in the nineteenth century for concentrating finely ground gold-bearing sulfides, as well as gold.

*Vein:* A tabular dikelike structure of quartz and other minerals, with well-defined walls, containing gold, silver, other valuable minerals, or some combination thereof.

*Whim:* A hoisting device, like a windlass, operated by a horse or other beast of burden, used for shallow or prospecting shafts.

*Windlass:* A hoisting device, generally made of a wooden drum, operated by hand crank or handle, used in small or prospecting shafts.

*Winze:* An underground shaft, sunk from a drift or crosscut.

Calaveras County has a rich mining history, but sources of information are spotty and often very narrowly focused. Production and development of the major mines are generally well covered in published professional bulletins from state and federal bureaus, but much harder to find are extensive and substantial treatments of smaller operations, personalities, labor issues, and comparative data showing Calaveras in the context of state or federal mining history. We have gleaned what we could from the sources available, but we have also had to rely on our own personal knowledge and experience to develop a comprehensive picture. Most of our sources are discussed in this essay, but we have not compiled a complete bibliography here since the notes contain full bibliographic references.

### MANUSCRIPT COLLECTIONS

David Clarence Demarest's unpublished manuscript, "California Gold," at the Holt-Atherton Library, University of the Pacific, is one of the most important comprehensive treatments of Calaveras mining, but it must be used with caution. Demarest was a Calaveras pioneer and an important figure in Calaveras mining for most of his life, but he had strong opinions, and his production figures and other data were often inaccurate. His lengthy study is based on firsthand knowledge of both the major mines and the men who ran the industry from the 1870s to the end of World War I.

Also at the Holt-Atherton is the Segerstrom Family Collection of correspondence, field notes, company files, clippings, professional reports, maps, and other materials. These are indispensable for the Carson Hill story after World War I. Other important manuscript collections in the Holt-Atherton Library include the Jeffrey Schweitzer Papers, the newspaper typescripts of Reatha Parcel Smith, the Archie Stevenot Papers, the Richard Coke Wood Papers, and the diary of John Wallis.

At San Andreas, the Calaveras County Historical Society has an important collection of mining history sources, including the Francis E. Bishop manuscript on the Union Water Company; Joseph Giovinco's study of ethnicity (1977); a notebook of data compiled by Edward Leonard; a copy of the Leonard Noyes manuscript; and maps, claim data, and a superb photograph collection. Also at San Andreas is the county archives depository, one of the best on the Lode, with a nearly complete documentary record of public transactions, assessment rolls, and court records, as well as an extensive collection of newspapers, maps, and photographs.

At nearby Sonora, the Tuolumne County Historical Society contains extensive newspapers and manuscripts bearing on Carson Hill, Angels Camp, and other Calaveras districts. In particular, the Wax Collection of William J. Loring Papers is essential, not only for understanding Carson Hill's development in the 1930s, but also for important corporate reports and other documents relating to Loring's work in Australia with the Bewick Moreing Company.

In Sacramento, various state agencies contain key Calaveras sources. In addition to the invaluable collection of newspapers on microfilm, the California State Library holds useful manuscript collections, including the William Higby Letters (1837–1859), the 1852 manuscript census for Calaveras County, and the annual reports of the California State Railroad Commission. The California State Archives are particularly useful for Public Utilities Commission records on Calaveras railroads, company annual reports filed with the PUC, and financial data for major mining companies in state incorporation records. The library of the Division of Mines and Geology contains agency records and photos, plus an extensive collection of books, maps, trade journals, and other technical publications.

Two other important primary sources for Calaveras mining history are the Bancroft Library at the University of California's Berkeley campus and the Huntington Library in San Marino. At the Bancroft, the most useful for Calaveras are the newspaper clippings in the county scrapbook collection; the De Laittre and Lascy Reminiscences; the correspondence of Alexander Dickinson, William Hanson, Alonzo W. Merrill, and James Megeath; the George Lundy Hunt diary; the Jonathan F. Locke Papers; the James Madison Grover Memoirs; and the Samuel Linus Prindle Records and Papers. At the Huntington, the John Brown, John Hovey, Elias Ketchum, Joseph Pownall, J. W. Stow, and Selim Woodworth Papers are especially valuable. Worth mentioning are Herbert Hoover's personal papers at the Hoover Presidential Library in West Branch, Iowa. They contain interesting correspondence relating to Hoover's brief interest in the Royal mine at Hodson.

NEWSPAPERS

Regional and local history would be barren indeed without newspaper files, often the only source of information on important but transitory personalities, events, and issues. Useful collections of Calaveras newspaper clippings are located in the Bancroft Library at the University of California at Berkeley, in the Holt-Atherton Library at the University of the Pacific, and in the Calaveras County Historical Society at San Andreas. The California State Library's microfilm collection is the best single source of historic papers containing Calaveras mining information. Among the most important newspapers are the *Alta California, Calaveras Chronicle, Calaveras Prospect, Copperopolis Courier, Los Angeles Times, Mining and Scientific Press, Sacramento Union, San Andreas Independent, San Andreas Register, San Francisco Bulletin, San Francisco Call, San Francisco Chronicle, San Joaquin Republican, Stockton Evening Mail, Stockton Independent, Stockton Record,* and *Wall Street Journal.*

ORAL HISTORY

Personal interviews should not be overlooked as a source of valuable information on the regional mining industry. Over a period of nearly twenty years, both authors have gleaned useful material from dozens of interviews with Calaveras miners or their family members, some of whom are no longer available to share their information with others. Among the most important were interviews with Ted Bird on Melones (1998), Samuel Bryan on the lumber industry (1979), Glen Bump on doodlebug operations (1979), Tony Dutil on mining in southeast Calaveras (1997), Emil Guidicy on Sheep Ranch mining (1987), Ella (McCarty) Hiatt on Copperopolis and Salt Spring Valley (1998); J. C. Kemp van Ee (1968), Mike Kizer on the Carson Hill mine (1987), Glen Nevens on dredging (1998), Duane Oneto on Calaveras water and power development (1998), Dick Rolleri on the Calaveras Central (1998), Vickie Saunders on Paloma (1998), Barden Stevenot on Carson Hill and other Calaveras economic developments (1998), Charlie Stone on Copperopolis (1991, 1997), Muriel Tiebolt on Mother Lode social history (1987), and Howard Tower (1989, 1998) on the Hodson district.

The Bancroft Library Oral History Project has an extensive collection of transcribed interviews of individuals important to western mining history. The reminiscences of Phil Bradley (1988) were particularly useful for this study.

GOVERNMENT PUBLICATIONS

Publications of federal agencies such as the Bureau of the Census, the Forest Service, the Geological Survey, and the Bureau of Mines have been very useful, particularly the U.S. Geological Survey Professional Papers 73 (Waldemar Lindgren's *Tertiary Gravels of the Sierra Nevada of California*) and 157 (Adolph Knopf's *Mother Lode System of California*). C. E. Julihn and F. W. Horton's report in the U.S. Bureau of Mines Bulletin 413 (*Mines of the Southern Mother Lode*) and the annual volumes of mineral statistics from 1867 through the 1940s were invaluable sources of production data and other information. The annual reports on *Mineral Resources of the States and Territories West of the Rocky Mountains,* published by the U.S. Treasury between 1867 and 1871, are filled with valuable information on early Calaveras County developments.

Equally important are the extensive publications of the California Mining Bureau, later called the Division of Mines, and still later the Division of Mines and Geology. The late-nineteenth-century bulletins and annual reports of the state mineralogist contain invaluable data, and more recent bulletins such as 57 (*Gold Dredging in California,* by Lewis E. Aubury); 108 (*The Mother Lode Gold Belt of California,* by Clarence A. Logan); 141 (*Geological Guidebook Along Highway 49,* ed. Olaf P. Jenkins); and William B. Clark and Philip A. Lydon's County Report 2, "Mines and Mineral Resources of Calaveras County, California," are indispensable descriptive narratives that incorporate many of the earlier data.

Other useful government publications include the early *Transactions of the California State Agricultural Society;* the decennial *California Statistical Abstracts;* William B. Clark's *Gold Districts of California* (1970); W. Turrentine Jackson's *Historical Survey of the New Melones Reservoir Project Area* (1976); and Dorothea J. Theodoratus's *Ethnographic Study of the New Melones Lake Project* (1976).

THESES AND DISSERTATIONS

Academic institutions across the United States and Canada contain rich but often underutilized resources having a mining and regional history focus. Our study utilized theses and dissertations by Jason Beck ("California Gold Rush Violence, 1849–1854: A Psychological Interpretation," 1978); Warren Ronald Blomquist ("Gold Dredging in California," 1973); Leigh Bristol-Kagan ("Chinese Migration to California, 1851–1882: Selected Industries of Work, the Chinese Institutions, and the Legislative Exclusion of a Temporary Labor Force," 1982); Albin J. Dahl ("British Investment in California Mining, 1870–1890," 1961); Lary Dilsaver ("From Boom to Bust: Post–Gold Rush Patterns of Adjustment in a California Mining Region," 1982); Rodney S. Ellsworth ("Discovery of the Big Trees of California, 1833–1852," 1933); Donald Ray Floyd ("The Role of Narrow-Gauge Railroads in California's Transportation Network," 1968); Philip Charles Habib ("Some Economic Aspects of the California Lumber Industry and Their Relation to Forest Use," 1952); Richard M. MacKinnon ("The Historical Geography of Settlement in the Foothills of Tuolumne County, California," 1967); Donald Macleod ("Miners, Mining Men, and Mining Reform: Changing the Technology of Nova Scotian Gold Mines and Collieries, 1858 to 1910," 1981); Ward McAfee ("Local Interests and Railroad Regulation in Nineteenth-Century California," 1966); Brian Eugene Roberts ("The California Gold Rush and the Making of the American Middle Class," 1995); and Kenneth Aubrey Smith ("California: The Wheat Decades," 1969).

BOOKS

Regional history is best if it is written from a broad perspective. Our efforts to place Calaveras mining history in the proper context began with what scholars and popular historians have said about mining and related economic development in western America. Especially helpful were books by Katherine M. Albert, William B. Hull, and Daniel M. Sprague (*The Dynamic West, a Region in Transition: A Guide for State Policy Makers on the Top Ten Trends Transforming the West*, 1989); Hubert H. Bancroft (*History of California, 1848–1859*, 1884, and *Chronicles of the Builders*, 1892); Gray Brechin (*Imperial San Francisco: Urban Power, Earthly Ruin*, 1999); John W. Caughey (*Gold Is the Cornerstone*, 1948); Mary Hill (*Gold: The California Story*, 1999); John S. Hittell (*The Resources of California, Comprising Agriculture, Mining, Geography, Climate, Commerce . . . and the Past and Future Development of the State*, 1866); Theodore H. Hittell (*History of California*, 2 vols., 1885–1898); Robert Evelyn Holmes (*The Southern Mines of California: Early Development of the Sonora Mining Region*, 1930); Patricia Nelson Limerick (*The Legacy of Conquest: The Unbroken Past of the American West*, 1987); Ralph Mann (*After the Gold Rush: Society in Grass Valley and Nevada City, California, 1849–1870*, 1982); Walter Nugent (*Into the West: The Story of Its People*, 1999); Watson Parker (*Gold in the Black Hills*, 1966); Rodman Paul (*California Gold*, 1947, and *Mining Frontiers of the Far West, 1848–1880*, 1963); Richard H. Peterson (*The Bonanza Kings: The Social Origins and Business Behavior of Western Mining Entrepreneurs, 1870–1900*, 1977); Thomas A. Rickard (*History of American Mining*, 1932, and *Retro-*

*spect: An Autobiography,* 1937); William Robbins (*Hard Times in Paradise: Coos Bay, Oregon, 1850–1986,* 1988); Josiah Royce (*California: From the Conquest in 1846 to the Second Vigilance Committee in San Francisco,* 1948 reprint of 1886 ed.); Charles Howard Shinn (*Mining Camps: A Study in American Frontier Government,* 1884); Clark D. Spence (*British Investments in the American Mining Frontier, 1860–1901,* 1958, and *The Lace-Boot Brigade: Mining Engineers in the American West,* 1970); Louis B. Wright (*Culture on the Moving Frontier,* 1955); Otis E. Young Jr. (*Western Mining: An Informal Account of Precious-Metals Prospecting, Placering, Lode Mining, and Milling on the American Frontier From Spanish Times to 1893,* 1970); and Sally Zanjani (*Goldfield: The Last Gold Rush on the Western Frontier,* 1992).

To commemorate the Gold Rush sesquicentennial, Gary Kurutz of the California State Library compiled an invaluable reference tool, *The California Gold Rush: A Descriptive Bibliography of Books and Pamphlets Covering the Years 1848–1853* (1997). The California Historical Society published a comprehensive series of articles beginning in 1998 that was simultaneously published in book form by the University of California Press under separate titles. We found most useful the second volume, *A Golden State: Mining and Economic Development in Gold Rush California* (1999), ed. James J. Rawls and Richard J. Orsi, containing twelve essays by specialists in banking, agriculture, transportation, technology, environmental history, law, labor, corporate development, and other important economic themes. Huntington Library made a substantial contribution to the sesquicentennial that same year by publishing Peter J. Blodgett's *Land of Golden Dreams: California in the Gold Rush Decade, 1848–1858,* a comprehensive illustrated history of the Gold Rush era.

An excellent specialized study of the mining machinery industry is Lynn R. Bailey's *Supplying the Mining World: The Mining Equipment Manufacturers of San Francisco, 1850–1900* (1996). Powell Greenland's *Hydraulic Mining in California: A Tarnished Legacy* (2001) is rich with technical details, but does not replace the classic work by Robert L. Kelley, *Gold vs. Grain: The Hydraulic Mining Controversy in California's Sacramento Valley, a Chapter in the Decline of the Concept of Laissez Faire* (1959). Also still useful is Philip R. May's *Origins of Hydraulic Mining in California* (1970). Contemporary machinery and tools are illustrated and described in *The Miner's Own Book, Containing Correct Illustrations and Descriptions of the Various Modes of California Mining, Including All the Improvements Introduced From the Earliest Day to the Present Time* (1949 facsimile of 1858 ed.). Julia Costello and Judith Cunningham's *History of Mining at Carson Hill* (1988) is a carefully documented compilation by a private consulting firm, as is *Madam Felix's Gold: The Story of the Madam Felix Mining District, Calaveras County, California* (1996), written by the same team with the addition of Willard P. Fuller Jr.

Other technical and specialized studies we found useful include Thomas Allsop's *California and Its Gold Mines* (1853); David T. Ansted's *Gold Seeker's Manual* (1849); Walter R. Crane's *Ore Mining Methods, Comprising Descriptions of Methods of Support in Extraction of Ore, Detailed Descriptions of Methods of Stoping and Mining in Narrow and Wide Veins and Bedded and Massive Deposits Including Stull and Square-Set Min-*

*ing, Filling and Caving Methods, Open-Cut Work, and a Discussion of Costs of Stoping* (1910); Emanuel Fritz's *Development of Industrial Forestry in California* (1960); Hank Johnston's *They Felled the Redwoods: A Saga of Flumes and Rails in the High Sierra* (1966); James F. Kemp's *Ore Deposits of the United States* (1895); Eliot Lord's *Comstock Mining and Miners* (1980 reprint of 1883 ed.); J. Arthur Phillips's *Mining and Metallurgy of Gold and Silver* (1867); Arnold R. Ross's *Twenty-five Years of Building the West* (195c); Grant H. Smith's *History of the Comstock Lode, 1850–1920* (1943); W. H. Storms's *Timbering and Mining: A Treatise on Practical American Methods* (1909); and James Ward's *History of Gold As a Commodity and As a Measure of Value: Its Fluctuations Both in Ancient and Modern Times, With an Estimate of the Probable Supplies From California and Australia* (1852).

For environmental issues related to mining, the best general treatment remains Duane Smith's *Mining America: The Industry and the Environment* (1987). Other works we found helpful include William Cronon's *Changes in the Land: Indians, Colonists, and the Ecology of New England* (1983) and his interpretive article in *Kennecott Journey: The Paths Out of Town* (1992), ed. W. Cronon, G. Miles, and J. Gitlin; Lary M. Dilsaver and William C. Tweed's *Challenge of the Big Trees of Sequoia and Kings Canyon National Parks* (1990); Francis P. Farquhar's *History of the Sierra Nevada* (1965); the Mineral Policy Center's *Golden Dreams, Poisoned Streams: How Reckless Mining Pollutes America's Waters, and How We Can Stop It* (1997); John Muir's *Our National Parks* (1981 reprint of 1901 ed.); David Stiller's *Wounding the West: Montana, Mining, and the Environment* (2000); and several recent U.S. Geological Survey studies of toxic substances.

Though Calaveras lacks a mine labor study, labor issues in the industry as a whole have been extensively covered in recent years. We found most useful books by Harry Braverman (*Labor and Monopoly Capital: The Degradation of Work in the Twentieth Century*, 1974); Joseph R. Conlin (*Big Bill Haywood and the Radical Union Movement*, 1969); Alan Derickson (*Workers' Health, Workers' Democracy: The Western Miners' Struggle, 1891–1925*, 1988); Jerry Dolph (*Fire in the Hole: The Untold Story of Hardrock Miners*, 1994); Richard Edwards (*Contested Terrain: The Transformation of the Workplace in the Twentieth Century*, 1979); Elizabeth Jameson (*All That Glitters: Class, Conflict, and Community in Cripple Creek*, 1998); Vernon H. Jensen (*Heritage of Conflict: Labor Relations in the Nonferrous Metals Industry up to 1930*, 1950); Larry Lankton (*Cradle to Grave: Life, Work, and Death at the Lake Superior Copper Mines*, 1991); Richard E. Lingenfelter (*The Hardrock Miners: A History of the Mining Labor Movement in the American West, 1863–1893*, 1974); David Noble (*America by Design: Science, Technology, and the Rise of Corporate Capitalism*, 1977); John Rowe (*The Hard-Rock Men: Cornish Immigrants and the North American Mining Frontier*, 1974); William Serrin (*Homestead: The Glory and Tragedy of an American Steel Town*, 1992); and Mark Wyman (*Hard Rock Epic: Western Miners and the Industrial Revolution, 1860–1910*, 1979).

For the early years of Stockton and its Mother Lode connections, we found most useful Frank T. Gilbert's *History of San Joaquin County, California, With Illustrations*

*Descriptive of Its Scenery, Residences, Public Buildings, Fine Blocks, and Manufactories, From Original Sketches by Artists of the Highest Ability* (1879); George P. Hammond's *Weber Era in Stockton History* (1982); George P. Hammond and Dale L. Morgan's *Captain Charles M. Weber: Pioneer of the San Joaquin and Founder of Stockton, California* (1966); Neal Harlow's *California Conquered: The Annexation of a Mexican Province, 1846–1850* (1982); George H. Tinkham's *History of Stockton* (1880); and George H. Tinkham, *History of Stanislaus County, California: With Biographical Sketches of the Leading Men and Women of the County Who Have Been Identified With Its Growth and Development From the Early Days to the Present* (1921). Mother Lode transportation history lacks comprehensive treatment, but various segments of the story can be found in Dorothy N. Deane's *Sierra Railway* (1960); David F. Myrick's *Railroads of Nevada and Eastern California,* vol. 1 (1962–1963); Jack R. Wagner's *Short Line Junction* (1956); and Oscar O. Winther's *Express and Stagecoach Days in California From the Gold Rush to the Civil War* (1936). A term paper by Larry Bradfield ("The Elephant at the Depot Door," 1979) was carefully done and especially useful.

No modern comprehensive history has been written for Calaveras. Richard Coke Wood's *Calaveras, the Land of Skulls* (1955) remains the best published work. Edna Bryan Buckbee's *Pioneer Days of Angels Camp* (1932) is useful for mining history anecdotes if used with caution. A few booklets and pamphlets are worth consulting, including *Calaveras County Illustrated and Described, Showing Its Advantages for Homes* (1976 reprint of 1885 ed.); Emmett P. Joy and Ellen H. Ladd's *Chronicles of San Andreas: One Town That Rose From a Golden Channel* (1972); editor Mark B. Kerr's *Mining Resources of Calaveras County, California* (1898); and Edward C. Leonard's *Brief History of Angel's Camp* (1973).

For Calaveras in the Gold Rush, a number of published reminiscences are available, including John Woodhouse Audubon's *Audubon's Western Journal, 1849–1850* (1969 reprint of 1906 ed.); James J. Ayers's *Gold and Sunshine: Reminiscences of Early California* (1922); Jacob Henry Bachman's "Diary of a Used-Up Miner," in *California Historical Society Quarterly* (1943); Milo Bird's *Melones Memories: Recollections of a Mother Lode Gold Mining Camp by a Latter-Day Tom Sawyer* (1985); John David Borthwick's *Three Years in California* (1948 reprint of 1857 ed., reprinted as *The Gold Hunters: A First-Hand Picture of Life in California Mining Camps in the Early Fifties,* 1917); James H. Carson's *Recollections of the California Mines* (1950 reprint of 1852 ed.); Walter Colton's *Three Years in California* (1949 reprint of 1851 ed.); Peter Y. Cool's "Goodness, Gold, and God: The California Mining Career of Peter Y. Cool, 1851–52, a Journal," in *Pacific Historian* (1966); Antonio F. Coronel's *Tales of Mexican California* (1994 reprint of 1848 ed.); editors and translators Maria del Carmen Ferreyra and David S. Reher's *Gold Rush Diary of Ramon Gil Navarro* (2000); Josiah Foster Flagg's "Diary of a Philadelphia Forty-Niner," in *Pennsylvania Magazine of History* (1946); John McHenry Hollingsworth's *Journal of Lieutenant John McHenry Hollingsworth of the First New York Volunteers, Stevenson's Regiment, September 1846–August 1849* (1923); Thomas Jefferson Matteson's *Diary* (1954); Byron Nathan McKinstry's *California Gold Rush Overland Diary of Byron N. McKinstry, 1850–1852* (1975); William

Perkins's *Three Years in California: William Perkins' Journal of Life at Sonora, 1849–1852* (1964); Thomas A. Rickard's *Interviews With Mining Engineers* (1922); William Redmond Ryan's *Personal Adventures in Upper and Lower California, in 1848–9* (1973 reprint of 1850 ed.); editor Joseph Schafer's *California Letters of Lucius Fairchild* (1931); Riley Senter's *Crossing the Continent to California Gold Fields* (1938); Pringle Shaw's *Ramblings in California: Containing a Description of the Country, Life at the Mines, State of Society, &c.* (1856); William Shaw's *Golden Dreams and Waking Realities: Being the Adventures of a Gold-Seeker in California and the Pacific Islands* (1851); Rhoda Stone and Charles A. Stone's *Tools Are on the Bar: The History of Copperopolis, Calaveras County, California* (1991); Bayard Taylor's *Eldorado; or, Adventures in the Path of Empire* (1949 reprint of 1850 ed.); George R. Underhill's *Voyage to California and Return: George R. Underhill, 1849–1852* (1980); Arthur R. Wilson's *My Life and Boyhood Days in West Point, 1881–1959* (1959); and Harvey Wood's *Personal Recollections* (1955 reprint of 1896 ed.).

Not specific to Calaveras, but still valuable as firsthand accounts are Henry William Bigler's *Bigler's Chronicle of the West: The Conquest of California, Discovery of Gold, and Mormon Settlement As Reflected in Henry William Bigler's Diaries* (1962 reprint of 1848 ed.); James Stephens Brown's *California Gold: An Authentic History of the First Find With the Names of Those Interested in the Discovery* (1894); Edward Gould Buffum's *Six Months in the Gold Mines, From a Journal of Three Years' Residence in Upper and Lower California, 1847-8-9* (1959 reprint of 1850 ed.); Charles William Churchill's *Fortunes Are for the Few: Letters of a Forty-Niner* (1977); Jesse L. Coffey with George Hoeper's *Bacon and Beans From a Gold Pan* (1972); Henry P. DeGroot's *Recollections of California Mining Life* (1884); Joseph Libbey Folsom's *Letter of Captain J. L. Folsom Reporting on Conditions in California in 1848* (1944); Frederich Gerstaecker's *California Gold Mines* (1946 reprint of 1942 translation of 1856 German ed.); Leonard Kip's *California Sketches, With Recollections of the Gold Mines* (1946 reprint of 1850 ed.); Samuel Frank Marryat's *Mountains and Molehills; or, Recollections of a Burnt Journal* (1952 facsimile of 1855 ed.); John Steele's *In Camp and Cabin: Mining Life and Adventure in California During 1850 and Later* (1928 reprint of 1901 ed.); and editor Walker D. Wyman's *California Immigrant Letters* (1952).

Readers interested in the social history of mining would also do well to start with contemporary journals and reminiscences. In addition to the volumes by John David Borthwick and William Colton, cited above, editor Charles L. Camp's *John Doble's Journal and Letters From the Mines: Mokelumne Hill, Jackson, Volcano, and San Francisco, 1851–1865* (1962) is an extremely valuable firsthand account, as is editor Walter Van Tilburg Clark's *Journals of Alfred Doten, 1849–1903* (1973). The best comprehensive social history of Gold Rush argonauts is Malcolm J. Rohrbough's *Days of Gold: The California Gold Rush and the American Nation* (1997). For the Southern Mines, Susan Lee Johnson's *Roaring Camp: The Social World of the California Gold Rush* (2000) is invaluable. Women are covered in JoAnn Levy's *They Saw the Elephant: Women in the California Gold Rush* (1990).

On Native Americans, the best general account is James J. Rawls's *Indians of California: The Changing Image* (1984), but see also Albert L. Hurtado's *Indian Survival on the California Frontier* (1988) as well as the work by Leonard Pitt, *The Decline of the Californios: A Social History of the Spanish-Speaking Californians, 1846–1890* (1966). For the Central Sierra Miwok, consult Eugene L. Conrotto's *Miwok Means People: The Life and Fate of the Native Inhabitants of the California Gold Rush Country* (1973) and James Gary Maniery's *Six Mile and Murphys Rancherias: An Ethnohistorical and Archaeological Study of the Two Central Sierra Miwok Village Sites* (1987). Robert F. Heizer's *Destruction of California Indians* (1993 reprint of 1974 ed.) provides contemporary examples of genocide, a subject David E. Stannard addresses more generally in *American Holocaust: Columbus and the Conquest of the New World* (1992). Less polemical is Shelburne F. Cook's *Conflict Between the California Indian and White Civilization* (1976). On the same subject, see also Wilbur R. Jacobs's *Fatal Confrontation: Historical Studies of American Indians, Environment, and Historians* (1996).

For other ethnic studies, see E. A. Bielharz and Carlos W. Lopez, eds. and trans., *We Were 49ers! Chilean Accounts of the California Gold Rush* (1976); Ping Chiu, *Chinese Labor in California, 1850–1880: An Economic Study* (1963); Rudolph M. Lapp, *Blacks in Gold Rush California* (1977); Robert E. Levinson, *Jews in the California Gold Rush* (1978); Jay Monaghan, *Chile, Peru, and the California Gold Rush of 1849* (1973); A. P. Nasatir, *A French Journalist in the California Gold Rush: The Letters of Etienne Derbec* (1964); Alexander Saxton, *The Indispensable Enemy: Labor and the Anti-Chinese Movement in California* (1995 reprint of 1971 ed.); Shih-Shan Henry Tsai, *The Chinese Experience in America* (1986); Stephen Williams, *The Chinese in the California Mines, 1848–1860* (1971 reprint of 1930 ed.); Rufus K. Wyllys, *The French in Sonora (1850–1854): The Story of French Adventurers From California Into Mexico* (1932); and Liping Zhu, *A Chinaman's Chance: The Chinese on the Rocky Mountain Mining Frontier* (1997).

Violence in the mining camps is discussed in David T. Courtwright's *Violent Land: Single Men and Social Disorder From the Frontier to the Inner City* (1996); Frank Latta's *Joaquin Murrieta and His Horse Gangs* (1980); Roger D. McGrath's *Gunfighters, Highwaymen, and Vigilantes: Violence on the Frontier* (1984); and Clare V. McKanna Jr.'s *Homicide, Race, and Justice in the American West, 1880–1920* (1997). Recently, the colorful history of the Calaveras County Bench and Bar from its beginnings to modern times has been chronicled in Michael B. Arkin and Franklin T. Laskin's *From the Depths of the Mines Came the Law* (2000).

Kevin Starr's *Americans and the California Dream* (1986 reprint of 1973 ed.) is an excellent introduction to Gold Rush literature and its enduring impact. Still useful is Franklin Walker's *San Francisco's Literary Frontier* (1939). For Mark Twain's and Bret Harte's Calaveras connections, consult editors Frederick Anderson, Michael B. Frank, and Kenneth M. Sanderson's *Mark Twain's Notebooks and Journals,* vol. 1 (1975); Margaret Duckett's *Mark Twain and Bret Hart* (1964); William R. Gillis's *Memories of Mark Twain and Steve Gillis* (1929); and Richard Coke Wood's *Tales of Old Calaveras* (1949).

PERIODICAL LITERATURE

Trade journals and commercial publications are helpful for production figures and financial data, and often contain useful descriptive articles and biographical sketches. For Calaveras, the most important are the transactions of the American Institute of Mining Engineers (AIME) and the *American Gold News,* successor to the *Western Mining and Industrial News;* the *California Mining Journal;* the *Commercial and Financial Chronicle;* the *Engineering and Mining Journal;* the *Industrial Labor Relations Review; Mining and Metallurgy;* and *Mining Engineering.* The Calaveras lumber industry immediately after World War II can be followed in *American Eagle,* the newsletter of the American Forest Products Corporation. The 1899 anthology published by the state's (AIME) affiliate titled *California Mines and Minerals* contains a number of noteworthy articles relating to Calaveras and the Mother Lode. By far, the most useful journal for western mining history is the *Mining and Scientific Press,* published weekly between 1860 and 1922.

Regional periodicals, published by professional organizations, historical societies, and other public and private agencies, are invaluable research tools for the regional historian. A wealth of information is contained in the Calaveras County Historical Society's bulletin, *Las Calaveras,* published quarterly since 1952. In addition, solid historical information on Calaveras mining and ancillary industries can be found in *California History,* the *Pacific Historian,* and the *San Joaquin Historian.* For the mining West in general, we found helpful articles in *Arizona and the West,* the *Australian Economic History Review,* the *East European Quarterly,* the *Huntington Library Quarterly,* the *Journal of the West, Marine Chemistry,* the *Mining History Journal, Montana,* the *Nevada Historical Society Quarterly,* the *Pacific Historical Review,* the *Southern California Historical Quarterly, Technology and Culture,* and the *Western Historical Quarterly.*